Textile Management: A Guide for Technicians

Arvind Kumar Upadhyay

CRC Press
Taylor & Francis Group
Boca Raton London New York

CRC Press is an imprint of the
Taylor & Francis Group, an **informa** business

WOODHEAD PUBLISHING INDIA PVT LTD

New Delhi

About the author

ARVIND KUMAR UPADHYAY

A competent and result oriented techno-commercial textile executive with over four decades of rich experience in worsted, synthetics & cotton industries with a tremendous track record. Through his rich career in the textile industry – having gone from a shop floor technical expert to a Business Head, with the largest and most prominent textile companies in India and Africa - the author has had special access to varying levels of information and critical vantage points in the Indian as well as international industry and has acquired a comprehensive professional expertise in managing fully integrated textile mills (from fiber to garment) with sharp problem-solving skills and utmost efficiency. He has been known to formulate and execute unique and ingenuous strategies for addressing a wide range of issues affecting textile industries globally, including crisis management, turning around lean businesses, meticulous quality and production management, process engineering, product development, preventive maintenance management, materials management, resource planning, cost efficiency, prudent administrative and team building skills, initiating new projects, brand building & marketing and creative solutions for overall growth of the business. A lot of these are covered in the book.

Companies served as senior management

India

- CEO, Uniworth Textile Ltd
- President & COO, Indoworth India Ltd
- President & COO, Digjam Ltd
- President, Reid & Taylor, Mysore
- Vice President (Manufacturing & Commercial), OCM India Ltd
- COO, CSM, a division of S. Kumars Nationwide Ltd (SKNL)
- DGM, Raymond Ltd
- Early years - Middle and lower management at Raymond, Digjam, Lal Imli & J K Cotton Mill

Africa

- Business Head - Ken Knit Group of Industries, Eldoret, Kenya
- Business Head - Bhojsons Industries PLC, Lagos, Nigeria
- Profit Centre - Head, International Textile Industries, Lagos, Nigeria

Awards and Recognition

- Honoured with Management Excellence Award by Hon' Minister of Industries, Govt of Punjab
- Published various articles and interviews in widely published technical magazines
- Ex-Chairman, Textile Association of India (MP), during which tenure organised grandly successful national textile seminars and conferences
- Actively associated with corporate advisory boards in various management and other institutes
- Participated in several international textile and machinery conferences, most recently as speaker on behalf of Technical Advisor, VOS Group
- Exchanged technical know-how with senior technicians of Marzotto & Piacenza of Italy, the famous worsted manufacturers in Italy.
- Past Technical Advisor, VOS Group, Austin, Texas, USA (leadership development and strategic advisory firm)

Dedicated to my mother
Late Smt. Ramvanti Devi Upadhyay
&
all textile technicians

First published 2024
by CRC Press
4 Park Square, Milton Park, Abingdon, Oxon, OX14 4RN

and by CRC Press
2385 NW Executive Center Drive, Suite 320, Boca Raton FL 33431

CRC Press is an imprint of Informa UK Limited

© 2024 Woodhead Publishing India Pvt. Ltd

British Library Cataloguing-in-Publication Data
A catalogue record for this book is available from the British Library

ISBN13: 9781032629964 (hbk)
ISBN13: 9781032629995 (pbk)
ISBN13: 9781032630007 (ebk)

DOI: 10.4324/9781032630007

Typeset in Times New Roman
by Bhumi Graphics, New Delhi

Contents

- o COMPANY SECRETARY
- o SALES
- o HRD AND ADMINISTRATION
- PRODUCTION REPORTS
  - o RAW MATERIAL PROCUREMENT
  - o DYEING
  - o SPINNING
  - o WEAVING
  - o FINISHING
  - o FOLDING
  - o WAREHOUSE
  - o PRODUCTION PLANNING AND CONTROL
  - o QUALITY ASSURANCE
  - o PROCUREMENT AND STORES
  - o ENGINEERING
- EXAMPLES OF FORMATS FOR COMMON REPORTS:
- DAILY PRODUCTION AND SALES REPORT
- DAILY SALES, EXPENSES AND TENTATIVE PROFIT & LOSS REPORT
- MONTHLY PERFORMANCE REPORT AND COMPARATIVE ANALYSIS OF A WEAVING & PROCESSING UNIT ALONG WITH A SMALL DIVISION OF PLASTIC ITEMS
- CASH FLOW FORMAT OF A COMPOSITE TEXTILE MILL
- BUDGET ANALYSIS FOR A FINANCIAL YEAR (APRIL TO MARCH)
- LIQUIDITY ASSETS, NET WORKING CAPITAL AND NET WORTH
- RATIO ANALYSIS
- ADMINISTRATIVE OVERHEADS
- ANNUAL PROFIT/LOSS AND BALANCE SHEET OF A COMPANY
- COMBINED EXPENSES STATEMENT FOR AND UPTO MARCH'03
- COMBINED OPERATING STATEMENT FOR AND UPTO MARCH'03
- COMBINED CONTRIBUTION STATEMENT FOR AND UPTO MARCH'03
- BALANCE SHEET
- STOCK SUMMARY
- SALES REGISTER
- DEPARTMENTS COMBINED PRODUCTION REPORT
- DYEING DEPARTMENT MONTHLY PERFORMANCE AND TENTATIVE EXPENSES
- MONTHLY PERFORMANCE REPORT (Worsted Dyeing Department)
- CATEGORY-WISE DYEING PRODUCTION
- CONSUMPTION OF DYES AND COST PER UNIT
- DAILY PERFORMANCE REPORT – WORSTED SPINNING DEPARTMENT
- SPINNING DAILY PERFORMANCE REPORT
- MONTHLY PERFORMANCE AND TENTATIVE EXPENSES OF A SPINNING DEPT (Worsted)
- END BREAKAGE STUDY IN SPINNING DEPARTMENT
- DEPARTMENT-WISE WASTE GENERATION REPORT (worsted)
- COTTON SPINNING AND WEAVING MILL DAILY WASTE GENERATION REPORT
- COTTON SPINNING MONTHLY WASTE RECORD

- ALLOCATION OF DEPRECIATION
- DEPARTMENT-WISE MONTHLY DIRECT AND ALLOCATED COST SUMMARY
- DEPARTMENT-WISE DIRECT AND INDIRECT COST SUMMARY
- COST CALCULATION GENERATOR SET POWER vs GOVT POWER SUPPLY COST
- CALCULATION OF STEAM COST
- ELEMENTS OF DIRECT COST OF PRODUCTION DEPARTMENTS
- ANNUAL COST ANALYSIS OF A COMPOSITE MILL
- COSTING OF COTTON YARN
- WORSTED SPINNING, SECTION-WISE COSTING
- WORSTED YARN COSTING-A
- WORSTED YARN COSTING -B
- CONVERSION COST OF WORSTED YARN BASED ON MONTH PRODUCTION
- ANALYSIS OF COST ELEMENTS OF WORSTED SPINNING PLANT
- MONTHLY DYES AND CHEMICAL COST
- ANNUAL COMPARATIVE DYEING COST
- FABRIC COSTING FOR EXPORT
- FABRIC COSTING FOR EXPORT (Summary)
- FABRIC COSTING OF LUXURY SUITINGS FOR DOMESTIC MARKET
- COTTON FABRIC COSTING
- COSTING OF BLANKETS
- COSTING OF GARMENT
- COST COMPARISON OF DIFFERENT OPTIONS OF WEAVING MACHINES
- CAPITAL COST AND COTTON YARN MANUFACTURING COST COMPARISON AT DIFFERENT RING FRAME SPINDLES AND SPEED

9. COST REDUCTION

- COST REDUCTION PROCESS
- MAJOR COST ELEMENTS AND WAYS TO CONTROL
- EXAMPLES OF COST REDUCTION
- COST REDUCTION REPORT FOR A WORSTED COMPOSITE PLANT
- COST OF REPROCESS OR REWORK
- EXAMPLE OF REPROCESSING COST IN FINISHING DEPARTMENT
- REPROCESSING COST IN DYEING DEPARTMENT
- EXCESS COST OF DYEING DUE TO UNDERUTILISATION OF DYEING M/CS CAPACITY
- DYEING RECIPES
- REDUCTION OF DYEING COST THROUGH ALTERNATE RECIPES
- COST SAVING BY REDUCING DEFECTIVE PRODUCTION

10. SALES AND MARKETING

- SALES AND MARKETING MANAGEMENT
- KEY AREA OF SALES STRATEGY
- ROLE OF SALES EXECUTIVE
- PRIMARY SALES
- SECONDARY SALES
- IN SHOP BRAND VISIBILITY

Preface

During my more than four decades of work experience in the international textile industry in various positions from 'Apprentice Supervisor' to 'Business Head', I have realised that a technologist or engineer cannot do justice to his/her duties as a complete manager without a thorough understanding of the commercial and financial aspects of the business. It is commonly observed that even after decades of working in the industry, most technicians remain unaware about essential features of a business such as basic accounting, budgeting, preparation of projects, costing of products, sales and marketing, auditing, profit and loss accounts, etc.

It is my humble but strong belief that a technician can never be a truly successful manager until all actions or decisions taken by them are keeping in mind its commercial and financial impact on the business.

In their daily working, technologists or engineers are expected to have deep knowledge of their core subjects and keep them updated with fast-changing technological upgradations in the industry. As a result, their focus is usually limited to technical aspects of the industry.

To refresh their knowledge of the latest technologies they are put through mandatory technical training programs conducted by reputed machinery manufacturers before and after selling their machineries, on specific demand or voluntarily from time-to-time. The companies also organise different management programs and courses related to motivation, change management, human relations, team building, etc. There are also several quality management programs conducted for system or quality improvements such as ISO, TQM, Six Sigma, etc. However, a similar emphasis is not given to the managerial, administrative or overall business aspects of the industry.

It is a common practice in the Indian textile industry to identify and select upcoming candidates based on their self-learning abilities and performance to elevate them systematically for managerial responsibilities upto senior most positions, either in the same organisation or elsewhere depending on their stability.

Manufacturing, being a human oriented industry where outcomes depend on the joint efforts of an entire team contributing at different stages of the production chain, a single mistake by an individual's negligence or lack of know-how may severely damage end results. However, usually it is the section in-charge or group leader that are held responsible for any failure - be it large or small, without any accountability or responsibility on the team.

In an open and market driven economy, businesses are getting highly competitive day by day. Customers demand the best quality at lowest prices, within a minimum lead-time. Therefore, management too needs to be competitive and focus on cost saving drives by adopting latest technologies, automation, generating out of the box ideas and innovative design and development strategies.

For optimum utilization of an organisation's technology and infrastructure, it is essential that the shop floor team members, section in-charge, and manager have basic commercial and financial understanding to manage their area of responsibilities efficiently and effectively and contribute to running of the business successfully. This is equally important for them to have clarity about the commercial and financial impacts of the activities carried out in their area, to assess their contribution in the business and present the financial gain of their special achievements, if any, to the management.

Keeping the above in mind, the objective of this book is to provide relevant information about critical business aspects of the textile industry and ancillary tools. I have tried to cover important concepts of business in a brief and simplified, yet practical way through examples in the form of reports, formats, figures related to respective subjects. However, the figures given in various tables on different topics are exemplary and are intended to serve as guidelines for understanding the various subjects and should not to be relied on as norms or industry standards.

The book provides readers with the necessary, basic tools to understand the overall management of a textile manufacturing company and help technicians to transform themselves as techno-commercial managers. It may not only help expedite the career development of students, individual technicians and managers working in the industry but may also contribute towards a strong apparatus for the larger benefits of the Indian textile industry as a whole.

Acknowledgements

This book is the outcome of the contribution of many textile units and professionals I have been fortunate enough to work with, in different capacities through my long and fulfilling career. More than four decades of this long journey have provided me tremendous opportunities of great value in writing this book. I gratefully acknowledge the profound intellectual debt that I owe to all the companies, my seniors and colleagues during the span of my career. My readers will also be deeply indebted to the companies that allowed access to their valuable information, used in this book as guidelines for the benefit of technicians and the textile industry as a whole.

I have always felt such a deep sense of obligation to the textile industry and believe that it is my moral duty to pay back adequately to this life resource in my limited capacity and this is my honest attempt in this direction. I have tried to share my limited experience honestly to the approaching generations through this book and sincerely hope my humble effort may be useful for sufficiently equipping students of textile, engineers and future managers in their respective professional journeys.

Outside of my professional world, I am foremost grateful to my late father Shree Jay Shankar Upadhyay who was responsible for initiating my journey in textiles and contributed greatly to my life, career and a creative academic pursuit.

I am also grateful to my late father-in-law late Shree Gyan Chandra Dwivedi for being a motivating presence through his actions all his life.

This book cannot be complete without the mention of my mentor, late Shree J. N. Singh former officiating Principal, Government Central Textile Institute, Kanpur who in his unique ways, gave me a great beginning to my career in the textile industry and was the source of many blessings and inspiration.

My deepest gratitude is to my wife Sushma who believed in me through thick and thin, has always been an endless spring of constant motivation, and encouraged me to write this book, always providing unfailing support.

I must acknowledge the encouragement, support, and motivation I have received from my sisters, brothers and all the youngsters in the entire family doing excellent work in their own professional careers.

I should not forget to acknowledge the contribution of my son Avijit Kumar and daughter Pankhuri Upadhyay who have always been with me as my moral support. I am most grateful to my daughter Pankhuri for being pro-actively involved in the publishing journey of this book and helping with the editing, book design and other critical aspects

I must express gratitude to my daughter-in-law Sasha for being part of our family, an inspiring entrepreneur and wonderful mother to our adorable granddaughter - the ultimate source of our collective joy, Arya Tara Upadhyay.

Last but not the least I am grateful to Woodhead Publishing India, who have excelled in ensuring elegant and expeditious publication of this book.

Types of Business

It is a statement of fact that no individual or country can be self-sufficient in its entire needs. Every country has its own strength depending on its natural resources, climatic conditions, available work force, and economic conditions. Some countries are technologically and scientifically advanced and produce high-end technological products, useful for humanity. Besides this, there is the theory of comparative advantage, which advocates specialization rather than duplication to bring about inter-dependence locally, nationally, and internationally. However, not every individual can engage in trade to meet his or her needs.

Usually, the basic requirements to start any business are:

1. Market

2. Know-how

3. Interest

4. Energy

5. Capital

DIFFERENT FORMS OF BUSINESS SETUPS

SOLE PROPRIETORSHIP BUSINESS

This is the oldest and most common business structure found in India where one person is the owner and is personally liable for all the debts and liabilities of the business. It is the simplest form of an entity with minimal compliance procedures. Most common businesses like your local grocer, the chemist, private- practicing doctors or lawyers are also the examples of sole proprietors. A sole proprietor is the only person who operates and manages the whole business, so 100% of the profits belong to him/her.

Legal compliances for a sole proprietorship are minimal. There is no provision for a Certificate of Incorporation or Registration Certificate. Therefore, compliances depend upon registrations or licenses taken by a particular sole proprietorship for conduct of business. For example, if a sole proprietorship registers itself under GST law, then it will have to comply with the GST return filing rules, etc. and a GST Registration Certificate will be the proof of business (as compared to a Certificate of Incorporation in the case of a company).

The assets and liabilities of the sole proprietorship are the assets and liabilities of the sole proprietor. A sole proprietor is required to file his/her usual income tax (IT) returns and show profits earned through the business within that return itself. Separate IT return is not required for the sole proprietorship firm. Tax is calculated at income tax slab rates applicable to an individual. Other tax liabilities like GST will depend upon the nature of business.

The financial reports of sole proprietorships remain private unlike the public or private limited companies where financial statements and audit reports are made public for the users.

Since, a sole proprietorship is managed and operated single-handedly there is no scope for any conflict of ideas or decisions. The sole proprietor has the freedom to make decisions independently.

Like mentioned earlier, in India there is no specific provision for registration under any law to set up a sole proprietorship. But one can apply for a few registrations or licenses under various laws for ease of conducting business - such as registration under the Shops and Establishments Act, or get an Udyog Aadhaar Certificate under the Ministry of MSME or get a GST Registration Certificate, if GST is applicable.

To register any of the above as a sole proprietorship, the documents required are Aadhaar card, PAN card, bank account and address proof of place of business.

The local laws prevailing in the area where sole proprietorship is located require registration under the Shops and Establishments Act. The following are examples of registrations / licenses that may be opted by a sole proprietor depending on the type of business:

- **Registration under the MSMED** (Micro, Small and Medium Enterprises Development) Act – it is required in case of micro, small, and medium enterprises.

- **Registration under GST** (Goods and Service Tax) – It is required in case the turnover of the business exceeds a specified limit.

- **Trademark registration** – It is required in case you want to trade your products or services with an exclusive name, brand or logo.

- **Registration for exports license** – It is required if you wish to export your products. Different licenses or certificates are required according to the nature of business.

ONE PERSON COMPANY (OPC): This concept was introduced by the new Companies Act, 2013. This provides for an option to register a company with just one member, instead of the usual requirement of minimum two members. In a way, an OPC is the corporatisation of a sole proprietorship. It is a separate legal entity and liability of the director is limited. Unlike a sole proprietorship, an OPC has to be compliant with Indian company laws and filings with the Registrar of Companies (ROC), including audit requirements. The income tax rates applicable are also those of a private limited company. Some entrepreneurs prefer incorporating an OPC for increasing legitimacy in the market and with clients.

PARTNERSHIP BUSINESS: The partnership business organisation evolved from the sole trader, arising from the need for people of different skills to work together and need to raise capital jointly. Just as in case of the sole trader, there is no distinction between the assets of the business and that of the partners. Thus, any asset can be put into any use or can be used to satisfy any debt. If the business is not able to pay its creditors all the property of the business will be sold to satisfy the creditors.

LIMITED LIABILITY PARTNERSHIP (LLP): A relatively newer concept and type of business structure is an LLP. This type of entity allows for a partnership structure where each partner's liabilities are limited to the amount of their contribution to the business. The LLP is a separate legal entity from it partners and is fully liable to the extent of its assets but the debt does not transfer to the partners, like in a partnership business. Also, by having business partners one shares risk, skills and labour. One partner is not liable for the default or unauthorised actions of another partner. Therefore, this form of corporate structure has the benefits of limited liability of a company (described below) and the flexibility of a partnership.

PRIVATE LIMITED COMPANY: A Private Limited Company is a business entity held by small group of people (joint stock company), incorporated under the Companies Act (earlier Act of 1957 and now Act of 2013) in India and governed by the Ministry of Corporate Affairs. It is registered for pre-defined objectives and owned by a group of members called shareholders.

The liability of each shareholder or member is limited. This means, if the company runs into a loss, the company shareholders are liable to sell their company shares to clear the debt or liability. The individual or personal assets of shareholders or members are not at risk.

The company restricts the transfer of shares and prohibits invitation to the public for the subscription of shares and debentures. It uses the term 'private limited' at the end of its name.

PUBLIC LIMITED COMPANY: A Public Limited Company or PLC is a joint stock company formed and registered under the Companies Act. There is no defined limit on the number of members. In addition, there is no restriction on the transferability of the shares. The company can invite the public for the subscription of shares or debentures, and that is why the term 'Public Limited' is added to its name. The liability of a public company is limited. No shareholder is individually liable for the payment. The public limited company is a separate legal entity, and each shareholder is a part of it.

COMPANY LIMITED BY SHARES: In this form of company, the capital is introduced in the form of Shares i.e., the capital of the company is divided into a small portion, known as shares. The shares are considered interest of the shareholders in the company. The number of equity shares held, measures the ownership of the shareholder in the company. If any requirement for capital arises in the company, the shares can be issued for subscription by shareholders. In this type of company, the liability of the members is limited up to the unpaid capital on the shares subscribed.

Further, this form of company can be registered as private limited company, One Person Company, or registered as a public limited company also, based on number of members and nature.

COMPANY LIMITED BY GUARANTEE: A company can be either a private limited company or a public limited company, where the capital is not divided into shares. Here, the capital to be introduced by the members, is in nature of guarantee and the percentage of ownership is based on the amount guaranteed. Whenever the requirement of capital arises, members introduce capital to the company. The liability of members is limited up to the amount of guarantee provided by them.

These companies can also issue shares, where the shareholders are also liable up to the amount unpaid on the shares as discussed above. However, shareholding is not the criteria for deciding the ownership.

UNLIMITED COMPANY: In this type of company, the liability of the members is not limited. In case of any debt, the liability of members does not limit to their contribution in the company, but extends to their personal assets also. In present times, this type of company is rarely chosen to be incorporated by entrepreneurs. The liability of members arises at the time of winding up or bankruptcy or otherwise, whenever the capital is to be raised or debt is to be paid.

Most popular type of company in most businesses is a company limited by shares.

DIFFERENCE BETWEEN PUBLIC LIMITED COMPANY AND PRIVATE LIMITED COMPANY

Criteria	Public Limited Company	Private Limited Company
Minimum number	Minimum 7 persons must be there to form a public limited company	Minimum 2 persons required to form a private limited company. One should be a citizen of India
Maximum number	No limit to the maximum number of shareholders	Maximum number of shareholders is limited to 200 in a private limited company excluding the past and present employees of the company
Certificate of Commencement of Business	Has to obtain the Certificate of Commencement of Business in addition to the Certificate of Incorporation in order to commence the business	Since 2019, has to obtain the Certificate of Commencement of Business in addition to the Certificate of Incorporation
Name of the company	Has to add word "Limited" at the end of its name	Has to add "Private Limited" at the end of its name
Minimum subscription	Has to secure minimum capital before allotting its shares	No such restriction for a private limited company and it can allot shares
Issue of prospectus	Can invite public to subscribe for its shares. It must issue a prospectus or file a statement in lieu of prospectus before issuing shares	As per the law, a private limited company has no rights to invite the public and as such cannot issue prospectus. They cannot get the public to subscribe for its share capital
Ownership	Regulation and ownership of shares can be sold to the public on an open market	Shares can be sold or transferred to other people by the choice of the owner. Shares of such companies are owned by founder management, or a group of private investors. Shares are not sold in the open market
Transfer of shares	Transfer of shares can be done easily in a public limited company	Rights of members to transfer their share are restricted in the Articles of Association in a private limited company

Statutory meeting	A public limited company must hold a statutory meeting within six months from the date of commencement of business	A private limited company need not hold any statutory meeting
Consent of Directors	Consent in writing is necessary	It is not necessary
Qualification Shares	Director needs to keep certain no of shares	No such condition applies
Minimum share Capital	A public company requires a minimum share capital of Rs. 5,00,000	There is no such minimum compulsion
Annual report	Has to file annual reports with the Registrar of Companies along with a declaration certifying that 1) No. of members does not increase to 50, 2) No share capital or debenture was raised from public, 3) Other member companies hold less than 25% of the shares	It is not necessary. Only an annual report needs to be filed. No such declaration is required
Inspection of annual accounts	Annual reports are public documents Any individual can see it. Confidential information such as executive compensation, legal settlements, and other essential information cannot be kept reserved in public companies	Such information is more secure in a private company. And available to members only
Special privilege	No privilege	Enjoys special privileges and exemptions
Meeting Quorum	5 Persons	2 Persons
Articles of Association (Document that specifics regulations for company's operations)	A public limited company can draw up its own Articles of Association or adopt Table F in the Companies Act, 2013	A Private limited company must draw up its own Article of association
Retirement of directors	Minimum two-third of the directors must retire from management by rotation	Rights of members to transfer their share are restricted in the Articles of Association in a private limited company

A MANUFACTURING FACTORY AND ITS BASIC ELEMENTS

A manufacturing process includes:

- Making, altering, repairing, ornamenting, finishing, packing, oiling, washing, cleaning, breaking up, demolishing or otherwise treating or adopting any article or substance with a view to its use, sale, transport, delivery or disposal;

- Pumping oil, water, sewage, or any other substance;

- Generating, transforming or transmitting power;

- Composing types for printing, printing by letter press, lithography, photogravure or other similar process or book-binding;

- Constructing, reconstructing, repairing, refitting, finishing or breaking up ships or vessels; or

- Preserving or storing any article in cold storage.

Men, Machines, Material, Money and Management are the key elements of a manufacturing organisation.

MANPOWER

Selection of the right person at the right place one of the most critical decisions for the success of any organization. Irrespective of technological advancement, the importance of work force will always be critical to operate and control on technology for its commercial success.

A company can buy a man's time, its physical presence at the workplace but it cannot buy its loyalty and his devotion to the work. Hence, it is important to adopt a systematic and scientific procedure for not only selection of workforce but also to focus continuously towards developing a positive work environment, happiness of employees and creating a feeling of belongingness.

Basic criteria for selection of manpower depends on:

- Nature of job
- Qualification (neither low nor overqualified) depending on specific job
- Work experience
- Age as suitable to job
- Specialization / expertise
- Languages known
- Native place/permanent and current address (to address local language and home sickness issues specially for lower and middle management)
- Extra-curriculum activities and hobbies
- Family background
- References, if any
- Previous employment details in case of an experienced candidate
- Special achievements in previous jobs, if any
- Comments from the previous management if necessary (track records)
- Medical fitness (height/weight/physic/eyesight/colour blindness /speaking/hearing, etc.)

Assessment Process

- To have a transparent recruitment policy
- Vetting through a group of experts depending on level of the candidate being interviewed
- Judging attitude and aptitude
- Entry in the interview room
- Confidence level
- Personality
- Dressing sense
- Face expression
- Style of standing/sitting/walking
- Body language
- Resume – qualifications, work experience vs. fresher, verification of facts presented
- Presentation (confidence, selection of words while speaking, speed, clarity about job applied for, understanding and communication skills, etc.)
- Knowledge of the subject
- Clarity about salary/incentives/terms & conditions/perks, etc.
- His/her opinion about the present and past employers
- Reason for leaving

- IQ Level (knowledge of other fields/subjects
- Leadership quality by group discussion
- Notice period to the previous employer
- Keep second candidate as reserve

Retaining employees: A good employer treats his employee as an asset for a company. Hiring an employee is just a start for creating a strong workforce. One of the primary objectives of an HR Department is also to retain qualified and committed people in the organisation.

For a successful business, the goal of the employer should be to decrease employee turnover, thereby decreasing training costs, recruitment costs and loss of talent and organizational knowledge. Employee retention means efforts of business organizations to maintain a working environment, which supports its present workforce in continuing with the company.

A good employee retention policy should aim at addressing the various needs of the employees to enhance their job satisfaction and reduce the costs involved in hiring and training new staff.

MACHINERY

Selection of Machinery and Equipment: Machineries and equipment are used to carry out a general or specific activity to manufacture the desired products. In a fast-changing competitive world, technology plays a leading role towards the success of a business. The selection of machines is a techno-commercial process and needs both comprehensive technical and commercial expertise.
Selection of machinery depends on:

- Type of products to be manufactured
- Quality of product to be manufactured and pricing segment
- Production capacity (workable speed and efficiency
- Make & manufacturer of m/c
- Cost of the machine
- Salient features of machines and its viability comparison with other manufacturers
- Payback period for required product with achievable production and efficiency
- Technology (user-friendly /complicated/advance/outdated, etc.
- Versatility (size/style, etc.) Should be able to switch over to produce other product
- Materialogy, usable life of machine
- Maintenance (cost)
- Electronics and computerisation
- Available spares and accessories with machine
- Erection and commissioning facility
- Cost of Energy
- Environment-friendly (sound/air, etc.)
- Training program and after sales service (updates/improvements/feedback, etc.)
- Spare parts availability
- Quality of instruction and maintenance manuals

MATERIAL

Raw material is the primary substance used as the input in a production process for subsequent modification and finally processed into a finished good. In a textile company, depending on the finished product, raw materials may be in different states such as fiber for spinning, yarn for weaving and fabric for apparel, dyes and chemical, etc. Raw material influences the quality of final product and decide its market value in textile manufacturing, since cost of the raw material is highest in total cost of product, its quality and price play a predominant role in the production of desired quality of product in a competitive price for a profitable business. Other than raw material, there are numbers of other inputs as spare parts, packaging materials lubricants, etc., which are equally important.

The selection of raw materials and other inputs and their procurement requires in depth technical knowledge and commercial expertise.

Selection of Materials (Raw Material)

- End products and their end-use

- Sales price of end products and market positioning

- Quality of end products required (as per prescribed norms domestic/international/institutional)

- Customer segments (low/medium/upper/premium/luxury/ defense /sensitive area/life savings, etc.

- Other than raw material for the product, there are several other materials used in a manufacturing company directly and indirectly including stores and spare parts, packaging materials stationary, computers, civil materials, and fuels, etc.

Procurement process of raw material and other inputs

- Design and follow a defined and transparent procurement policy and prepare SOP

- Selection of vendors – their understanding about the material, capability to produce or supply credibility, commitment, etc.

- Retain reliable vendors and help them to develop

- Introduction of new vendors for alternate arrangement/back-up and healthy competition

- Long term agreements wherever required and possible

- Quotation or tender system and techno-commercial comparison

- Criteria for decision-making should not be based on lowest price but it should be based on the quality of product (including materialogy, dimensions, finish, etc.), capability, capacity, and credibility of supplier. Procurement policy should have a system of marking for individual aspects of material to arrive on conclusion.

- Scientific system (data-analysis) of price negotiation - cost of material, cost of fabrication, cost of technology, etc.

- Knowledge of materials/workmanship/fabrication, national/international price/forecast, etc.

- Use testing facility in-house or outsourced to verify the quality of materials supplied

- Quality approval by competent person of concerned department

- Historical consumption data and life of individual items/spares

- Maintain optimum inventory level (limit) and review regularly with concerned users

- Monthly report of consumable and slow-moving inventory and discuss with management for corrective measures

Storage

Proper storage of specific materials is important to avoid damage. Location must be easily traceable and easy to handle. Each item must be stored under a proper temperature and humidity according to its physical and chemical properties such as:

- Storage of stationeries requires dust free, dry and optimum temperatures
- Storage of leathers/plastics in an appropriate temperature to avoid brittleness
- Storage of inflammable item
- Storage of health hazardous items
- Storage of electronic / electrical items
- Storage of food items / drugs (climatic conditions required / date of expiry)
- Record keeping
- Safety of inventory
- Inventory alerts
- Consumption details

MANAGEMENT

Management is the administration of an organisation. Management's role is the overall coordination and administration of tasks to achieve a goal. It includes strategic planning, decision making, organizing, leading, motivating, and controlling the financial and human resources of an organization to achieve its goals efficiently and effectively.

There are several levels within management according to the size of an organisation. Level refers to a line of demarcation between various managerial positions in an organisation. The level of management determines a chain of command, the amount of authority & status enjoyed by any managerial position. The levels of management can be classified in three broad categories:

Top-Level or Senior Management: Board of directors, chief executive, or managing director. The top management is the ultimate source of authority, and it manages goals and policies for an enterprise. High-level managers tend to have a substantial amount of experience, ideally across a wide variety of functions. Many high-level managers become part of an executive team by mastering their functional disciplines across various roles, becoming the Chief Operations Officer (COO), Chief Marketing Officer (CMO), Chief Technology Officer (CIO or CTO), Chief Financial Officer (CFO) or Chief Executive Officer (CEO).

The primary role of the top-level managers is to derive broad strategic plans for Company's policies execution, financial investments, strategic alliances, discussions with the board, and stakeholders.

The role of the top management can be summarised as follows:

- Decides the objectives and broad policies of the enterprise
- Issues necessary instructions for preparation of department budgets, procedures, schedules, etc.
- Prepares strategic plans & policies for the enterprise
- Appoints executives for middle-level management i.e. departmental managers
- Controls & coordinates the activities of all the departments.
- Responsible for maintaining relevant networks and contact with the outside world
- Provides guidance and direction
- Responsible towards shareholders for the performance of the enterprise

Mid-level Management: Middle management is the intermediate management level accountable to top management and responsible for leading lower level managers.

The branch managers and departmental managers constitute middle-level. They are responsible to the top management for the functioning of their department. They devote more time to organisational and directional functions.

In small organisation, there is only one layer of mid-level of management but in big enterprises, there may be senior and junior mid-level management. Middle-level managers can include general managers, branch managers, and department managers. A middle manager's role may have emphasis on:

- Executing the plans of the organisation in accordance with the policies and directives of the top management
- Making plans for the sub-units of the organisation
- Participating in employment & training of lower-level management
- Interpreting and explaining policies from top-level management to lower-level
- Responsible for coordinating the activities within the division or department
- Providing important reports and other important data to top-level management
- Evaluating performance of junior managers
- Inspiring lower-level managers towards better performance

Lower-level Management or Front-line Management: At the front-line, managers are often highly skilled and even functional specialists. Front-line managers are best positioned when they focus on controlling and directing specific employees (thinking in terms of supervisors, team leaders, line managers, and project managers).

A frontline manager needs to have two distinctive skill sets: interpersonal skills to manage people as well as the technical expertise to be among the front-lines actively executing functional tasks. As a result, frontline managers are often highly valuable team members with the versatility to contribute in various ways.

Lower level is also known as supervisory / operative level of management. It consists of supervisors, foreman, section officers, superintendent, etc. Their activities include:

- Assigning of jobs and tasks to various workers
- To guide and instruct workers in day-to-day activities
- Responsible for the quality as well as quantity of production
- Entrusted with the responsibility of maintaining good relations within the organisation
- Communicate workers' problems, suggestions, and recommendatory appeals, etc. to the higher-level and higher-level goals and objectives to the workers
- Help to solve the grievances of the workers
- Supervise & guide the subordinates
- Responsible for providing training to the workers
- Arrange necessary materials, machines, tools, etc. for getting the things done
- Prepare periodical reports about the performance of the workers
- Ensure discipline in the enterprise
- Motivate workers
- Image builders of the enterprise because they are in direct contact with the workers

Board of Directors and Chairman: A Board of Directors is an appointed or elected group of individuals that has overall responsibility for the management of an organization. The board is typically the governing body of an organisation and seeks to ensure the best interest of all stakeholders in all types of management decisions. The board of directors typically includes key individuals involved with the management of an organisation. Other individuals may be appointed or elected based on their expertise and experience in areas pertaining to the management of the organisation. The board will often have a mix of internal and external trustees. The board of directors is elected by the shareholders of a company and is usually composed of both inside directors, who are senior officers of the company, and outside directors, who are individuals not employed by the company.

Chief Executive Office (CEO), Chief Operating Office (COO), President: A CEO is considered the highest-ranking officer in a company. However, in corporate governance and structure, several permutations can take shape. So, the roles of CEO, COO and President may be different depending on the company. In some corporations and organizations, the president is the leader of the company's executive group. In the corporate world, however, the president often refers to someone who is the leader of a segment or critical part of the overall company, rather than the leader of the overall company.

In some instances, the President is also the CEO. In small businesses, the President might also be the owner of the company. In an organisation or company where a CEO is already in-charge, the President is the second in command.

In the corporate world, presidents often hold the position of Chief Operating Officer (COO). The COO, responsible for day-to-day operations, has vice presidents for different parts of the company reporting to him or her.

Chief Financial Office (CFO): A Chief Financial Officer (CFO) is the senior executive responsible for managing the financial actions of a company and analysing its financial strengths and weaknesses and proposing timely corrective measures. Their duties include managing coordinating with financial institutions, maintaining account through accountant, cash flow and financial planning, preparing Profit and Loss account (P&L), Balance sheet, Management information system MIS, conducting audit, etc.

Chief Marketing Office / Chief Sales Office (CMO/CSO, CSMO): A Chief Marketing Officer (CMO) is responsible for overseeing the planning, development and execution of an organization's marketing and advertising initiatives and continuous development of company's brand image. Usually, a CMO reports directly to the CEO and is primarily responsible for the business growth through successful marketing for the entire organization, using market research, correct pricing, product marketing, brand positioning, marketing communications, advertising, and public relations. In many cases, the CMO role is expanded to include sales management, new business development, product development, distribution channel management, and customer service. The CMO ensures the organization's message is distributed across channels and to targeted audiences in order to meet sales objectives.

The CSO or Chief Sales Officer is one of the key personnel in a business. He/she is in-charge of the management of the entire sales department and overseeing all sales-related activities. The CSO is responsible for making sales strategies, document sales policy and implementing it. Their other important responsibilities are to achieve daily sales targets, maintain credit facilities, revenue collection to maintain smooth cash flow for the smooth functioning of the business. They study sales numbers to assess how successful sales strategies are in the business. The CSO is answerable directly to the business stakeholders. They provide leadership, direction, and resources to the sales department and is accountable for the overall sales department performance, the achievement of sales department goals and targets, and the alignment of the business's strategy. The Chief Sales Officer is also charged with ensuring sustainable revenue growth by constantly maximizing market penetration.

Design and Development Officer In a competitive market, a company is required to meet fast changing customer choices at a rapid speed. Innovation and product development is an essential part of a manufacturing company for its presence in the market and to achieve sustainable growth. The Design and Product Development Officer is responsible for developing new products by innovative imagination, getting regular feedback from the trade, coordinating with different markets, attending trade exhibitions, fashion shows, study fashion magazines, etc. to update himself and keeping abreast with trends. He/she needs to conceptualise various ideas into products and develop samples for trade approval before bulk production in case of premium and luxury products and support the sales team on any customized requirements for customers.

Production Planning and Control (PPC): This is also known as system chain management (SCM). The responsibility of a PPC head is to coordinate with production, sales and finance team to prepare annual production and sales plan, ensure its execution, arrange raw material, maintain raw material and WIP inventory level and ensure timely delivery of material to meet up sales requirement.

Quality Control and Quality Assurance Head: The terms quality control (QC) and quality assurance (QA) are often used synonymously and may be considered interchangeable, but there are distinct differences, and both are critical to the success of a company.

Quality assurance and quality control personnel and their responsibilities overlap, which lends itself to confusion that could affect the total quality system. Therefore, it is important for decision makers and managers, to understand the differences between both and their function. Both work together as an integral part of a quality management system (QMS).

QMS contributes in establishing the quality of a product and helps in its acceptability and positioning in the market and also in maximizing its profit. It collects and analyses relevant statistical data to assist in decision making for improvements in the quality of a product.

QC personnel perform the testing of materials at different stages and compare the results with the standard specification. However, they have limitations to analyse it and conclude the reasons if the testing results are out of the norms. They may or may not have authority to take decision depending on whether they are the owner or a third party.

QA personnel check the test procedures used by quality control personnel and verify the authenticity of the data They analyse the test results and identify the specific reasons of faults related to material, specific process or machines and manual malfunctioning. They are ultimately responsible for online assurance of the quality from raw material to the finished products and final packaging. Their responsibility is to monitor the test results at every stage, suggest appropriate and timely corrective measures to achieve the desired quality. A quality assurance head prepares the Quality Policy and SOPs in consultation with concerned manufacturing heads, sales and marketing team, commercial heads, HRD In-charge and gets it approved by the business head. He/she organises various quality assurance programs to create awareness about the precautionary measures for achieving quality as per pre-determined norms. Thus, a QA head helps in improving the organisation's profitability by achieving the maximum yield of raw material, controlling waste generation and production of defective material, and minimising rework.

Manufacturing Department Heads: Manufacturing activities are key areas in a manufacturing company, with all the rest in place as support and servicing departments. Depending on the type of product, a manufacturing company has different sections and processes, which need expert guidance and supervision. Department heads are experts of their area and are responsible for conversion from inputs into outputs, as per plan. A department head should have excellent leadership qualities, good communication skills and must be able to take ownership of their respective department. They are directly responsible for all activities within their department, which includes achieving production and quality targets, maintenance of their machines and equipment, maintaining optimum inventory levels, minimizing manufacturing costs and ensuring timely delivery with assortments and maintaining internal discipline.

Commercial Head or Chief Commercial Office (CCO): A company's Chief Commercial Officer is a key position and plays a direct role in controlling a major part of expenses. He is responsible for preparing the policy, overseeing objectives and implementation of all commercial strategies including procurement of raw material, machines, equipment, spare parts and its storage. He/she is not only supposed to have a deep knowledge of the material and scientific approach for procurement but also have logical and convincing negotiation skills. They are also responsible for coordinating with user departments, monitor and control the consumption and maintenance of inventory level at stores. They should be aware and updated about all applicable government, industrial and trade policies and ensure the company gets maximum benefits

Company Secretary (CS): A Company Secretary is an important managerial and administrative part of a company. He/she is one of the legal representatives of a company who performs and manages various statutory and regulatory compliance functions like incorporation of the firm, preparation and audit of business reports, filing annual returns, ensuring compliance with amended regulations on a regular basis and also makes sure that decisions of the board of directors are implemented. A CS also plays a definite role as a business advisory to the board of directors of the company guiding them on corporate laws, corporate governance, strategic management, project planning, capital markets & securities laws and contracts. In short, a Company Secretary works as an in-house legal expert and compliance officer with the company. The Companies Act provides specific rights to a company secretary to exercise promotion and incorporation of companies to handle company's audit and certification services, to sign annual returns, to handle corporate restructuring and takeovers, to scrutinize reports and voting procedures in a transparent manner. Other tasks include overseeing delivery details of share certificates, allotments, keeping records of share warrant holders to comply, releasing statutory declaration to receive commencement certificate, managing statutory books, giving meeting updates and notices of general meetings to every member, signing and authenticating financial reports and statements (Balance Sheet & Profit & loss statements) in cases of non-banking financial companies, filing resolution with registrar and preparing minutes of all General and Board Meetings within 30 days. A CS also submits reports of different committees such as Audit committee, Nomination and Remuneration Committee (NRC), Stakeholders' Relationship Committee (SRC), Corporate Social Responsibility (CSR) Committee and Risk Management Committee (RMC) in the board meetings. A Company Secretary is legally bound by restrictions and policies defined by ICSI and various regulatory authorities.

Human Resource Development Head: An HR manager plays a pivotal role in achieving the objectives of an organisation. It's the human resource/work force of a company that perform tasks and achieve company goals. An HR head is a strong bridge between an employer and employees. The HR department's specific duties include recruitment of right candidates at the right time and preparing various plans to retain such employees, conduct motivational and skill development programs throughout the calendar year, personal record keeping of individual work force, implement vetting and assessment processes, preparing growth and succession plans for the company, implementing performance incentives and welfare activities.

Specific responsibilities of HRD Head are to prepare and implement the following:

- Recruitment Policy
- Prepare Training Programs and yearly schedules
- Prepare Schemes to retain the manpower (human assets)
- Prepare escalation scheme/promotions/incentives, etc.
- Identify future leaders and prepare succession plans
- Prepare welfare activities
- Culture building exercises
- Coordination between management & work force to maintain cordial relations
- Prepare HR or Company manual

Administrative Head: The duty of an administrative head is to supervise day-to-day operations of the administrative department and staff members, developing, reviewing, and improving administrative systems, policies, and procedures.

Duties of an Administration Head is to ensure the following work is performed by specific section in-charge(s):

- Personal record keeping of individual work force, office times for daily attendance record of entire work force, preparation of wages and salary and distribution
- Record keeping of leaves/medical/LTA/advances, performance incentives, bonus, etc.

- Cooperative society

- Vehicle and fuel record/parking arrangement

- Maintaining Company's internal and external communication systems

- Monitoring and record keeping of visitors and their welfare

- Coordination with Workers Union

- Overall discipline within the premises of the company

- Securities arrangement for the Company's work force and assets

- Liaisoning with local police and administration authorities and maintain cordial relation

- Coordination between management & workers

- Liaisoning with local police and administration authorities, associations, government agencies, local leaders, social activists, etc. and maintain cordial relation

- Address the company's legal issues / court cases of the company through legal cell

Occupier: The Indian Factories Act, 1948 states that every factory should have an 'occupier' and defines an 'occupier' as the person who has ultimate control over the affairs of the factory. Any director can be the occupier of a factory.

General Duties of an Occupier

- Ensuring health, safety and welfare of all workers while they are at work in the factory

- Provision and maintenance of plant and systems of work in the factory that are safe and without risks to health

- Appointment of Safety Officers and formulating his duties

- Provisions of such information, instruction, training and supervision as are necessary to ensure the health and safety of all workers at work

- Provision, maintenance or monitoring of such working environment in the factory for the workers that is safe, without risks to health and adequate as regards facilities and arrangements for their welfare at work

- Compulsory disclosure of information by the Occupier

- Specific responsibility of the Occupier in relation to Hazardous Process

- Worker's participation in safety management

- Enforcing right of workers to be warned about imminent danger

- Facilities for sitting and canteens

- Annual leave with wage

- Safety and Occupational Health Survey

Factory Manager/Labour Officer A Factory Manager should have the capability to demonstrate outstanding problem-solving skills and resolve any issues that may derail production. Ultimately, a top-performing labour officer should have exceptional communication, management, and leadership skills to ensure that factory operations run smoothly.

According to the Industrial Disputes Act of India, every employer should form a works committee if they have more than 100 employees in the organisation. The works committee members should be comprised of equal number of workmen (employees) and individuals representing employers. The employer should select the employees in consultation with the Union (if already formed in the organisation).

Duties of Factory Manager

- Right of workers to be warned about imminent danger

- Notice of periods of work for adults

- Register of adult workers

- Notice of certain dangerous occurrences to the appropriate authority

- Notice of certain disease to the appropriate authority

- Safety and occupational health survey

- Notice of certain accidents to the appropriate authority

- General procedure as to filing of cases in case of factory accidents

The role and responsibility of a factory manager is to ensure safety of workmen in the factory and provide a safe and hygienic atmosphere at workplace. They also oversee necessary arrangement for safe handling of hazardous operations, health check-ups, canteen management, medical check-up of factory employees, maintain harmonious relations with labour union and labourers, coordinate between workers and management, attend issues related to labour courts, etc. Further, compliance of the company with the Indian Factories Act, 1948 and labour laws is the top most priority of a factory manager.

LABOUR LAWS

Factories Act: The main objective of the Indian Factories Act is to regulate the working conditions in factories - to regulate health, safety, welfare, and annual leave of workmen and includes special provisions in respect of young persons, women, and children who work in factories. Every company has a Labour Officer and Welfare Officer to ensure compliance with such laws. However, it is important that the lower, middle and senior management staff who directly or indirectly interact with workers must also have some basic knowledge about rules and regulations as per the Factories Act related to the following points to avoid any exploitation knowingly or unknowingly for or by them.

- **Working Hours:** According to the provisions, every adult (18 years or more) cannot work for more than 48 hours in a week and 9 hours in a day. A weekly holiday is also prescribed in the Act. There is also a restriction on employment of women for working between 7:00 pm and 6:00 am (which can be relaxed by a Chief Inspector, but not beyond 10:00 pm and 5:00 am).

- **Health:** There is an Act for protecting the health of workers as factory cleaning and all necessary precautions to be taken in this regard, proper drainage system, emergency exit of adequate size in case of emergency, etc.

- **Welfare:** Act for the welfare of the workers regarding adequate and suitable facilities for washing for the use of workers, facilities for storing and drying clothing, facilities for sitting, first-aid appliances and availability of shelters, rest rooms and lunchrooms, creches, etc.

- **Penalties:** The employer or employee nay be liable for penalties If any law or any rules made under the Act, or any order given in writing under the Act is violated by employer and If a worker misuses an appliance related to welfare, safety and health of workers, or in relation to discharge of his duties.

THE NEW LABOUR CODES

In India, since independence, several legislations have been enacted to promote the condition of labour. There have been 44 labour-related statutes promulgated by the Central Government over the years. In 2020, the government proposed the amalgamation of 29 erstwhile labour legislations into four comprehensive labour codes:

- Code on Social Security, 2020,

- Code on Wages, 2019

- Industrial Relation Code, 2020,

- Occupational Safety, Health, and Working Conditions Code, 2020

The Codes deal with, among other things, minimum wages, accidental and social security benefits, occupational safety and health, conditions of employment, disciplinary actions, formation of trade unions, and industrial relations.

As of September' 2022, the provisions of Section 142 of the Code on Social Security, 2020 and the provisions related to the Central Advisory Board as specified under Section 42 and 67 of the Code on Wages, 2019 had come into effect. The President had given his assent to the Central Government for all the four Codes, but the Central Government was yet to announce an 'effective date' for their implementation.

The new Labour Codes are expected to simplify compliance by ensuring single licensing mechanism for industries. Previously, the industries were required to apply for licenses under different laws. The introduction of technology in creating national databases has ensured workforce formalization, which is a primary step toward providing welfare benefits. Further, the scope and ambit of the workmen/employees has been widened to cover all those who were earlier left out from the ambit of labour laws. Uniform definitions for wages are provided such that there is no confusion in its interpretation. The statutory benefits o fer more coverage that includes informal workers. Also, the ease of doing business and faster dispute resolution mechanism is beneficial to the businesses. Usage of technology for efficacious implementation and ease of compliance is a positive step envisaged by the labour codes.

MAJOR POINTS TO REMEMBER BEFORE SETTING A MANUFACTURING PLANT

Products to be Manufactured	Selection of Land
Market Survey / Competitors	Industrial Park
Market size - Domestic and Export	Private / Agriculture Land
Targeted Share / Timing	Government Land
Conduct SWAT Analysis	Developed
Selection of Place	Or Under Development
Geographical Location	Road Conditions
Connectivity – Road/Railway/Air	Plot on the main road or side road / Road Width
Distance from Port	Plot Size
Climatic Conditions	Shape of plot
Availability of Raw Materials	Rate of Land
Availability of Workers	Total Amount for Land
Policy for wages	Lease or Free Hold
Energy	Selection and Recruitment for Key Positions
Availability of Power	Project Report
Voltage	Arrangement of Finance
Frequency	Selection and finalisation of Machines
Power Tariff / Unit	Work out Viability, Pay Back and ROI
Quality of Power Supply	Order for Machines keeping in mind the lead time
If Generator sets are required/Its viability	Building Construction
Availability of Water	Factory Layout / Building Area
Terrif of Water /KL	Admin Block / Area
Quality of Water	Machines lay out

Water Treatment Plant	Water Lines - lay out - Pipes Diameter/ Length
Bore hole allowed or not	Steam Lines - lay out - Pipes Diameter / Length
Availability of Fuels	Electrical Supply - lay out - Cable/Wires Size/length
Gas	Air Supply lay out - Sizes of Pipes
LPFO / Coal	Cost of Construction
Rates of Gas / Fuel	Construction of Reserve Water Tank -Capacity
Energy Cost / Unit of Production	Construction of Reserve Oil / Gas Tank-Capacity
Import/Export Policy	Construction of Roads within Premises - Area
Duty for Machines and Equipment	Accommodation for Staff Members, if necessary
Import Duty for Spare Parts	Entertainment facilities
Import Duty for Raw Materials	Guest House if necessary
Export Duty for Finished Products, if any	Few Houses within Co. Premises for Essential Services
Export Incentives	Local Hotel Tariffs
Govt Industrial Policies / Permissions	Accommodation for Workers
Central and state Government trade policies	Local Transport for Staff and workers
Policy for FDI	
Incorporation of Company -No of Days	
Permission for Expatriates Quotas -Nos	
Resident Permit	
Import/Export Licenses	
GST Registration	
Labour/Factory Acts	
Company Acts	
Permission for Power	
Permission for Water	
Environmental Policy	
Policy for Boiler	
Policy for Weight & Measures	
Other Govt Departments Permission	
Detail of Central and Local Taxes	

'SWAT' ANALYSIS OF A MANUFACTURING COMPANY

A company should conduct a "SWAT" analysis before launching and also at regular intervals later, to understand its strength, weakness, opportunity, and threat to the business.

Examples:

STRENGTH

- Brand awareness across consumer segments
- Wide distribution network with pan-India and international presence
- Wide product portfolio with good quality perception
- Experienced Technical & Product development team
- Integrated manufacturing complex with fully balanced facilities from fiber to finished product

- State of the arts technology and agile flexibility
- Geographical location favouring the business
- Availability of manpower
- Availability of uninterrupted power supply and source of good water
- Favourable climatic conditions

WEAKNESS

- Long working capital cycle
- Aged Equipment (in case of an old company)
- Not perceived as the first choice among the relevant audience
- Unfavourable geographical location of the company (higher freight, connectivity issues, etc.)
- Seasonal market demand

OPPORTUNITY

- Growing disposal income of the working class
- Consumer preference shifting towards branded quality fabrics
- Identity among major brands in the market as competitor
- Export potential to developed countries
- Potential to develop RMG and Institutional Sectors
- Availability of raw material and skilled workforce
- Favourable government policy
- Growing global demand

THREAT

- Vertical expansion of decentralized sectors with competitively priced products
- Dumping of goods by other countries
- Shift in Youngster choice towards the material out of company profile
- Increased input cost
- Inconsistent raw material availability and prices
- Short term trade policy (if any).

2

Types of Management Styles

A management style is the way an enterprise chooses to function in order to achieve their targets. It includes the way it plans, organises, makes decisions, delegates, and manages the workforce.

Management styles may vary widely based on the type of industry, size, culture, and location. An effective management is flexible and adjusts its style in response to different situations while keeping its focus on smooth functioning and achieving targets successfully.

These styles are influenced by both internal and external factors. Internal factors include overall culture of the company, its policies, priorities, employee engagement and staff skill levels. External factors include employment laws, the economy, competitors, suppliers, and consumers.

There are three broad categories of management styles: Autocratic, democratic, and laissez-faire.

A. **Autocratic management style:** In this style of management, employees are expected to follow orders, not question the authority of management, and perform their tasks the same way each time. In this style, managers dictate exactly what they require from their subordinates and punish those who do not comply.

There is a lack of trust or confidence that their employees can achieve their goals without direct and constant supervision.

This system allows quick decision-making and creates clearly defined roles of employees and management expectations. Setting clear goals can allow workers to operate without uncertainty. Productivity may increase, but it may need continuous supervision.

However, this style is prone to increase in dissatisfaction among employees, which leads to higher turnover, resentment and a lack of belongingness. There is no scope for participation and innovative ideas from employees.

Management may at times use their persuasive skills to convince employees for their participation and unilateral decisions for the benefit of the organisation.

However, employees still feel dissatisfied due to their lower position and limited participation in policy matters.

B. **Democratic management style:** With this style, management encourages employees to participate and share their ideas, and give inputs during the decision-making process, but ultimately takes responsibility for making final decisions.

Communication goes both ways, top-down and bottom-up, and team cohesiveness is increased. In this process diverse opinions, skills, and ideas are used for decision making and solving the problems.

It helps in promoting a deeper bond and trust between staff and management. There is more opportunity for learning from the ideas, opinions, and experience of the employees for the progress of the enterprise. This may also help in smooth implementation of major changes in an organisation.

At times management creates an open forum for ideas to discuss extensively before making decisions based on a majority rule. Staff is empowered to take ownership of outcomes, which can lead to increased engagement, innovation, and creativity.

Staff feels trusted, valued and heard by all levels of their management team. They are inspired to put forth their best work, find collaborative solutions to problems, and engage completely with the process. Open communication means workplace conflicts are often solved before real issues arise. Turnover is increased when employees are engaged, and diverse voices often lead to better solutions and outcomes.

However, consulting with several staff members can be labour and time consuming. If there is a lack of transparency and doubt of favouritism among the staff members or management, members do not encourage and give weightage to their suggestions, it may discourage and demotivate employees and adversely affect the overall work atmosphere. Also, sharing of trade secrets and access to sensitive information can be risky for the management.

C. **Laissez-faire management style:** 'Laissez–faire' is a French term, meaning 'Let it be' or 'Allow to do'.

This is directly opposite to the autocratic style of management. Instead of a single leader taking all decisions for the organisation, management makes few decisions and allows staff members to choose appropriate ones for the workplace.

Staff are given liberty towards decision making, problem solving and trusted to do their work without much supervision. Management delegates power and expects the staff to control their workflow and outcomes. Management's presence is to help or assist staff on their requests.

Since staff are given space to address their own issues, it strengthens teamwork. Job satisfaction is increased due to autonomy.

However, if team members are not motivated, highly skilled and experts in their field, they may lose focus and results may suffer without proper direction, leadership, and supervision. Poorly managed conflicts may flare up and breed resentment. Some staff may feel that the management is not contributing enough towards the team's success and become resentful.

MANAGEMENT STYLE IN THE INDIAN TEXTILE INDUSTRY

More than 90% of the textile business is in decentralised sectors. There are a number of small and medium sized, independent spinning, weaving factories, process houses, and apparel manufacturing factories. They are small, medium and large sized in the categories of sole proprietorship, private limited etc. There are also a few big sized fully integrated textile companies but they are more or less a family business.

Despite modernisation, the textile, and apparel industry remains labour oriented. A majority of workers are trained on the job and most lower, and middle staff members are not professionally qualified but accrue knowledge of the jobs themselves by experience at the workplace. In different states, different local cultures, languages and financial conditions influence the functioning of management.

Indian textile companies usually have a flexible approach and adopt a combination of all the above three types of management styles.

The role of a leader in management is of utmost importance. The smooth functioning and progress of an organisation depends on personal skill, expertise, vision, attitude and approach towards problems. *The basic qualities of a successful team leader are to be honest, fair, unbiased, transparent and a good listener. He must never use his power to satisfy his personal ego at the cost of the organisation.*

PYRAMID AND INVERTED PYRAMID SHAPE MANAGEMENT

Traditionally, organisations follow a pyramid structure, with the person in-charge at the top. This pyramid usually has many layers in the middle management, experts in different fields for supervision, and all frontline workers are at the bottom. This structure is usually associated with a large number of employees and the lowest number of leaders. The power and authority are located at the top of the pyramid. They make policy decisions to move the company forward. In some poorly-managed organisations, middle or lower management members not allowed to use their brain or participate in important issues even when they may give their best suggestions.

In most hierarchical organisations, there's one CEO, one COO, President, CFO, CMSO, few Vice Presidents and General Managers collectively managing a large number of frontline workers. A very wide base and a narrow top creates a pyramid. Leadership styles are often influenced by individual personalities as they grow and transform as leaders accrue knowledge and experience.

Problems with the Pyramid Structure: Leadership styles being influenced by individual personalities can be harmful for the business if the person is not capable enough.

There is often a compensation disparity between the top and bottom of the organisation and the justification for this is often questioned. Is the CEO of a company worth multiple times compared to the bottom layer of the organisation even though the workers at the bottom perform allocated jobs and do the hardest work? Their contribution is equally important for efficiently executing allocated jobs.

It also creates an obstacle in upward mobility. There is a long queue at the lower and middle-levels of the organisation, waiting to be promoted into the next tier. As the number of higher positions reduce, only few get the chance to be upgraded. Presumably, many members may be equal in terms of doing a good job, but only a few of them will have a chance to be promoted to the next level. Despite performance-based upliftment processes, there is dissatisfaction among members. At times it may even feel like a chance at winning the lottery from one level to the next. However, in reality there may not be enough quality candidates available for each tier.

Therefore, the question remains why the majority of the corporate compensation should go to few senior management members, when a business is built on the shoulders of many others. How to save middle class, and frontline workers? This is the basic thinking among the middle and lower workforce in an organisation. There is a serious challenge before management gurus to reinvent organisational design for a modern 21st century business to address dissatisfaction and make middle and lower workforces happy.

INVERTING THE PYRAMID

Considering the problems with the pyramid style of management, a question is often raised - what may be the result if a traditional employment pyramid model is reversed? Would an opposite approach be functional? The reversed or inverted employment pyramid model is at the heart and soul of servant leadership and promises many benefits for companies seeking to leverage the progression of success in their business endeavours by creating a company culture of human compassion, understanding, and community.

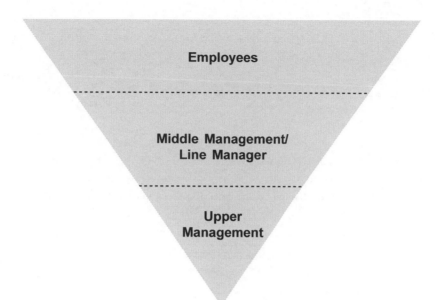

In a lean organisational model, the mentality is very different. The management's role is only to prepare policies and give directions. They are there to support employees in their daily work. Knowing their work and the direction to take, employees are able to propose solutions and improvements to make their jobs as well as the organisation sustainable.

However, it is slightly difficult to manage this kind of a management system compared to a pyramid model. If the management does not involve employees and does not provide clear directions and instructions, it can lead to confusion and can result in a sort of anarchy within the organisation. For successful implementation, sufficient authority is to be given to employees, and management has to support them in achieving their targets.

They must accompany their employees in their empowerment, guide them, and give them a chance to solve their problems themselves, and focus on strategic direction and skill development. This is not an easy task. There arises a need for conducting various training programs to educate employees and a lot of support is needed to implement this model. It demands a change in the work culture at every level of the organisation, which is a time-consuming process.

3

Risk Management

Today, advanced technology plays a vital role within the textile industry. Any manufacturing unit employs five M's - Men, Machine, Material, Management and Money. For organisational success, managers need to focus on synchronising all these factors and developing synergies within and outside an organisational operation. Many textile companies are leveraging technological power to add value to their business.

The twentieth century gave us an abundance of what may be called a 'miracle' in the global apparel and textile industries. Textile scientists did not just imitate products of nature, but improved on them. This success was derived through imaginative and innovative use of low cost feed stocks and efficient manufacturing processes to achieve consumer driven and functional characteristics of comfort and aesthetics.

In the past decade, the textile industry has undergone enormous changes. Both China and India have emerged as global economic powerhouses in the field of apparel & textile after quota elimination. The concept of disruptive innovation is the route for future growth and success. Future product and process discoveries will be built on multi-disciplinary technology platforms. The borders between material, biological and information science erode and become seamless; and the areas where they interact or overlap will catalyse the re-conceptualisation of tomorrow's textiles with disruptive technologies as the hallmark. The world has become a small village and no country can remain unaffected with the ups & down of each other.

While the open global market has created tremendous opportunities for the entire world, it has also increased high risk factors for businesses due to dependency on each other.

There is always an inherent risk attached to any business. A textile business needs to implement a continuous monitoring mechanism to deal with risks on an ongoing basis.

To avoid any unpleasant surprises, a company must lay down risk assessment and minimisation procedures. These procedures must be periodically reviewed to ensure that executive management controls risks through means of a properly defined framework.

Details of various initiatives towards achieving this objective are provided below:

Strategic Planning

Senior management must meet periodically for a detailed strategic and operational review of the business, taking into account existing and prospective business environments.

Communication & Reporting

Members of the core management team must review the implementation of these strategies, and ensure that adequate attempts are made to mitigate the risks perceived.

Actual performances of budgets should be reviewed by the management on a monthly or quarterly basis and should be presented at board meetings. The monthly and quarterly MIS should be designed to ensure timely dissemination of information and risks of non-achievement of business objectives to key management.

Operational Initiatives for Managing Risks

To strengthen internal controls over business processes, and thereby reduce the risks of manual intervention and error, the company should document important policies and procedures and circulate across the organisation such as approval limits (authorisation matrix document specifying the financial powers for every nature of expenses and every executive).

RISK IDENTIFICATION & MITIGATION

Risks can be identified in the following categories:

1. **Business Risks:** These strategic risks may threaten the existence or smooth running of the business in the future.

2. **Operational Risks:** These are risks encountered in the day-to-day running of the business and mainly mitigated by putting strong business processes with internal controls in place. These risks are segregated function-wise.

3. **Financial Risks:** Finance functions including borrowing forex management and hedging, etc are centralized for the company, and hence these risks are for the central finance department.

4. **IT Risks:** Risk of disruption in business due to failure of IT systems within the company.

5. **Legal Risks:** These include legal risks whilst ensuring compliance with applicable laws in a company, but the mitigation procedures are fairly decentralized depending on the compliance requirements.

Following are some insights into the risk management framework for a textile company.

1. AREA OF BUSINESS RISKS AND THEIR MANAGEMENT

I. Raw Material

Area: Availability of raw materials at competitive prices

Risks:

- Volatility in input prices and constraints in availability of raw materials

- Increase in prices of essential materials (Cotton, wool, polyester, acrylic etc.)

- Imported raw materials may be adversely impacted due to decline in bilateral relations, war, trade embargo, etc.

Controls/Procedures:

- The Company should cover its RM requirements for a defined period. This ensures availability as well as price competitiveness of materials.

- Management should continuously monitors events affecting raw material availability and suppliers.

- The division constantly explores alternatives for the existing supplier.

- The division also looks at captive processing units to lower dependence on suppliers.

II. Product Portfolio

Risks:

- Changes in demand or market trends and product concentration or dependence

- Primary markets may no longer accept the existing products due to new fashion trends, low economy, changing age group of major consumers, etc.

- Changes in demand patterns may reduce demand for the company's existing product

- Changes in preferences may affect sale of a particular category of products and hence affect profitability (e.g, shift from woollen to PV & acrylic products, 100% cotton to P/C, 100% polyester, etc. affects margins.)

- At times, a certain product range dominates the company's product range. Significant revenues or profits of the company flow from these products and hence any adverse impact on the sale of few products could impact the business severely.

Controls/Procedures:

- The company must monitor even subtle changes in the industry and customer preferences through market surveys, particularly in the end application population for any indicators that could change or affect demands of the company's product.

- The company may encounter the risk of shift to ready-mades by two prong method of supplying to institutional ready-made manufacturers as well itself having a presence in the ready-made garments industry.

- The company must host dealer meets and increase interactions with agents to gauge customer trends.

- Data captured by the POS (point of sale) of retail stores should be collected and analyzed regularly to determine customer-buying trends. Company must focus of its research & development consistently on topmost priority.

- The company should monitor demand forecasts and threats to products closely. New developments in the product area (like new fibers, finish & fashion developments) are to be closely watched and the company should aim to be the leader in adopting changes to the product range.

- The company should also continuously undertake new product development initiatives or spin-offs from existing products that will reduce the dependency on a small number of products.

III. Customer

Risks

- Credit risks and customer concentration

- Risk of customers not meeting payment obligations

- Just a few customers should not account for the bulk of sales or profits, to avoid over dependency on a small customer base

- These customers could dictate terms of sales and loss of a single customer may adversely affect the business.

Controls/Procedures:

- Strong credit control and monitoring mechanisms must be in place to restrict undue exposure.

- The credit policy should include other mitigating actions like making agents responsible for collections and accepting deposits from customers.

- Further receivables position should be reviewed and closely monitored daily, weekly and in monthly management meetings.

- The company should continuously appoint new dealers and retail shops which achieve the dual purpose of increasing reach as well as broadening the customer base to reduce customer concentration.

- In addition, activities of large customers should be closely watched, at all times avoiding the misconception that due to long association with such customers, risk may be low.

IV. Competition

Risks: Loss of market share to competitors and global competition, e.g. availability of foreign goods

Controls / Procedures:

- The company should focus on increasing the domestic market as well as export market share. Marketing strategies should be continuously aimed towards improving market share by meeting existing and emerging customer demands and providing incentives to customers (rewards, loyalty programs).
- Competitor's moves should be tracked closely for any major developments.
- The current distribution network should be strengthened. The company should also expand into retailing.
- Continuous improvement programs should be initiated to ensure that high product quality standards are maintained. Focus should be on brand building & its recognition.

V. Technology

Risks:

- The present technologies or manufacturing processes may become obsolete under influence of emerging technologies.
- Availability of alternate manufacturing processes to manufacture products at lower costs.
- Upgradation of manufacturing facilities in time.
- Inadequate resources for product development and research in emerging technologies.
- Inability to correctly assess impact of new technology on existing business model (e.g., Internet and–e-Commerce.

Controls / Procedures:

- A company should consciously upgrade its manufacturing facilities to adopt the latest technologies.
- It should also take steps to retain or reduce costs of production by initiatives, which may include shifting the manufacturing location to lower costs area.
- There should be ongoing efforts for development and research of new products and new technologies.
- On the selling front as well, the company should adopt the latest channels of promotions and selling to the extent feasible (e-commerce, online sales,.ctc.),

VI. Economy & Market Risks

Risks:

- Slow economic growth/recession, and inadequate market analysis.
- Reduction in buying power of the end customer base and inability to foresee demand patterns.

Controls/Procedures:

- The company should track economic and buying trends and accordingly adapt product mix to suit current market demands (producing lower value fabrics during recession and high value finer fabrics during growth times).
- The company should spend significant efforts on demand forecasting.
- The company may install common software at all sale points through which information on sales trends is available. This information is used to forecast future demands.
- Other means like inputs from market research agencies, media, customer surveys, sales and marketing teams are also be taken to constantly gauge customer needs and behaviour.
- Regular analysis of the individual components that are built into the product cost and ways to reduce one or many cost components. Simultaneous market analysis to determine optimum setting price.

VII. Regulatory Environment

Risks: Adverse changes in local government policies or regulations leading to adverse impact on business environment.

Controls/Procedures: Government policies are to be closely monitored. The company should continuously make representations to the authorities through trade associations for changes that would favourably affect the industry and the company.

2. AREAS OF OPERATIONAL RISKS AND THEIR MANAGEMENT

I. Human Resources

Risks:

- Employee integrity, industrial relations and possible shortage of skilled work force and risk of high staff turnover
- Compromise of employee integrity could lead to fraud or pilferage
- Industrial unrest could lead to strike or disruptions in activities
- Unavailability of new talent or inability to retain existing talent
- High staff turnover could affect continuity and growth of the business

Controls / Procedures:

- The code of conduct must be communicated and mandated by a company, requiring high standards of integrity and ethical behaviour. The screening process for senior management positions should also include a thorough background check to avoid any such issues.
- A strong and robust internal control framework ensures transparency and early detection of any deviation.
- There must be regular efforts to maintain cordial relations with the unions & management. Company should periodically sign a memorandum of settlement with the worker unions.
- The company must have good HR strategies to attract and retain talent.
- Compensation should not be less than the industry norms and profit-sharing policy should be adopted.

II. Procurement

Risks:

- Non-continuity of supplies, raw material and other input material
- Unreliable suppliers or frequent disruption of supplies could pose risk of disruption in operations.
- Dependence on single source suppliers or service provider leading to business partner risks.

Controls / Procedures:

- Vendor selection process ensures reliability of suppliers.
- There should be continuous exercise to develop alternative vendors for Class B items.
- Alternatives should be developed for imported items.
- For key materials, contracts should be in place with the main vendors.
- Terms with certain suppliers should also include penalties for delays or failure to supply which also acts as a mitigating factor.

III. Logistics & Supply chain

Risks:

- Non-optimal inventories and late delivery of finished goods.

- Risks of overstocking leading to blockage of working capital and obsolescence of material and risk of under-stocking leading to loss of sales. Company's inability to deliver on time to meet peak season demands or delivery commitment.

Controls/Procedures:

- Inventory management controls, supported by IT systems should be in place.

- Targets in terms of number of days, sales or consumption should be defined for all items.

- Inventories at all levels should be monitored regularly, reported and discussed in monthly management meetings.

IV. Production

Risks:

- Risk of rejections, disruptions in resource supply and shop floor hazards.

- High rejection rates for finished goods could result in business and reputation loss.

- Inadequate resources for manufacturing like power, water, fuel, etc. Failure to take adequate safety precautions may lead to accidents in manufacturing facilities

Controls/Procedures:

- All the divisions of the company must follow stringent quality assurance & inspection norms. The final product must be properly inspected & confirmed for its quality before dispatch to customer. Company must recognize quality of its product as its strategic strength and should have controls in manufacturing processes to ensure the continuity of this factor.

- A company plant should be setup considering availability of resources as a key criterion. The continuity of these resources should be monitored regularly.

- A special arrangement may be made to ensure availability of these resources in an uninterrupted manner such as special water pipelines, captive power plants, etc.

- Safety requirements must be specified as per norms.

V. Sales & Marketing

Risks:

- Incorrect pricing and high number of customer claims.

- Over or under pricing the products.

- Large number of customer claims for failure in fulfilling commitments or delivering the right product.

Controls / Procedures:

- Introduction of proper sales policy.

- Continuous feedback from markets.

- Regular follow up with production planning to ensure timely delivery.

VI. Commercial

Risks:

- Loss of benefits and business partner risks.

- Inability or failure to claim available benefits e.g., DEPB, EPC), etc.) for either lack of knowledge or lack of compliance could lead to financial loss.

- Failure of a business partner / counter partner such as vendor, service provider, dealer to meet obligations or service, results in disruptions in business operations.

- Dependence on single service provider for services leading to business partner risks.

Controls/Procedures:

- The company should deal with required suppliers and service providers. Contractual safeguards and SLAs (service level agreement) should be included in the form of contracts with each business partner.

- The company should define selection criteria for service providers. Contracts with service providers should contain penalty clauses in cases of unsatisfactory service.

- Specific SLAs and penalty clauses should be in place for key service providers (e.g., maintenance contracts with software vendors for key business systems).

VII. Othes operational areas

Risks:

- Inadequate internal controls and damages to property & asset.

- Inadequate or weak internal controls across functions.

- Operations (manufacturing facilities, warehouses, corp. offices, etc.) may be adversely impacted due to natural disasters / acts of God.

- Damages due to industrial espionage, sabotage, etc.

- Damages due to carelessness or lack of safety measures

Controls/Procedures:

- The company should appoint internal auditors to regularly audit the status of internal controls.

- Audit reports and actions on recommendations should be reviewed by the Audit Committee of the Board every month/quarter.

- Insurance functions should be in place to review insurance coverage by each division.

- The company must have adequate insurance cover for various heads including acts of God as well as accidents or sabotage.

- Regular training to be provided to workers on safety and precautionary measures.

3. AREAS OF FINANCIAL RISKS AND THEIR MANAGEMENT

Financial risk management is the practice of economic values in a firm by using financial instruments to manage exposure to risk, particularly credit risk and market risk. Other types include foreign exchange, volatility, liquidity, inflation risks, etc. Similar to general risk management, financial risk management requires identifying its sources, measuring it, and plans to address them.

Financial risk management can be qualitative and quantitative. As a specialization of risk management, financial risk management focuses on when and how to hedge using financial instruments to manage costly exposures to risk.

Management should prepare budgets and operating plans for the business and every cost centre (specifying both revenues and capex). The following policies & procedures' manuals must be prepared & followed strictly:

- Capital expenditure
- Purchasing
- Travel policy
- Accounting policies and procedures
- Bill passing manuals
- Retail shops manual
- Insurance policy
- HR policies

A firm of Chartered Accountants, whose appointment is approved by the audit committee, should conduct regular internal audit reviews. The observations and recommendations should be reviewed and discussed with top management and the implementation status should be reviewed regularly. These observations may be presented every quarter to the Audit committee (AC). The AC should also be reviewed by the management to assess the observations and recommendations made by the auditors.

4. AREAS OF IT RISKS AND THEIR MANAGEMENT

IT risk management is the application of risk management principles to the context of information technology in order to manage IT risk. IT risk provides an end-to-end, comprehensive view of all risks related to the use of IT. It can be considered as a component of a wider enterprise risk management system.

It encompasses not only the negative impact of operations and service delivery, which can bring destruction, or reduction in the value of the organization, but also the risk associated with missing opportunities to use technology so as to enable or enhance business with IT project management for aspects like overspending or late delivery with adverse business impact.

5. AREAS OF LEGAL RISKS AND THEIR MANAGEMENT

In today's globalised business environment, companies face a complex assortment of new and often contradictory laws and regulations. High-profile corporate scandals involving compliance failures teach us that loss of reputation can have a significant, if not fatal, effect on a company.

International companies recognise this and invest heavily in systems designed to detect and prevent compliance breaches. However, such systems and controls cannot succeed without the development of a strong compliance culture that secures buy-in from executives, managers, employees, contractors and business partners at all levels.

A company must have experts for innovative expertise and guidance for the development and management of a sophisticated legal risk management and compliance operation. While identifying risks and regulatory challenges, the management should also explore how professionals can manage processes, implement changes, and track issues.

Management should also focus on stopping the loss of potential clients, partners, employees, contractors and implement appropriate remedial measures, form professional board structures, follow corporate governance to stop fraud and bribery in the organisation.

4

Accounting

Accounting is said to be the language of business because it is used in the business world to describe the transactions entered into by all kinds of organisations. It is defined as the act of recording, analysing, and interpreting business transactions in terms of money. The record making phase of accounting is usually called book-keeping. Most businesses have two types of transactions - receiving of benefits and giving of benefits. Accounting is the proper record keeping of both types of transactions.

A company enters into hundreds of transactions every day, while dealing with people outside the company as well as within the company (employees).

There are also transactions between the organisation and suppliers to purchase materials or services, transactions between organisation and customers to sell their goods, and transactions within different sections or departments of the same company. For example, material moving from a raw material godown to the production department or inter-department transactions during production or finished goods being transferred from production floor to the finished goods store, a machine being serviced by maintenance department, HRD department organising training programs, electricity, steam, air-conditioning, air supply related services being supplied by the engineering department to the production departments, etc.

A transaction may also be agreed to orally in case of an individual buying an item from a shop. However, all transactions in a formal business are recorded. The dual nature of a transaction are given specific symbols.

The giver of a benefit is credited: Cr

The receiver of a benefit is debited: Dr

The accountant records the sale invoice amount as a debit or receivable and records the same amount as credit or amount to be received. An account is a page of a book divided into two equal sides

Dr	Cr
Debit Side	Credit Side

The left-hand side is debit and right-hand side is credit. According to the nature of the transaction, a name or code is given to an individual account, e.g., sales account, purchase account, wages and salary account, energy accounts, etc.

Each account has a date, value, and types of transactions. Every transaction includes two accounts, an account, which receives a benefit, is debited and that which gives benefit is credited with the monetary value of the benefit. The result of the application of this system is that for every debit entry, there must be a corresponding credit entry.

ACCOUNTING DOCUMENTATION AND RECORDS

Business transactions when put into formal record keeping are called accounting documents. These documents consist of three main areas:

1. Information about the parties to the transactions

2. The nature, type and the amount involved in the business transactions

3. The areas that show the parties to the transactions, process, the necessary authorities and intended formal business relationships

Good accounting documentation must have the information listed below:

- Name of the company/department issuing the document and initiating the business relationship
- Name of the company/department receiving the document and agreeing to business relationship
- Date of business relationship
- Serial/Reference no. of document for ease of identification
- Appropriate name of the document such as invoice, local purchase order (LPO, receipts, etc.
- Detail of the transactions as to the number of the items, unit cost, quantity and value
- Particulars of person issuing the documents
- Particulars of the person authorising the documents
- Particulars of other party to the transaction

An accounting document is usually the only prima facie evidence of transaction and the design of its documentation varies from company to company. The ultimate aim is that the design lends itself to easy data collection and analysis.

For the purpose of documentation, business activities are divided into cycles namely:

1. *Production Cycle:* For all activities involved in the process from raw material to the finished products. E.g., Store issue note, stock return and transfer note, stock taking sheet, stock ledger cards, stock records (bin cards), etc.

2. *Payment Cycle:* For all those activities involved in the payment of money. E.g., Local and overseas purchase orders (LPO/OPO), Goods received notes (GRN), Supplier invoices, Payment vouchers, Cheques, Debit note/ Credit note, Petty cash vouchers, etc.

3. *Revenue Cycle:* For all activities involved in the receipt of income and money. E.g., Sales Invoices, Waybills, Receipts, Debit notes, Credit notes, Bank tellers, etc.

The documents mentioned above must be handled with utmost care to avoid errors which can at times result in huge financial losses.

PREPARATION OF FINANCIAL STATEMENTS

After proper documentation of business transactions in ledgers, the next step is to balance all the accounts. This is the difference between all the debit entries and credit entries in cash account title.

Balance B/D and Balance C/D: In bookkeeping, Balance B/D and Balance C/D are terms used for balancing and closing of ledger accounts from the current period to the following period.

Balance B/D: is the balance brought down as opening balance of a ledger pulled from the previous accounting period.

Balance C/D: Is the balance carried down as the closing balance of a ledger pushed to the next accounting period.

If Debit side > Credit side, it is called Debit Balance

If Credit side > Debit side, it is called Credit Balance

Example – To Balance C/D and By Balance B/D

To Balance C/D: In a ledger account when Credit side > Debit side the difference in balance is inserted on the debit side to balance the account, the differential amount is denoted as "To Balance C/D".

By Balance B/D: In the following accounting period closing credit balance of previous period (To Balance C/D) is brought down to the credit side of ledger account, this amount is the opening balance of next period and is denoted as "By Balance B/D".

Dr	Cash Account		Q 4 Previous period		Cr	
Date	Particulars	Amount		Date	Particulars	Amount
	To Capital A/C	50000			By Bank A/C	20000
	To Bank A/C	10000			By Purchase A/C	40000
	To Sales A/C	25000			By Salary A/C	30000
	To Balance CD	5000				
Total		90000				90000
	Cash Account		Q 1 Current Period			
Date	Particulars	Amount		Date	Particulars	Amount
	To Capital A/C	50000			By Balance B/D	5000
	To Bank A/C	10000			By Purchase A/C	20000
	To Sales A/C	30000			By Salary A/C	40000
					By Bank A/C	25000
Total		90000				90000

Example – By Balance C/D and To Balance B/D

By Balance C/D: In a ledger account when *Debit side > Credit side the difference in balance is inserted on the credit side* to balance the account, the differential amount is denoted as "By Balance C/D"

To Balance B/D: In the next accounting period closing debit balance of previous period (By Balance C/D) is brought down to the debit side of ledger account, this amount is the opening balance of next period and is denoted as "To Balance B/D".

Dr	Cash Account		Q 4 Previous Period		Cr	
Date	Particulars	Amount		Date	Particulars	Amount
	To Capital A/C	50000			By Bank A/C	20000
	To Bank A/C	10000			By Purchase A/C	40000
	To Sales A/C	30000			By Salary A/C	25000
					By Balance CD	5000
Total		90000				90000
	Cash Account		Q 1 Current Period			
Date	Particulars	Amount		Date	Particulars	Amount
	To Balance B/D	5000			By Bank A/C	20000
	To Capital A/C	50000			By Purchase A/C	40000
	To Bank A/C	10000			By Salary A/C	30000
	To Sales A/C	25000				
Total		90000				90000

Thus, in accounts department similar exercises are carried out on all the accounts at the end of each accounting period, which for most companies is monthly. Therefore, each month there will be new opening balance for each account and this exercise is repeated monthly until the end of the financial year.

It is to be noted that all the information collected in the accounts department did not actually originate there. The account department is only a service department and not an operational department.

BASIS OF PREPARATION OF ACCOUNT

Accounts are prepared on an accrual basis, which is an accounting convention. The accrual basis can be defined as a recognition of the fact that income and expenditure cannot be fully determined by receipt and payment of cash. Income is earned when a third party receives a product or service. Income may be understated when an invoice has not been raised for a service rendered, probably due to local customs or practices of the trade. For example, in professions like accountancy, engineering, law firms, etc. one may be at various stages of the clients' work at the end of an accounting period, this does not mean that you have not earned the fees because the work has not yet been completed.

In all cases, an estimate should be made of the value of the work that has been done but has not been invoiced and this must be included in your income account via a journal entry.

Journal Voucher:

Journal Voucher	Dr	Cr
Customer name/Account	X	
Income account (Sales)		X

The same thing applies to expenditure. Sometimes you enjoy a service for which you have received no invoice e.g., electricity, telephone, and water bill, etc. The cost of these services should be estimated and added to the relevant accounts to have a true picture of the result of your business activities. This is similarly done through a journal voucher entry:

Journal Voucher	Dr	Cr
Expenditure name/Account	X	
Supplier name		X

Therefore, with accrued income, you have debtors in the form of customers and with accrued expenses, you have creditors in the form of suppliers. Accruals are necessarily estimates and when the actual income or expenditure is determined, there may be need for further adjustment to these accounts arising from over or under accruals.

Another entry from the accounting department is that of pre-payment.

Pre-payment: It means payment in advance for goods or services yet to be received / enjoyed. It is the opposite of accruals. At times when demand for a product is higher, customers deposit an advance amount to ensure future allocation of the required goods. This advance payment is also called Deposit. The important thing to recognise is that income has not yet been earned and if by mistake it is included in the income, the accountant at the end of each period will separate it via a journal voucher entry. This will ensure that income is not overstated.

Journal Voucher	Dr	Cr
Income (Sales)/Account	X	
Customers Deposit		X

On the other side of the income, one also may pay in advance for services not yet received. The most significant are two to three years advance payments for few services as per the practice of that trade.

E.g., rent for company houses, real estate, advance payment for machineries, etc.

The original transaction is recorded as:

Debit - Rent Account

Credit - Bank Account

It will not be correct to charge the entire advance amount to one single year as the service may be enjoyed through other years too. So, via a journal voucher entry, the amount for the portion which has not been enjoyed will be taken out of the rent account into another account called pre-paid account which is also similar to a deposit and included amongst the company's assets.

Journal Voucher	Dr	Cr
Prepaid rent account	X	
Rent account		X

This adjustment will make the financial statement to be more realistic.

Depreciation: The account department also addresses the issue of use of assets. With each day's use, there is wear and tear which cannot be compensated by repair and maintenance but ultimately by replacement of the assets. Past experience, actual use or the machine manufacturer's recommendation could disclose the number of years for which assets can be used.

The assets will thus be written off, meaning charged to the income of each period for the use of the assets during that period. Even though no money is paid out of these uses, it is provided for its eventual replacement and this provision is kept in an account called "Provision for depreciation account' and actual usage is called 'Depreciation' and kept under this head.

Depending on the internal policy of the individual company, some tentative rates for depreciation used for various assets are:

Land: 0 %

Factory Building: 2-5 %

Plant and Machinery: 4-10 %

Furniture and fixture   10-12 %

Motor vehicle: 25-50 %

So, the term depreciation refers to allocation of cost of a tangible asset to expense to the periods in which the asset is expected to be used to obtain monetary benefit. For example, a company purchases a machine for Rs 100,000 and it is estimated that it will be used for a period of 10 years. The cost of the machine (i.e., Rs. 100,000) will be allocated to each of the 10 years, using some systematic and rational allocation method.

The cost of an asset is initially recorded as an asset in accounting records because it is supposed to be used for several years in the future. The portion of cost is allocated to the period of its expected life, removed from the total cost of the asset and becomes the expense of that particular period and is matched against revenue like any other expense.

There are different methods of calculating depreciation in the industry: Any of the following depreciation policies may be chosen by an organisation depending on the circumstances:

- Straight line method
- Depletion method
- Revaluation method
- Declining balance method
- Sum of years digit method
- Activity method

Straight-line method of depreciation: In this method of depreciation, cost is allotted to each period of the asset's useful life. This method assumes that depreciation is a function of the passage of time rather than the actual productive use of the asset. Under this method, the depreciation expense for a period is calculated by dividing the depreciable cost of the asset by the years of its useful life. Depreciable cost is arrived at by deducting salvage or residual value from the original cost of the asset. For example, if an asset is purchased for Rs 100,000 and its salvage value is expected to be Rs 50,000 at the end of its useful life, the depreciable cost would be calculated as follows:

Depreciable cost = Rs 100,000 – Rs 50,000 = Rs 50,000

Formula:

The following formula is used to calculate depreciation under straight-line method:

Depreciation Expenses = (Cost – Salvage Value) / No. of Years of Useful life

Revaluation method of depreciation: This is one of the easiest ways of calculating depreciation on fixed assets. The value of the asset on which depreciation charge is to be calculated is assessed both at the start and at the end of the year and any revaluation losses arising during the year are considered as the depreciation charge. The valuation is usually performed by either an internal competent member or an external professional valuer.

Revaluation method of depreciation is mostly used for the calculation of depreciation of trivial, inexpensive, and small fixed assets that are normally accounted as a collective unit. This method makes the calculations easier, because it becomes complicated to assess depreciation of each such asset separately.

Formula:

Depreciation expense = Value of asset at the start of the year + Additions during the year – Deductions during the year – Value of asset at the end of the Year

ACCOUNTING CONCEPTS

Some fundamental accounting concepts are described below:

1. **Entity:** Every economic unit, regardless of its legal form of existence is treated as a separate entity (in accounting) from parties having proprietary or economic interest, in it.

2. **Going Concern:** The assumption is that the business unit will operate in perpetuity and is not expected to be liquidated in the near future. A business is considered a going concern if it is capable of earning a reasonable net income and there is no intention or threat from any source to curtail significantly its line of business in the near future.

3. **Periodicity:** Although the result of a business unit cannot be determined with precision until its final liquidation, the business community and users of financial statements require that the business be divided into accounting periods (usually one year) and that changes in position are measured over these periods.

4. **Monetary Unit:** Accounting simply means the recording of financial transactions in monetary terms . Money serves as a standard for deferred payment and measure of value. Therefore, all transactions made on credit will be recorded in monetary form. The use of money is very important in accounting as recording of figure may require addition and subtraction.

5. **Matching Concept:** One of the most important purposes of financial accounting is to calculate income resulting from transactions. This means that quantifying the date of the transaction and setting off against those days the expenses related to that transaction. Therefore, these concepts help the accountant to prepare income statement by matching revenue with the expenses, so the revenue of the previous year and the current period will be taken into consideration.

BASIC CONVENTIONS IN ACCOUNTING

Following are some basic protocols to be followed in accounting:

1. **Fairness:** In view of the fact that there are many users of accounting information, all having different needs, the fairness principle requires that accounting reports should be prepared not to favour any group or segment of society.

2. **Prudence:** This convention demands that great care should be exercised in the recognition of profit whilst all known losses are adequately provided for.

3. **Conservatism:** The convention of conservatism states that while the accountant is contemplating on which figure to choose for a given item, he will select the one, which will understate the profit, rather than the one, which will overstate it.

4. **Materiality:** The treatment of an item as being material or immaterial varies from one company to another. Some firms may fix an item under Rs 5,000 as being material, which may be immaterial to another firm – depending on the size or capital of the firm. Therefore, any material items are capitalised and called capital expenditure, while immaterial items are recorded into expenses account.

5. **Industry Practice:** Based on the way and manner the industries practice, their policies vary. Therefore, one must go by such practice in a particular industry to which one belongs and stick to the principle governing that industry.

COMMON TERMS USED IN BALANCE SHEETS

1. **Debit:** A debit is an expense, or amount of money paid from an account, that results in the increase of an asset or a decrease in a liability or owner equity on the balance sheet. Debit balance refers to the balance that remains after one or a series of book-keeping entries. This amount represents an asset or an expense of the entity.

2. **Credit:** Credit is generally defined as a contractual agreement in which a borrower receives something of value now and agrees to repay the lender later as a payable.

   For a general ledger to be balanced, credits and debits must be equal. **Debits** increase asset, expense, and dividend accounts, while **credits** decrease them. **Credits** increase liability, revenue, and equity accounts, while **debits** decrease them.

   A debit entry in an account represents a transfer of value to that account, and a credit entry represents a transfer from the account.

   Example of Debit and Credit: Company "A" sells 10 Items to a customer for Rs 1 lac. The bill is paid immediately, in cash. Company "A" deposits the money directly into his company's business account. Now it is time to update the company's online accounting information.

   Company "A" goes online. For this transaction, he records a debit to his cash account (under "Assets") of Rs 1 lac. His sales (under "Revenue") are credited Rs 1 lac.

   A few weeks later company "A" takes out a loan of Rs 3 Lacs for some upgrades to his store. He will then debit his loans payable account (under "Liabilities") for Rs 3 lacs and credit his cash account (under "Assets") for the same amount.

3. **Assets:** These are economic resources or valuable possessions owned by the firm, capable of being measured in monetary terms and from which future benefits accrue to the firm. Assets are classified into:

   (a) Fixed Assets

   (b) Current assets

   Fixed Assets: These can be defined as machineries and equipment necessary to carry out jobs that a company may be engaged in. They are permanent in nature with useful life of more than an accounting year. There may be tangible fixed assets or intangible fixed assets.

   - Tangible fixed assets:
   - Land and building
   - Plant and machineries
   - Office and business equipment
   - Office furniture and fitting
   - Motor vehicles

The cost of tangible assets is allocated over their useful lives. The amount so allocated each year is called depreciation.

Intangible fixed assets: Represents the firm s rights and intellectual property and includes patents, copyrights, franchises, trademarks, trade name, trade dress, trade secrets, goodwill, etc.

Current assets: These are resources of the firm which are either held in the form of cash or expected to be converted into cash within the accounting period or the operating cycle of the firm, whichever is greater. These can be defined as assets the company trades in. They keep changing and are not permanent in nature.

Examples:

- Stocks:
  - Raw material stock
  - Finished goods stock
  - Packaging material stock
  - Engineering stock
  - Stationary stock
- Work in progress (WIP)
- Debtors (receivable) and prepayments or advances
- Cash in bank
- Cash in hand

4. **Liabilities:** These are debts payable in the future by the firm to creditors. They represent economic obligations to pay cash or provide goods or services within fixed periods.

Sometimes, the capital of the shareholders may not be sufficient to procure required assets and borrow from individuals and or from financial Institutions. Any type of borrowing is referred to as a liability because it will have to be paid back. A liability could be of a long or short term in nature.

Like assets, liabilities can be classified into current liabilities and long-term liabilities.

Current Liabilities: These are debts payable within an accounting period e.g., amount owed to trade creditors, amount owed to associated companies, bank over drafts, etc.

Long-term Liabilities: These are obligations or debts payable in a period greater than the accounting period. Loans which are payable in long terms, debentures and preference shares are some examples of long-term liabilities.

Debentures: At times, the money raised through the issue of shares is not inadequate to meet the growing financial requirements of business. Hence, to meet financial requirements, a company resorts to short term or long-term borrowings. Short-term borrowings are overdraft, bills payable etc. Long term borrowings may be debentures, bonds, term loan from banks, public deposits etc. Issue of debentures is very common source to raise borrowings. The word 'debenture' is used to signify – 'A written acknowledgement of a debt by a company under its seal, and generally containing a provision as to payment of interest and repayment of principal.' Debentures carry interest at a certain percent. As it is a loan taken by the company, it is repaid after a specified period or at the option of the company as per terms of the issue. The company may or may not give a charge on the assets of the company as security for the loan. Therefore, the biggest advantages of debentures is that the company may get its required funds without diluting equity. Since debentures are a form of debt, the equity of the company remains unchanged.

Preference Shares: Preference share is a kind of participation in the ownership of a company which also secures a return of the amount invested regardless of whether the company makes a profit or loss. Ownership of preference shares is by a certificate and inclusion in a register of owners of such shares.

- Other types of long-term borrowings are stocks and bonds.

- Business Equation - In the event of borrowing to augment the capital of a company the business equation changes to:

$$Capital\ +\ Liability\ =\ Assets$$

5. **Top Line and Bottom Line**

   The P&L statement provides the top and bottom line for a company. It begins with an entry for revenue, known as the top line and subtracts the costs of doing business, including the cost of goods sold, operating expenses, tax expenses, interest expenses, and any other expenses sometimes referred to as "extraordinary" or "one-time" expenses. The difference, known as the **bottom line** is net income, also referred to as profit or earning.

6. **Revenue:** Revenue is used to describe the increase in business capital that occurs whenever there is sale of goods or discharge of services and this is equal monetarily to the amount of cash and receivables obtained in compensation for the goods delivered or service rendered.

7. **Expenses:** These are amounts spent to generate income (revenues). In essence, they represent a decrease in business capital (cash) caused by the revenue producing operation of a business. Expenses are not incurred solely for its purpose but rather as a set of payments for an item or the cost of merchandise, which will be resold rather than being consumed, the sale of the merchandise in turn, brings about the generation of revenue.

8. **Gross Profit** This represents the total income accruing to the business from all sources after deducting the cost directly incurred to generate this income.

9. **Net Profit:** This represents the balance of gross profit after deducting all non-directly traceable expenses incurred by the business and netted off.

10. **EBITDA:** Earnings before interest, taxes, depreciation, and amortization. It measures a company's profitability and expresses the net income before deductions of above elements that are often considered irrelevant in the decision-making process. Adding these expenses back into net income allows us to analyse and compare the true operating cash flows of the businesses.

$$EBITDA\ =\ net\ profit\ +\ interest\ +\ taxes\ +\ depreciation\ and\ amortization$$

11. **PBDT:** Profit before depreciation and taxes

12. **PBT:** Profit before taxes

13. **Earnings:** It is the net profit or simply net income. This is the bottom-line profit for the company mentioned at the bottom of the income statement.

14. **Interest:** Depends on the financial structure of a company. It depends on the amount of funds a company has borrowed to support its business activities. Different companies have different capital structure resulting in different interest expenses. Hence, it is easier to compare the relative performance of companies by adding interest amounts.

15. **Taxes:** Tax expense depends on the type of business and changes from year to year. This often depends on the industry, location, and size of the company. They are a function of tax rules, which are not really a part of assessing a management team's performance and, thus, many financial analysts prefer to add them back when comparing businesses.

16. **Break-Even Point (BEP):** The break-even point is the production level at which total revenues generated for a product equal total expense. When a business reaches the break-even point, the total sales equal the total expenses. At the break-even point, a business does not profit or generate a loss. Break-even analysis indicates the no. of units of a product that must be sold to cover the fixed and variable costs of production.

In accounting, the break-even point formula is determined by dividing the total fixed costs associated with production by the revenue per individual unit minus the variable costs per unit. It is important to analyse the break-even point figures of the business to make action plan to increase sales prices, reduce expenses, or both to cross the break-even point and make profit

$$Break\text{-}even\ point\ in\ units = Fixed\ Cost\ /\ (sales\ price\ per\ unit - variable\ cost\ per\ unit)$$

17. **Fixed Costs:** Cost that do not change in relation to production volume such as depreciation, insurance, advertisement, etc.

18. **Sales Price per Unit:** This is the amount, a company charges to consumers for one unit of products.

19. **Variable Costs per Unit:** Variable costs are costs directly proportional to the volume of production. Variable costs often fluctuate, and are typically a company's largest expense.

20. **Depreciation and Amortization:**

    Depreciation and amortization depend on the historical investments a company has made and not on the current operating performance of the business. Depreciation and amortisation refers to any investment that loses its value over a period of time. When it comes to depreciation, this means tangible fixed assets like machines and equipment, vehicles, or a building. For amortisation, it means intangible assets like patents.

    Both are non-cash expenses, showing the deteriorating value of these assets. In the case of a vehicle, for example, a business owner may expect it to be on the road for five years. So, every month until the end of its useful life, the resale value on that vehicle will go down. The same goes for intangible assets. While deterioration of a patent is not visible but its value deteriorates on its expiry.

    Determining depreciation expenses and amortisation expenses is subjective. Tentative workable life of any machines or equipment is decided at the time of its procurement and their salvage value is determined when they are not viable. Because of their subjective nature, they are also removed from consideration in the EBITDA metric.

21. **Equity:**

    Equity is the amount of capital invested or owned by the owner of a company. The equity is evaluated by the difference between liabilities and assets recorded on the balance sheet of a company. The worthiness of equity is based on the present share price, or a value regulated by the valuation professionals or investors. This account is also known as owners or stockholders or shareholders equity.

    Equity Formula: The accounting equation is

$$Assets - Liabilities = Equity$$

22. **AR and AP Outstanding:** Generally, **Accounts Receivable** (AR) is the amount of money owed to the company by buyers for goods and services rendered. The receivables should not be confused with **Accounts Payable** (AP). While AP is the debt a company owes to its suppliers or vendors, AR is the debt of the buyers to the company.

PROFIT AND LOSS ACCOUNT

The profit and loss (P&L statement is a financial statement that summarizes the revenues, costs, and expenses incurred during a specified period, usually a fiscal quarter or year. The P&L statement is synonymous with the income statement. These records provide information about a company's ability or inability to generate profit by increasing revenue, reducing costs, or both. Some refer to the P&L statement as a statement of profit and loss, income statement, statement of operations, statement of financial results or income, earnings statement or expense statement.

The profit and loss account presents the summary of revenues, expenses, and net income of a firm for a specific period. Thus, it serves as a measure of profitability. P&L management refers to how a company handles its P&L statement through revenue and cost management.

It is important to compare P&L statements from different accounting periods, as the changes in revenues, operating costs, R&D spending, and net earnings over a time are more meaningful than the numbers themselves.

Together with the balance sheet and cash flow statement, the P&L statement provides an in-depth look at a company's financial performance.

THE BALANCE SHEET

The balance sheet is a statement of the financial status f a firm at a particular time. It is a summary of the effect of all transactions, on a firm s financial statement - income statement, and fund flow statements, which are two derivatives of various accounts, maintained within the accounting system of a firm. In accounting language, the balance sheet communicates information about the assets, liabilities, and shareholders' equity on a specific date. It provides a basis for computing rates of return and evaluating the company's capital structure. This financial statement provides a snapshot of what a company owns and owes, as well as the amount invested by shareholders.

The balance sheet shows a company's resources or assets, and how those assets are financed, whether through debt under liabilities or by issuing equity as shown in shareholder equity. It provides both investors and creditors with a report of how effectively a company's management uses its resources. Just like the other financial statements, the balance sheet is used to conduct financial analysis and to calculate financial ratios.

TRIAL BALANCE vs BALANCE SHEET

It is important to note that a trial balance is different from the balance sheet. This is an internal report that stays in the accounting department. The balance sheet, on the other hand, is a financial statement distributed to other departments, investors, and lenders.

The trial balance provides financial information at the account level, such as general ledger accounts, and is therefore more granular. Eventually, the information in the trial balance is used to prepare the financial statements for the period.

In contrast, the balance sheet aggregates multiple accounts, summing up the number of assets, liabilities and shareholder equity in the accounting records at a specific time. The balance sheet includes outstanding expenses, accrued income, and the value of the closing stock, whereas the trial balance does not.

RELATIONSHIP BETWEEN PROFIT AND LOSS ACCOUNT AND BALANCE SHEET

Although the balance sheet and the P&L statement contain some of the same financial information including revenues, expenses and profits - there are important differences between them. Main difference is that the balance sheet reports the assets, liabilities, and shareholder equity at a specific point in time, while a P&L statement summarizes a company's revenues, costs, and expenses during a specific period.

The P&L statement requires accountants to add up the company's revenue on one portion and add up all of its expenses on another. The total amount of expenses is subtracted from the total revenue, resulting in a profit or loss. The balance sheet has a few different calculations that are all performed as representations of one basic formula:

$$Assets = Liabilities + Owner's\ Equity$$

Example: A private company is formed by three members who have contributed Rs 1 lac as their contribution. These three members A, B and C are now promoters or co-founders of the company and money now belongs to the company, say ABC Private Ltd. and it will not be returned to them unless the company is wound up. The company formed acknowledges this indebtedness to its members by keeping these in an account in a bank, called share capital account record.

Thus, there is a primary equation of a company i.e., Assets = Capital

The money now in the possession of the company is called an asset and since it was contributed by company's promoters, it is acknowledged as capital. Thus, the primary liability (indebtedness) of a company is to the members who have contributed money to kickstart it.

Equity Shares: Equity Shares are the main source of raising funds for a company. All equity shareholders are jointly owners of the company, and they have the authority to control the affairs of the business. It is a form of partial or part ownership in the company in which shareholders bear the highest business risk. Ownership in the company depends on the %age of shares they hold. Equity shares are also called **ordinary shares**. A share represents a unit of ownership in a company and has an exchangeable value that is influenced by market forces.

Preference Shares: Preference shares, as the name suggests are the shares in which shareholders share in the profits of the company in the form of dividends before equity shareholders at a fixed dividend rate. Money or capital that a company raises through the issue of preference shares is called preference share capital. Preference shareholders do not have the authority to control the affairs of the company. In case of insolvency issues, preference shareholders are paid first from company assets.

PRACTICE EXCERSISE

A & B Limited is a manufacturing company with an authorised share capital of 100,000 Rs 1.00 Ordinary Shares and 60,000 Rs 1.00 Preference Shares

Trial Balance as of 31st March, 201X was as follows:

Company - A & B Limited		
	(Amount, Rs 000)	(Amount, Rs 000)
Issued Share Capital		
80,000 Rs 1.00 Ordinary Shares		80000
40,000 Rs 1.00 Preference Shares		40000
Profit and Loss account balance b/f		1020
Stock: (Opening)		
Raw Materials	5226	
Work in progress	10117	
Finished goods	6121	
Factory rent	1417	
Factory lighting and power	3820	
Factory wages	6808	
Administrative Expenses	4752	
Raw materials purchases	24592	
Wages	26261	
Sales		90814
Plant and Machinery (at cost)	60000	
Plant and Machinery		
Depreciation provision		17540
Fixtures (at Cost)	7500	
Fixtures depreciation provision		2500
Premises (at Cost)	76800	

Debtors	42000	
Cash at bank	10000	
Creditors		51620
Bad debts Provision		1920
	285414	285414

The following adjustments are required:

1. Amount due for factory power Rs 285

2. Factory rent prepaid Rs 142

3. Depreciation provision for the year to be 10 % on the reducing balance method for plant and machinery and fixtures

4. Provision for doubtful debts to be 5 per cent of total debtors.

5. Stock as of 31st March 201X:

> Raw Materials   Rs 8,134
>
> Work in progress   Rs 7,115
>
> Finished goods   Rs 6,002

Exercise:

1. To prepare manufacturing, trading and profit and loss account for the year

2. To prepare Balance Sheet as of 31st March 201X

Answer for Exercise No. 1

A & B Limited			
BALANCE SHEET AS OF 31ST MARCH 201X			
FIXED ASSETS	**COST**	**Accumulated**	**NBV**
		Depreciation	
	Rs 000	Rs 000	Rs 000
Land	76800		76800
Plant & Machinery	60000	21786	38214
Fixtures	7500	3000	4500
	144300	24786	119514
CURRENT ASSETS			
Stock-Raw Materials	8134		
Work-in-progress	7115		
Finished goods	6002	21251	
Debtors (42000 - 2100)		39900	
Prepaid Rent		142	
Cash at Bank		10000	
		71293	
Less: **CURRENT LIABILITIES**			
Creditors		51620	
Accrued Factory power		285	
		51905	

NET CURRENT ASSETS (71293-51905)			19388
TOTAL ASSETS (119514+19388)			138902
FINANCED BY:			
ISSUED SHARE CAPITAL			
80,000, Rs 1.00 Ordinary Shares			80000
40,000, Rs 1.00 Preference Shares			40000
Profit & Loss Account b/f		1020	
Profit & Loss Account			
Current year		17882	18902
			138902

Answer for Exercise No. 2

A & B LIMITED		
PROFIT AND LOSS ACCOUNT		
FOR THE YEAR ENDED 31ST MARCH, 201X		
	Rs 000	Rs 000
Raw materials (Opening)	5226	
Add Purchases	24592	
	29818	
Less Raw Material (Closing)	8134	
	21684	
Add Factory Overheads		
Factory wages	6808	
Rent (1417-142)	1275	
Lighting & power (3820 + 285)	4105	
Depreciation - Plant & Machinery	4246	
	38118	
Add Work-in-progress (Opening)	10117	
	48235	
Less Work-in-progress (Closing)	7115	
Cost of goods manufactured		
Transferred to Trading Account	41120	
Sales		90814
Less Cost of goods sold	6121	
Finished goods (Opening)	41120	
Cost of goods manufactured	47241	
Less Finished goods (Closing)	6002	
GROSS PROFIT		41239
		49575
Less Expenses		
Administrative Expenses	4752	
Wages	26261	
Depreciation - Fixtures	500	

Provision for doubtful		
debts (increase) 2100-1920	180	31693
NET PROFIT		17882

ANALYSIS OF FINANCIAL STATEMENTS

Financial statements are prepared primarily for ease in the decision-making process.

There is a basic limitation in traditional financial statements comprising the balance sheet and profit and loss account, as they do not give all the information related to the financial operation of a firm. Nevertheless, they provide some extremely useful information to the extent that the balance sheet mirrors the financial position on a particular date in terms of the structure of assets, liability, owner's equity and so on. In addition, the profit and loss account shows the result of operations during a certain period in terms of revenues obtained and cost incurred during the month/ quarter or year. Thus, the financial statements provide a summarised view of the financial position and operations of a firm

Bank, shareholders, potential investors, government agencies, etc., also require special information necessary for specific purposes to understand more about a company from a careful examination and analysis of its financial statements for the use of decision-making.

FINANCIAL ANALYSIS

Financial analysis is a systematic examination and interpretation of available information to assess a firm s past performance and its present conditions and its future viability. The analysis attempts to pin-point unique characteristics of a firm that determine its success or failure. It presents a clear picture of the management's control in operating the business by examining certain aspects of the business as follows:

Financial Structure: The assets the company maintains and the liabilities it has incurred to acquire and keep these assets;

Operating Cycle: Lead time for converting raw material to the product for the market;

Trend: The direction in which different aspects of its operations are heading; and

Flexibility: Its capacity to deal with both planned and unexpected changes.

METHOD OF INTERPRETING FINANCIAL STATEMENTS

Practically all interpretations involve comparisons with previous periods and budgets for the same company and comparison with corresponding figures for other companies in the same trade. It is necessary to use ratios or percentages rather than absolute figures so that companies of different sizes can be meaningfully and effectively compared.

It is obvious that considerable skill and judgement is required to appreciate and unveil the realities underlying a set of final accounts but basically the financial analyst has at his disposal three main techniques:

- Ratio Analysis
- Straight forward criticism
- Movement of fund statements

Ratio Analysis: Accounting ratio is an interpretation technique that facilitates comparison of significant figure by expressing their relationship in the form of ratios or percentages thereby bringing into focus, salient features in a given financial statement. Ratios are useful for the higher management which is responsible for increasing profits and planning for future prospects of the company.

While accounting, ratios employed to denote past trends may give an indication as to future trends and thus act as signpost for plans and policies. The ability to make plans and take action, which will produce desired changes in a certain ratio, will depend on a full appreciation of those factors that can cause the ratio itself to change. On this basis, a target ratio, if properly understood and used, may therefore be computed as concomitant to budgetary control.

Financial statement ratio analysis focuses on three key aspects of a business: liquidity, profitability, and solvency.

RATIOS

Basically, there are many classifications of ratios, however different classifications exist for different purpose.

I. LIQUIDITY RATIOS

Liquidity ratios measure the ability of a company to repay its short term debts and meet unexpected cash needs.

A. Current Ratio:

The current ratio is also called the working capital ratio, as working capital is the difference between current assets and current liabilities. This ratio measures the ability of a company to pay its current obligations using current assets. The current ratio is calculated by dividing current assets by current liabilities.

Current ratio = Current Assets / Current liabilities

Example:

Particulars	Firm A	Firm B
Current Assets	Rs 180,000	Rs 30,000
Current Liabilities	Rs 120,000	Rs 10,000
	= 3:2 (1.5:1)	3:1

The higher the current ratio, the higher is the rupees available per rupee of current liability and more is the firm s ability to meet current obligations and greater is the safety of funds of short-term creditors.

In the above example, current ratio of Firm A is 1.5:1 which means for every one rupee of current liability, current ratio of 1.5 rupees is available. It indicates that current assets are one and half times the current liabilities.

The current ratio of 3:1 in case of Firm B indicates that current assets are three times than its short-term liabilities.

From the above, liquidity position of Firm B is better than Firm A.

It is also important to note that a very high ratio of current assets to current liabilities may also be indicative of slack management practices. The higher ratio may be a result of excessive inventory levels to meet current requirements, poor credit management i.e. extended days of amount receivable and the company may not be fully utilising its current borrowing capacity.

Different industries have different levels of expected liquidity. Whether the ratio is considered adequate coverage depends on the type of business, the components of its current assets, and the ability of the company to generate cash from its receivables and by selling inventory. In textile field also, it varies between independent spinning, weaving, processing, knitting and garmenting plants and a fully integrated composite textile company due to difference in cycle time, trade sales policy (payment terms) capital and working capital involved, etc.

B. Acid-Test Ratio:

The acid-test ratio is also called the quick asset ratio. Quick assets are defined as cash, marketable (or short term) securities, and accounts receivable and net of the allowances for doubtful accounts. These assets are quickly available liquid (easy to obtain cash from the assets) and therefore, available for immediate use to pay obligations. The acid test ratio is calculated by dividing quick assets by current liabilities.

Acid Test Ratio = Quick Assets (Current assets less Stock) / Current Liabilities

Example:

A-Cash	Rs 4000.00
B-Debtors	4000.00
C-Inventory	24000.00
D-Total Current assets (A+B+C)	32000.00
E-Total current liabilities (actual)	16000.00
1. Current Ratio (D/E)	32000/16000 = 2:1
2. Acid Test Ratio (D-C) / E	(32000-24000) = 8000 / 16000 = 0.5:1

In the above example, current ratio is 2:1 which can be considered satisfactory but the acid test ratio is not and it needs improvement. Acid test ratio of 0.5:1 is less than the norms and cannot be considered satisfactory. It means the firm is not able to meet with requirements of clearing its current claims immediately. It indicates that large parts of current assets of the firm are blocked in slow moving or unusable/dead inventories, it's paying its debts slow and finding it difficult to clear its current liabilities.

The traditional rule of thumb for this ratio has been 1:1. Anything below this level requires further analysis of receivables to understand how often the company turns them into cash. It may also indicate the company needs to establish a line of credit with a financial institution to ensure the company has access to cash when it needs to pay its obligations.

II. EFFICIENCY RATIOS

Efficiency ratios, also known as activity financial ratios, are used to measure how well a company is utilizing its assets and resources.

A. Receivables Turnover Ratio or Debtors Turnover Ratio

The receivable turnover ratio calculates the number of times in an operating cycle (normally one year) the company collects its receivable balance. It is calculated by dividing net credit sales by the average net receivables. Net credit sales is net sales less cash sales and return goods, if any. If cash sales are unknown, use net sales. Average net receivables are usually the balance of net receivables at the beginning of the year plus the balance of net receivables at the end of the year divided by two. If the company is cyclical, an average is calculated on a reasonable basis for the company's operations such as monthly or quarterly.

Receivables Turnover Ratio = Net Credit Sales / Average Net Receivables

Example:

Credit sale during the year = Rs 480,000

Outstanding amount of Debtors at the beginning of the year = Rs. 55000

Outstanding amount of debtors at the end of the year = Rs. 65000

Average Debtors = (55000 + 65000)/2 = Rs. 60000

Receivable turnover ratio = 480000 / 60000 = 8 times /year

Debtors or receivable Collection Period = 12 months / 8 = 1.5 months

Average collection period: The average collection period is the amount of time it takes for a business to receive payments owed by its clients in terms of account receivable (AR). Companies calculate the average collection period to make sure they have enough cash on hand to meet their financial obligations. The average collection period (also known as day›s sales outstanding) is a variation of receivables turnover. It calculates the number of days it will take to collect the average receivables balance. It is often used to evaluate the effectiveness of a company's credit and collection policies. A rule of thumb is the average collection period should not be significantly greater than a company's credit term period. The average collection period is calculated by dividing 365 by the receivable's turnover ratio.

Average collection period = 365 Days/ Receivables turnover ratio

Example:

Let's say a company has an average accounts receivable balance for the year of Rs 10, 0000. The total net sales the company recorded during this period was 12,00000. So, to calculate the average collection period, we use the following formula:

$$(10, 0000 \div 12,0\ 0000) \times 365) = 30.41 \text{ Days}$$

The average collection period, therefore, would be 30.41 days

B. **Asset Turnover:**

The asset turnover ratio measures how efficiently a company is using its assets. The turnover value varies by industry. It is calculated by dividing net sales by average total assets.

Assets Turnover = Net Sales / Average Total Assets

C. **Inventory Turnover ratio**

It measures the number of times the company sells its inventory during the period (one year). It is calculated by dividing the cost of goods sold by average inventory. Average inventory is calculated by adding beginning inventory and ending inventory and dividing by 2. If the company is cyclical, an average calculated on a reasonable basis for the company's operations should be used such as monthly or quarterly. Cost of goods sold means sales minus gross profit

Inventory turnover ratio = Cost of Goods Sold / Average Inventory

The ratio indicates how fast inventory is sold. A high ratio is good from the viewpoint of liquidity and vice-versa. The low ratio means inventory is not selling fast and laying in the warehouse for a longer time.

Example:

Sale of goods − Rs 300,000	
Gross profit margin	= 20% (60,000)
Inventory at the beginning	= Rs 35,000
Inventory at the end of the year	= Rs 45,000
Average Inventory	= (35,000+ 45,000)/2 = 40,000
Inventory turnover Ratio	= (300,000- 60,000) / 40000 = 6 Times
Inventory holding Period	= 12 months / Inventory Turnover Ratio (6) = 2 months

Day's sales on hand: Day's sales on hand is a variation of the inventory turnover. It calculates the number of day's sales being carried in inventory. It is calculated by dividing 365 days by the inventory turnover ratio.

Day's sales on hand = 365 Days/ Inventory Turnover

Stock ratio indicates efficiency in the control of stock

Raw material stock / Purchase OR

WIP / cost of Product OR

F G Stock / Total Turnover

III. PROFITABILITY RATIOS

Profitability ratios measure a company's operating efficiency, including its ability to generate income and therefore, cash flow. Cash flow affects the company's ability to obtain debt and equity financing

A. **Profit margin:** The profit margin ratio, also known as the operating performance ratio, measures the company's ability to turn its sales into net income. To evaluate the profit margin, it must be compared to competitors and industry statistics. It is calculated by dividing net income by net sales.

$$Profit\ margin = Net\ Income\ /\ Net\ Sales$$

$$Gross\ Profit\ \% = (Gross\ profit\ /Sales)\ x\ 100/1$$

$$Net\ Profit\ \% = (Net\ Profit\ /Sales)\ x\ 100/1$$

$$Working\ Capital\ Turnover = Sales\ /\ Working\ Capital$$

B. **Return on Assets:**

The return on assets ratio (ROA) is considered an overall measure of profitability. It measures how much net income was generated for each Rs 100 of assets the company has. ROA is a combination of the profit margin ratio and the asset turnover ratio. It can be calculated separately by dividing net income by average total assets or by multiplying the profit margin ratio to the asset turnover ratio.

$$Return\ on\ Assets = Net\ Income\ /\ Average\ Total\ Assets\ OR$$

$$Return\ on\ Assets = Profit\ Margins\ X\ Asset\ Turnover$$

$$Net\ Income\ /\ Average\ Total\ Asset = (Net\ Income\ /\ Net\ sales)\ x\ (Net\ sale\ /\ Average\ Total\ Assets)\ OR$$

$$Return\ on\ Capital\ Employed\ (ROCE) = (Net\ Profit\ before\ Tax\ /\ Capital\ Employed) \times 100/1$$

C. **Creditors Turnover ratio**

It is a ratio between net credit purchases and the average amount of creditors outstanding during the year.

$$Creditors\ or\ payable\ turnover\ ratio = Net\ credit\ purchase\ /\ average\ creditors$$

Net credit purchase is gross credit purchase less return to suppliers, if any.

Average Creditors payable is the average of payable outstanding at the beginning and at the end of the year.

A low turnover ratio shows credit terms granted by suppliers are liberal and high ratio indicates credit terms are short and firm has to settle the suppliers account within a shorter period.

After analysing the credit turnover ratio, it can be determined if a firm can reduce its requirement of current assets by increasing credit terms with his suppliers for his business.

Example:

Credit purchase	= 360000
Amount payable at the beginning of the year	= 85000
Amount payable at the end of the year	= 95000
Average Payable	= (85000 +95000) / 2 = 90000
Creditors turnover ratio	= 360000 / 90000 = 4 times
Creditors payment period	= 12 months / Current turnover ratio (4) = 3 months

IV. EARNINGS RATIOS

A. **Earnings per share:** Earnings per share (EPS) represents the net income earned for each share of outstanding common stock. In a simple capital structure, it is calculated by dividing net income by the number of weighted average common shares outstanding

Earnings Per Share= Net Income / Weighted Average Common Shares Outstanding

B. **Return on common stockholders' Equity:** The return on common stockholders' equity (ROE) measures how much net income was earned relative to each Rs. of common stockholders' equity. It is calculated by dividing net income by average common stockholders' equity. In a simple capital structure (only common stock outstanding), average common stockholders' equity is the average of the beginning and ending stockholders' equity.

Return on Common Stockholders' equity = Net Income / Average Common Stockholders' Equity

In a complex capital structure, net income is adjusted by subtracting the preferred dividend requirement, and common stockholders' equity is calculated by subtracting the par value (or call price, if applicable) of the preferred stock from total stockholders' equity.

Return on Common Stakeholders Equity = Net Income – Preferred Dividends / Average Common Stakeholders Equity

C. **Price and Earnings ratio:**

The price earnings ratio (P/E) represents the investors' expectations for the stock

Price Earnings Ratio = Market value per share / Earning per share

C. **Dividend Yield**

Another indicator of how a company performed is the dividend yield. It measures the return in cash dividends earned by an investor on one share of the company's stock. It is calculated by dividing dividends paid per share by the market price of one common share at the end of the period.

Dividend Yield = Dividend paid per share / market price of one share

V. SHAREHOLDER'S RATIOS

Capital Ratio = Capital Employed/ Total Indebtedness OR

Capital Employed /Fixed Assets OR

Fixed Assets / Capital Employed

Capital Gearing Ratio = Fixed Interest Debt/ Equity OR

Profit before Tax /Debenture Interest

VI. SOLVENCY RATIOS

Solvency ratios are used to measure long term risk and are of interest to long term creditors and stockholders.

A. **Debt to total assets ratio:** The debt to total assets ratio calculates the percent of assets provided by creditors. It is calculated by dividing total debt by total assets. Total debt is the same as total liabilities

Debt to total Assets Ratio = Total debt / Total assets

B. **Times interest earned ratio:** The times interest earned ratio is an indicator of the company's ability to pay interest as it comes due. It is calculated by dividing earnings before interest and taxes (EBIT) by interest expense.

Time Interest Earned = Income before Interest and Tax Expenses / Interest expenses

A times interest earned ratio of 2 to 3 or more indicates that interest expense should reasonably be covered. If the times interest earned ratio is less than 2 it will be difficult to find a bank to loan money to the business.

VII. PRODUCTION RATIOS

Factory cost / Sales OR Admin Cost / Sales OR Selling Cost / Sales

In summary, ratio analysis can be based on different purposes it intends to serve, as follows:

For Profitability and Efficiency

- Gross Profitability %
- Expenses %
- Net profit to Sales
- Sales to Total Assets
- Net Profit to Capital Employed or Total Assets

For Potential and Actual Growth:

- Earnings per Share
- Dividend per Share
- Price Earnings Ratio
- Earning Yield
- Dividend Yield

Short term Solvency and Liquidity:

- Stock turnover
- Current Ratio, Liquidity Ratio
- Debtors' collection period

EVALUATION OF COMPANY PERFORMANCE

UNDER-TRADING COMPANIES

A company is said to be under-trading or being over capitalized if the following features characterize it:

i) Its capital is locked up in assets, which cannot be remuneratively employed.

ii) There is an unprofitable utilization of the whole of its capital employed.

Causes: A situation where capital is represented by assets which cannot be profitably employed might arise as a result of a company purchasing a business where excessive values have been placed on tangible assets for which own shares are to be issued. The result would be that value of shares would fall just as earnings per share too will be falling.

The other situation where the company is unable to profitably utilize the whole of its capital employed might arise as a result of the company raising funds without regards to its level of production and its working capital requirements. It might also arise because of large retention of profit and build-up of reserves.

Consequences

i) Fall in share value of the company

ii) Inability of the company to earn sufficient revenue to justify its capital employed

iii. Capital reduction

OVERTRADING COMPANIES

A company is said to be overtrading or being undercapitalized when it endeavours to increase its level of activity at a rate or operate at a scale in excess of what can be financed by its working capital. The company is thus characterized by inadequate financial backing.

Causes:

i) Inflationary burdens

ii) Excessive burden of tax

A company might find it difficult to finance its activities where the company hopes to pass these burdens on to customers without necessarily considering the need to finance. For example, more stock quantities to correspond with its working capital.

Indicators

Listed below are signals to indicate that a particular company is over-trading:

 i) Rapid increase in the volume of assets which are financed substantially by credit especially through short term bank borrowing and use of trade credit with increase in average credit period

ii) Decrease in average credit period granted which means a fall in average debt collection period

iii) Increase in stock without corresponding increase in turnover

iv) Change in certain accounting ratios:

Debt Ratios: The proportion of total assets financed by shareholders funds will fall whilst the proportion of total assets financed by short-term credits will rise.

Liquidity Ratios: Current and quick ratios will decrease.

Profitability Ratios:

   a) The ratio of sales to working capital will increase

   b) Gross profit ratio might fall as a result of stock piling which might lead to selling off some slow-moving stocks at low prices or granting of discounts to dispose such goods

   c) Net Profit Ratio might fall because of substantial increase in overhead costs

INSOLVENT COMPANIES

A company is said to be insolvent if after a period of overtrading it is forced to recognize that it cannot pay its debts and continue operations.

The following are the major features inherent in an insolvent company:

i) Where it's a going concern and its ability is threatened, i.e. it is not certain that the company will continue to be in operational existence in the near future and if it has the intention to curtail significantly, its line of business, then the company can be said to be tending towards being insolvent.

ii) Inability to meet financial obligations and settle debts as and when they fall due.

Indicators

Adverse key financial ratios

Liquidity Crisis: non-availability of liquid funds to meet pressing financial needs.

5
Budget

A budget or business plan is a financial plan indicating the estimation of revenues and expenses over a specified future period, which includes - planned sales plans and revenue generation, production plans, raw material requirements, cost & expenses, profit and loss accounts, cash flows and other financial statements. A budget may be surplus, balanced or deficit. A **surplus budget** means the anticipated profit, **a balanced budget** is no profit and no loss and a **deficit budget** means loss i.e., expenses are more than the revenues.

Normally, a budget is prepared annually for a financial year & reviewed monthly or quarterly. However, it is appropriate to monitor it daily, weekly, & monthly to initiate timely corrective measures and arrest slippages if any, and avoid any surprises at the end.

It is usually the responsibility of the MIS (management information system) section to prepare the annual budget of a company in consultation with functional heads. The MIS head coordinates with all concerned department heads to collect the following information for preparing the budget:

- Forecast of monthly sales volume, sales prices (price list) and revenue generation considering the market situation
- Manufacturing capacity, utilisation, departments production, waste generation, rework, ratio of sound & defection percentage and monthly production plan
- Raw material procurement to revenue realisation cycle to assess the inventory level at different stages
- Raw materials requirement, availability, and prices to prepare monthly purchase and consumption plan
- Details of utilities such as power, fuel, and water consumption per unit production and expenses
- Department wise manpower budget, their numbers, expenses and provision for increments
- Expenses for sales promotion activities including advertisements, tours and travels, sales commissions, etc.
- Expenses for packaging materials
- Expenses for repair and maintenance
- Expenses for capital items (Capex)
- Expenses for training programs, seminars, conferences, etc.
- Expenses for design and development
- Expenses for logistics
- Export and import tariffs for the concerned items
- Expenses for transport
- Expenses for information system (software and hardware)
- Expenses for stationary
- Expenses for communications
- Expenses for legal issues
- Expenses for corporate affairs

- Expenses for miscellaneous heads

- Source, types and amount of funding and finance charges (rate of interest)

- Various taxes applicable to central, states, and local government bodies

- Estimated foreign currency conversion rates to consider export and import realisation keeping in mind the expected currency fluctuation.

- Rate of depreciation

- Monthly cash flow

Once the annual budget draft is prepared in consultation with all concerned departments, it is presented to the top management and board of directors for their comments and approval.

After board approval, copies are provided to individual department heads for implementation and monitoring.

EXAMPLES OF ANNUAL BUDGETS IN TEXTILE MILLS

This chapter contains examples of annual budgets of different sectors of textile companies of various sizes - for a basic understanding of budget formats and information usually inducted in the annual budget.

Important: The figures are imaginary, for illustrative purposes only and not to be relied on, considered as actual or as a norm.

A. BUDGET OF A MEDIUM SIZE COTTON SPINNING AND WEAVING MILL

PLATINUM TEXTILE LIMITED, ANNUAL BUDGET FOR THE YEAR 2015-16

INDEX

Summary of Financial Projection

Budget Assumptions

Monthly Business Statement - EBITDA

Monthly Business Statement – Spinning

Monthly Business Statement – Weaving

Monthly Business statement (Stock Tracking - Process Cloth

Spinning & Weaving Division Projection (Combined) Profit and Loss (P&L)

Monthly P & L (Overall)

Monthly P & L (Raw material procurement to Yarn Sale

Monthly P & L (Purchase yarn to Grey Cloth & Group Company Job work

Monthly P & L (Grey Cloth to Processed Fab. Sale

Spinning Division Projection (PV Yarn & Cotton Yarn)

P & L Account (PV Yarn + Cotton Yarn)

Raw Material Costing & Sale Plan

Manufacturing division P & L Account

Spinning Plan

Weaving Division Projection

P & L - Weaving. to Processed fabric Sale - Overall

Weaving Plan

Grey Cloth - Job Transfer to Group company Processed

Fabric Sales Plan

Manufacturing Division. P & L Account

Break-up of Manufacturing Expenses

Current Assets & Current Liabilities (CA/CL)

Provisional & Budgeted CA / CL

Details of should be stock – Yarn

Details of should be stock – Fabric

Cash Flow Plan

Month-wise overall summary

Cash Flow Plan – Spinning Division

Cash Flow Plan – Weaving Division

Cash Flow Plan – Backlog

Administration Overheads

Administration. & Factory Overheads

Capital Expenses (Capex)

Capex Requirement – Spinning Division

Capex Requirement – Weaving Division

Capex Requirement - General & Civil Work

Organisation Chart

SUMMARY OF FINANCIAL PROJECTIONS YEAR 2015-16			
SR. NO.	PARTICULARS	QUANTITY Kgs / Mtrs	AMOUNT IN LACS
A.	A. QUANTITY OF PRODUCTION		
1	Production of yarn MT	3019.2	
2	Production of Cloth Lacs Mtrs	105.72	
B.	QUANTITY OF SALES		
1	Yarn sales MT	3019.2	6312.64
2	Finished Fabric Sale Lacs Mtrs	16.34	3413.07
3	Conversation Charges – Group Company	81.96	756.24
4	Second sales (Waste)		26
C.	GROSS TURNOVER		10507.95
D.	NET TURNOVER		10507.95
E.	EBITDA		
1	Spinning Division		(3.32)

2	Yarn Sale Division				344.46
3	Weaving Division				237.66
4	Finished Fabric Sale				500.58
					1079.38
F.	**FIXED OVERHEAD**				2005.41
G.	**PUBLICITY PROVISION**				NIL
H.	**PROVISIONAL CA / CL AS ON 31.03.2015**				(506.60)
I.	**BUDGET CA / CL AS ON 31.03.2016**				1509.1
J.	**CORPORATE FUND REQUIREMENT**				
	For Sinning Division				
1	Working Capital				690
2	Capex Requirement				366
3	Capex Requirement General, Civil & Computer				347
	TOTAL				**1403**
	For Weaving Division				
1	Working Capital				1009
2	Capex Requirement				330
					1339
1	Backlog Payment (Old Dues from corporate)				204
	GRAND TOTAL				**2946**
K.	**PAY BACK PERIOD**				16 Month
L.	**SURPLUS FUND REFUND TO CORPORATE**				967
M.	**CAPEX REQUIREMENT**				
1	Sinning Division				366
2	Weaving Division				330
3	General, Civil & Computer				347
	TOTAL				**1043**

ASSUMPTIONS

SPINNING		APR-MAY	JUN-MAR	For The Year
No of Ring Frames /Spindles		44/19442	44/19442	
Average Count Ne		30.03	27.07	
Average Speed		12295	13578	
Utilisation %		96	96	
Efficiency %		90	90	
Production/ Day - MT		6.52	8.76	
Production/ Month - MT		195.6	262.8	**3019.2**
Considered Conversion Charges Per /Count				
Up to 39s K	In Rs.	1.10	1.10	
40s K & Above	In Rs.	1.15	1.15	
PV Grey	In Rs.	1.10	1.10	
Doubling	In Rs.	0.30	0.30	
TFO Outside	In Rs.	0.50	0.50	
Poly Grey / Kgs	In Rs.		130.0	

Viscose Grey / Kg	In Rs.		170.0	
Cotton P/Candy (355.62 Kgs)	In Rs.	48000	50000	
Average Conversion Cost / Count	In Rs.	1.24	1.37	
Realisation / P. Count		-0.35	0.07	
Yarn Sales - Avg Cost / Kg	In Rs.	186.27	199.36	
Average Yarn Sale Rate / Kg	In Rs.	187.54	212.29	
Contribution in Sales / Kg	In Rs.	1.27	12.92	**12.92**
Contribution in Yarn Sales	In Lacs Rs.	2.48	33.95	**344.46**
Contribution in Mfg	In Lacs Rs.	-20.35	5.19	**-3.32**
Total Contribution	In Lacs Rs.	-17.86	39.14	**341.14**
Overall % of Profit (Mfg + Mktg)		-4.87	7.02	**4.81**
WEAVING				
Looms Working				
Sulzer (2 width)		32	32	
Airjet		48	48	
Total		80	80	
Average Loom Speed		675/250	675/250	
Utilisation %		96	96	
Efficiency %		85/75	85/75	
Average Pick		54.76	54.76	
Prod./Day – Group Co & Others (Lac Mtrs)		0.36	0.36	
Prod./Month –Group Co &others (Lac Mtrs)		6.83	6.83	
Prod. /Month – For processed Sale Lac Mtrs		1.98	1.98	
Conversion charges Paisa /Pick		19.32	19.32	
Conversion Cost Paise/ Pick		15.11	15.11	
Realisation Paise /Pick		4.21	4.21	
Contribution in Manufacturing Rs. In Lacs		20.33	20.33	**187.76**
Conversation Charges				
Sulzer – QN-A With Sizing		0.14	0.14	
– QN-B With Sizing		0.12	0.12	
Airjet – QN-C – with Dobby Name Writer		0.24	0.24	
QN-D with Dobby Name Writer		0.22	0.22	
QN-E -do-		0.18	0.18	
Sizing Charges – For Job work Per Kgs In Rs		15	15	
Processing Charges				
Avg Process Cost Per Kg	In Rs	80	80	
Total Fabric Processed Mtrs	In Lacs	1.98	1.98	
Avg Process cost Per Mtr	In Rs	35.13	35.13	
Fabric Marketing				
Processed Fabric Sales – Avg Cost Per Mtr		163.44	163.44	
Avg Sales Rate Per Mtr		191.53	191.53	
Contribution Per Mtr – In Rs.		28.09	28.09	
Total Contribution Per Month Rs In Lacs		55.62	55.62	
EBITDA (Earnings before interest, tax and depreciation)				
Spinning Division – Manufacturing		-20.35	5.2	-3.32
– Yarn Sales		2.48	33.95	344.46
Weaving Division – Manufacturing		20.33	20.33	237.66
– Finished Fabric Division		55.62	55.62	500.58
TOTAL		**58.08**	**115.1**	**1079.38**

Note: Weaving Conversion Charge- 19.32 (Paisa /Pick) X 54.76 (Ave Picks/Inch) = 1057.96/100 = 10.57 Rs/Mtr.

The company has spinning and weaving facilities. Dyeing and finishing is outsourced. Spinning and weaving divisions are considered as individual profit centers. Profit and loss of manufacturing and sales activities are calculated separately as separate profit centers. Therefore, conversion charges for manufacturing in spinning and weaving division are considered as per market rates to determine their profit and loss up to manufacturing stage and assess the viability of the manufacturing unit. Yarn will be sold to weaving division as well as in the market. Weaving will do the job for the group company as well as sell the finished fabric in the open market. Spinning machineries are old so ring frame speed is lower, resulting in loss in spinning manufacturing.

Since it is a loss-making company, annual plan is prepared to revive it. Funds are to be injected for working capital, for machines upgradation (Capex) and to cover backlog of past years. The effect of new funds will be effective after two months because of increasing WIP and arranging spares, etc.

Monthly Statement

EBITDA

Month	EBITDA (Rs.in lacs)					
	Spg. Mfg.	Wvg. Mfg.	Yarn Sale	Grey Fabric Sale	Processed fabric sale	Total
Apr	(20.34)	20.33	2.48	-	-	2.47
May	(20.34)	20.33	2.48	-	-	2.47
Jun	5.20	20.33	33.95	-	-	59.48
Jul	4.80	20.25	33.95	-	55.62	114.62
Aug	4.80	20.25	33.95	-	55.62	114.62
Sep	4.80	20.25	33.95	-	55.62	114.62
Oct	3.16	19.38	33.95	-	55.62	112.11
Nov	3.16	19.38	33.95	-	55.62	112.11
Dec	3.16	19.38	33.95	-	55.62	112.11
Jan	2.76	19.26	33.95	-	55.62	111.59
Feb	2.76	19.26	33.95	-	55.62	111.59
Mar	2.76	19.26	33.95	-	55.62	111.59
Total	(3.32)	237.66	344.46		500.58	1079.38

EBITDA: Earnings before interest, taxes, depreciation, and amortization. It measures a company's profitability and expresses the net income before deductions of above elements that are often considered irrelevant in the decision-making process. Adding these expenses back into net income allows us to analyse and compare the true operating cash flows of the businesses.

$$EBITDA = net\ profit + interest + taxes + depreciation\ and\ amortisation$$

Earnings: This is the net profit or simply net income. It is the bottom-line profit for the company mentioned at the bottom of the income statement.

Interest: Depends on the financial structure of a company and the amount of funds company has borrowed to support its business activities. Different companies have different capital structure resulting in different interest expenses. Hence, it is easier to compare the relative performance of companies by adding interest amount.

Taxes: Tax expenses depend on the type of business and changes from year to year. This often depends on the industry, location, and size of the company. They are a function of tax rules, which are not really part of assessing a management team's performance and, thus, many financial analysts prefer to add them back when comparing businesses.

Depreciation and Amortization: Depreciation and amortization depend on the historical investments the company has made and not on the current operating performance of the business. Depreciation and amortisation refers to any investment that loses its value over a period of time. When it comes to depreciation, it means tangible fixed assets like machines and equipment, vehicles, or a building. For amortization, it means intangible assets like patents.

Both are non-cash expenses, showing the deteriorating value of these assets. In the case of a vehicle, for example, a business owner may expect it to be on the road for five years. So, every month until the end of its useful life, the resale value on that vehicle will go down. The same goes for intangible assets. While a deterioration of a patent is not visible but its value deteriorates on its expiry.

Determining depreciation expenses and amortization expenses is subjective. Tentative workable life of any machine or equipment is decided at the time of its procurement and their salvage value is determined when they are not viable. Because of their subjective nature, they are also removed from consideration in the EBITDA metric.

SPINNING – Monthly Contribution of Manufacturing & Yarn Sales

Month	Spinning Manufacturing						Yarn Sales				Overall			
	No. of Ring Frames	Avg Count Ne	Kgs / Month (M T)	Mfg. Expenses (Rs in Lacs)	Mfg. Charges (Rs in lacs)	Contribution (Rs in lacs)	Kgs (M T)	Cost Value (Rs. In lacs)	Sale Value (Rs. In lacs)	Contribution (Rs. In lacs)	Cost / Expenses (Rs in lacs)	Earnings (Rs in lacs)	Contribution (Rs in lacs)	Contribution%
Apr	44.00	30.03	195.60	93.50	73.16	(20.34)	195.60	364.34	366.82	2.48	457.84	439.98	(17.86)	(3.90)
may	44.00	30.03	195.60	93.50	73.16	(20.34)	195.60	364.34	366.82	2.48	457.84	439.98	(17.86)	(3.90)
Jun	44.00	27.07	262.80	92.28	97.48	5.20	262.80	523.95	557.90	33.95	616.23	655.38	39.15	6.35
Jul	44.00	27.07	262.80	92.68	97.48	4.80	262.80	523.95	557.90	33.95	616.63	655.38	38.75	6.28
Aug	44.00	27.07	262.80	92.68	97.48	4.80	262.80	523.95	557.90	33.95	616.63	655.38	38.75	6.28
Sep	44.00	27.07	262.80	92.68	97.48	4.80	262.80	523.95	557.90	33.95	616.63	655.38	38.75	6.28
Oct	44.00	27.07	262.80	94.32	97.48	3.16	262.80	523.95	557.90	33.95	618.27	655.38	37.11	6.00
Nov	44.00	27.07	262.80	94.32	97.48	3.16	262.80	523.95	557.90	33.95	618.27	655.38	37.11	6.00
Dec	44.00	27.07	262.80	94.32	97.48	3.16	262.80	523.95	557.90	33.95	618.27	655.38	37.11	6.00
Jan	44.00	27.07	262.80	94.72	97.48	2.76	262.80	523.95	557.90	33.95	618.67	655.38	36.71	5.93
Feb	44.00	27.07	262.80	94.72	97.48	2.76	262.80	523.95	557.90	33.95	618.67	655.38	36.71	5.93
Mar	44.00	27.07	262.80	94.72	97.48	2.76	262.80	523.95	557.90	33.95	618.67	655.38	36.71	5.93
Total			3019.20	1124.44	1121.12	(3.32)	3019.20	5968.18	6312.64	344.46	7092.62	7433.76	341.14	4.81

WEAVING – Monthly Contribution of Manufacturing & Fabric Sales (Grey & Processed)

	Meters in lacs
Processed Sale	1.98
Conversion for group Co.	6.83
Total Mtrs	**8.81**

Month	Weaving Manufacturing							Fabric Sales				Overall			
	Sulzer / Air Jet Nos	Total	Avg. Picks	Meters / Month (lacs)	Mfg. Expenses (Rs in lacs)	Mfg. Charges (Rs in lacs)	Contribution (Rs in lacs)	Meters (lacs)	Cost Value Rs. In lacs	Sale Value Rs. In lacs	Contribution Rs. In lacs	Cost / Expenses (Rs in lacs)	Earnings (Rs in lacs)	Profit / Loss (Rs in lacs)	Profit / Loss %
Apr	32 / 48	80	54.76	8.81	72.89	93.22	20.33					72.89	93.22	20.33	27.89
may	32 / 48	80	54.76	8.81	72.89	93.22	20.33					72.89	93.22	20.33	27.89
Jun	32 / 48	80	54.76	8.81	72.89	93.22	20.33					72.89	93.22	20.33	27.89
Jul	32 / 48	80	54.76	8.81	72.97	93.22	20.25	1.98	323.61	379.23	55.62	396.58	472.45	75.87	19.13
Aug	32 / 48	80	54.76	8.81	72.97	93.22	20.25	1.98	323.61	379.23	55.62	396.58	472.45	75.87	19.13
Sep	32 / 48	80	54.76	8.81	72.97	93.22	20.25	1.98	323.61	379.23	55.62	396.58	472.45	75.87	19.13
Oct	32 / 48	80	54.70	8.81	73.84	93.22	19.38	1.98	323.61	379.23	55.62	397.45	472.45	75.00	18.87
Nov	32 / 48	80	54.76	8.81	73.84	93.22	19.38	1.98	323.61	379.23	55.62	397.45	472.45	75.00	18.87
Dec	32 / 48	80	54.76	8.81	73.84	93.22	19.38	1.98	323.61	379.23	55.62	397.45	472.45	75.00	18.87
Jan	32 / 48	80	54.76	8.81	73.96	93.22	19.26	1.98	323.61	379.23	55.62	397.57	472.45	74.88	18.83
Feb	32 / 48	80	54.76	8.81	73.96	93.22	19.26	1.98	323.61	379.23	55.62	397.57	472.45	74.88	18.83
Mar	32 / 48	80	54.76	8.81	73.96	93.22	19.26	1.98	323.61	379.23	55.62	397.57	472.45	74.88	18.03
Total				105.72	880.98	1118.64	237.66	17.82	2912.49	3413.07	500.58	3793.47	4531.71	738.24	19.46

MONTHLY STATEMENT (Stock Tracking)

	April to March/Month
Processed Sale-	1.98
Conversion for Group company-	6.83
Total Mtrs in lacs	8.81

Month	Weaving Prod. Mtrs lacs	Transfer to Process House %	Transfer to Process House Mtrs. Lacs	Processed Fabric Sale Mtrs lacs	Month
Apr	1.98				
May	1.98	50	0.99		
Jun	1.98	75	1.49		
Jul	1.98	100	1.98	0.99	
Aug	1.98	100	1.98	1.49	
Sep	1.98	100	1.98	1.98	
Oct	1.98	100	1.98	1.98	
Nov	1.98	100	1.98	1.98	
Dec	1.98	100	1.98	1.98	
Jan	1.98	100	1.98	1.98	
Feb	1.98	100	1.98	1.98	
Mar	1.98	100	1.98	1.98	
Total	23.76		20.30	16.34	

Summary

Production for Process	Mtrs.in lacs	Stock lacs mtrs	Remark
Total Production	23.76	3.47	OWN
Transfer to Process House	20.30	3.96	Process House
Processed Fabric Sale	16.34		
Total		7.43	

MONTHLY PROFIT & LOSS STATEMENT

Spinning + Weaving Mfg + Cotton Fabric Sales

PARTICULARS (Rs in Lacs)	Apr	May	Jun	Jul	Aug	Sep	Oct	Nov	Dec	Jan	Feb	Mar	Total For the Year
INCOME													
– Yarn Sales	366.82	366.82	557.9	557.90	557.90	557.90	557.90	557.90	557.90	557.90	557.90	557.90	6312.64
– Grey Cloth Sales													
– Grey Cloth – Transfer to Processor	229.54	229.54	229.54	229.54	229.54	229.54	229.54	229.54	229.54	229.54	229.54	229.54	2754.54
– Process Fabric Sales				379.23	379.23	379.23	379.23	379.23	379.23	379.23	379.23	379.23	3413.07
– Waste Sales	3.00	3.00	2.00	2.00	2.00	2.00	2.00	2.00	2.00	2.00	2.00	2.00	26.00
– Weaving - Job Work Charges	63.02	63.02	63.02	63.02	63.02	63.02	63.02	63.02	63.02	63.02	63.02	63.02	756.24
TOTAL INCOME	662.38	662.38	852.46	1231.69	1231.69	1231.69	1231.69	1231.69	1231.69	1231.69	1231.69	1231.69	13262.44
EXPENDITURE													
Raw Materials - Cotton, Polyester, Viscose	282.82	282.82	419.58	419.58	419.58	419.58	419.58	419.58	419.58	419.58	419.58	419.58	4761.44
Raw Materials - Purchase Yarn	199.34	199.34	199.34	199.34	199.34	199.34	199.34	199.34	199.34	199.34	199.34	199.34	2392.08
Grey Cloth Cost - Transfer to Processor				229.54	229.54	229.54	229.54	229.54	229.54	229.54	229.54	229.54	2065.87
Processing Charges				69.55	69.55	69.55	69.55	69.55	69.55	69.55	69.55	69.55	625.95
MANUFACTURING AND ADMINISTRATIVE OVERHEADS													
Wages & Salaries	52.00	52.00	49.61	50.09	50.09	50.09	50.70	50.70	50.70	51.22	51.22	51.22	609.65
Salaries	17.29	17.29	17.29	17.29	17.29	17.29	19.19	19.19	19.19	19.19	19.19	19.19	218.89
Power & Fuel	61.84	61.84	60.38	60.38	60.38	60.38	60.38	60.38	60.38	60.38	60.38	60.38	727.48
Coal	4.35	4.35	4.35	4.35	4.35	4.35	4.35	4.35	4.35	4.35	4.35	4.35	52.20
Stores & Spares	10.09	10.09	9.82	9.82	9.82	9.82	9.82	9.82	9.82	9.82	9.82	9.82	118.38
Packing Materials	4.55	4.55	7.46	7.46	7.46	7.46	7.46	7.46	7.46	7.46	7.46	7.46	83.70
Sizing Materials	5.14	5.14	5.14	5.14	5.14	5.14	5.14	5.14	5.14	5.14	5.14	5.14	61.68
Administrative Expenses	6.00	6.00	6.00	6.00	6.00	6.00	6.00	6.00	6.00	6.00	6.00	6.00	72.00
Year End Expenses	5.12	5.12	5.12	5.12	5.12	5.12	5.12	5.12	5.12	5.12	5.12	5.12	61.44
SELLING EXPENSES													
Brokerage & Commission	8.08	8.08	6.41	13.99	13.99	13.99	13.99	13.99	13.99	13.99	13.99	13.99	148.50
Freight Charges	3.28	3.28	2.48	4.46	4.46	4.46	4.46	4.46	4.46	4.46	4.46	4.46	49.20
Value Loss / Shrinkage				14.95	14.95	14.95	14.95	14.95	14.95	14.95	14.95	14.95	134.57
TOTAL EXPENDITURE	659.91	659.91	792.98	1117.07	1117.07	1117.07	1119.58	1119.58	1119.58	1120.10	1120.10	1120.10	12183.02
PBDIT	**2.47**	**2.47**	**59.48**	**114.62**	**114.62**	**114.62**	**112.11**	**112.11**	**112.11**	**111.59**	**111.59**	**111.59**	**1079.38**

Spinning Division (Raw Material Procurement to Yarn Sales)

PARTICULARS	Apr	May	Jun	Jul	Aug	Sep	Oct	Nov	Dec	Jan	Feb	Mar	Total For the Year
INCOME													
– Yarn Sales	366.82	366.82	557.90	557.90	557.90	557.90	557.90	557.90	557.90	557.90	557.90	557.90	6312.64
– Waste Sales	3.00	3.00	2.00	2.00	2.00	2.00	2.00	2.00	2.00	2.00	2.00	2.00	26.00
TOTAL INCOME	369.82	369.82	559.90	559.90	559.90	559.90	559.90	559.90	559.90	559.90	559.90	559.90	6338.64
EXPENDITURE													
Raw Materials - Cotton, Polyester & Viscose	282.82	282.82	419.58	419.58	419.58	419.58	419.58	419.58	419.58	419.58	419.58	419.58	4761.44
MANUFACTURING AND ADMINISTRATIVE OVERHEADS													
Wages	35.05	35.05	32.66	33.06	33.06	33.06	33.50	33.50	33.50	33.90	33.90	33.90	404.15
Salaries	10.89	10.89	10.89	10.89	10.89	10.89	12.09	12.09	12.09	12.09	12.09	12.09	137.89
Power & Fuel	32.74	32.74	31.28	31.28	31.28	31.28	31.28	31.28	31.28	31.28	31.28	31.28	378.28
Stores & Spares	4.76	4.76	4.49	4.49	4.49	4.49	4.49	4.49	4.49	4.49	4.49	4.49	54.42
Packing Materials	2.35	2.35	5.26	5.26	5.26	5.26	5.26	5.26	5.26	5.26	5.26	5.26	57.30
Administrative Expenses	4.20	4.20	4.20	4.20	4.20	4.20	4.20	4.20	4.20	4.20	4.20	4.20	50.40
Year-End Expenses	3.50	3.50	3.50	3.50	3.50	3.50	3.50	3.50	3.50	3.50	3.50	3.50	42.00
SELLING EXPENSES													
Brokerage & Commission	8.08	8.08	6.41	6.41	6.41	6.41	6.41	6.41	6.41	6.41	6.41	6.41	80.26
Freight Charges	3.28	3.28	2.48	2.48	2.48	2.48	2.48	2.48	2.48	2.48	2.48	2.48	31.36
TOTAL EXPENDITURE	387.68	387.68	520.75	521.15	521.15	521.15	522.79	522.79	522.79	523.19	523.19	523.19	5997.50
PBDIT (Profit before depreciation, interest, tax)	(17.86)	(17.86)	39.15	38.75	38.75	38.75	37.11	37.11	37.11	36.71	36.71	36.71	341.14

Weaving Division (Yarn purchase to Grey Cloth production)

PARTICULARS	Apr	May	Jun	Jul	Aug	Sep	Oct	Nov	Dec	Jan	Feb	Mar	Total
INCOME													
– Weaving Charges – For Job work	93.22	93.22	93.22	93.22	93.22	93.22	93.22	93.22	93.22	93.22	93.22	93.22	1118.64
TOTAL INCOME	93.22	93.22	93.22	93.22	93.22	93.22	93.22	93.22	93.22	93.22	93.22	93.22	1118.64
EXPENDITURE													
MANUFACTURING AND ADMINISTRATIVE OVERHEADS													
Wages	16.95	16.95	16.95	17.03	17.03	17.03	17.20	17.20	17.20	17.32	17.32	17.32	205.50
Salaries	6.40	6.40	6.40	6.40	6.40	6.40	7.10	7.10	7.10	7.10	7.10	7.10	81.00
Power & Fuel	29.10	29.10	29.10	29.10	29.10	29.10	29.10	29.10	29.10	29.10	29.10	29.10	349.20
Coal	4.35	4.35	4.35	4.35	4.35	4.35	4.35	4.35	4.35	4.35	4.35	4.35	52.20
Stores & Spares	5.33	5.33	5.33	5.33	5.33	5.33	5.33	5.33	5.33	5.33	5.33	5.33	63.96
Packing Materials													
Sizing Materials	5.14	5.14	5.14	5.14	5.14	5.14	5.14	5.14	5.14	5.14	5.14	5.14	61.68
Packing Materials	2.20	2.20	2.20	2.20	2.20	2.20	2.20	2.20	2.20	2.20	2.20	2.20	26.40
Administrative Expenses	1.80	1.80	1.80	1.80	1.80	1.80	1.80	1.80	1.80	1.80	1.80	1.80	21.60
Year-End Expenses	1.62	1.62	1.62	1.62	1.62	1.62	1.62	1.62	1.62	1.62	1.62	1.62	19.44
TOTAL EXPENDITURE	72.89	72.89	72.89	72.97	72.97	72.97	73.84	73.84	73.84	73.96	73.96	73.96	880.98
PBDIT	20.33	20.33	20.33	20.25	20.25	20.25	19.38	19.38	19.38	19.26	19.26	19.26	237.66

Weaving division - (Grey Cloth to Processed Fabric Sales)

PARTICULARS	Apr-11	May-11	Jun-11	Jul-11	Aug-11	Sep-11	Oct-11	Nov-11	Dec-11	Jan-12	Feb-12	Mar-12	Total For the Year
INCOME													
– Grey Cloth Sales													
– Grey Cloth (Job Weaving for Group Co)	63.02	63.02	63.02	63.02	63.02	63.02	63.02	63.02	63.02	63.02	63.02	63.02	756.24
– Grey Cloth - Transfer to Processor	229.54	229.54	229.54	229.54	229.54	229.54	229.54	229.54	229.54	229.54	229.54	229.54	2754.49
– Process Fabric Sales				379.23	379.23	379.23	379.23	379.23	379.23	379.23	379.23	379.23	3413.07
TOTAL INCOME	292.56	292.56	292.56	671.79	671.79	671.79	671.79	671.79	671.79	671.79	671.79	671.79	6923.80
EXPENDITURE													
Raw Materials to Purchase Yarn	199.34	199.34	199.34	199.34	199.34	199.34	199.34	199.34	199.34	199.34	199.34	199.34	2392.08
Grey Cloth Cost to Transfer to Processor				229.54	229.54	229.54	229.54	229.54	229.54	229.54	229.54	229.54	2065.87
MANUFACTURING EXPENSES													
Processing Charges				69.55	69.55	69.55	69.55	69.55	69.55	69.55	69.55	69.55	625.95
Weaving Charges	93.22	93.22	93.22	93.22	93.22	93.22	93.22	93.22	93.22	93.22	93.22	93.22	1118.64
SELLING EXPENSES													
Brokerage & Commission				7.58	7.58	7.58	7.58	7.58	7.58	7.58	7.58	7.58	68.24
Freight Charges				1.98	1.98	1.98	1.98	1.98	1.98	1.98	1.98	1.98	17.84
Value Loss / Shrinkage				14.95	14.95	14.95	14.95	14.95	14.95	14.95	14.95	14.95	134.57
TOTAL EXPENDITURE	292.56	292.56	292.56	616.17	616.17	616.17	616.17	616.17	616.17	616.17	616.17	616.17	6423.18
PBDIT				55.62	55.62	55.62	55.62	55.62	55.62	55.62	55.62	55.62	500.58

PROFIT & LOSS ACCOUNT

SPINNING DIVISION (Manufacturing + Yarn Sale)

(June- March) - 29.17 Days Each Month (PV Yarn +Cotton yarn)

PARTICULARS		Rs / Kg
Working. Days/month	29.17	
No. of Ring frames	44	
Avg. Spindle Speed	13578	
Efficiency / Utilisation %	90/96	
Avg. Count	27.07	
MT / Day	8.76	
YARN PROD./ MONTH MT	**262.80**	
SALE VALUE YARN / MONTH /(Rs.in lacs)	**557.90**	**212.29**
TOTAL Rs.in lacs	**557.90**	**212.29**
Expenses		
MIXING	408.73	155.53
SPINNING / T F O CHARGES	97.48	37.09
VAT	8.85	3.37
TRANSPORT	2.48	0.94
SALES OHD'S (COMM. ETC.)	6.41	2.44
TOTAL EXPENSES Rs.in lacs	**523.94**	**199.37**
CASH GEN. Rs.in Lacs	**33.95**	**12.92**

SPINNING MANUFACTURING. DIVISION.	5.19	
YARN SALES DIVISION.	33.95	
TOTAL (Rs.in lacs)	**39.14**	
EBITDA	**39.14**	

Remark

Count conversion Charges in (Rs./Count)	
Up to 39s	1.10
40s & above	1.15
PV Grey	1.10
Add. For Doubling	0.30
Add. For TFO outside	0.50
Rate Poly. fibre Grey / kg Rs.	1.30
Rate Viscose fibre Grey / kg Rs.	170
Cotton rate / Candy - Rs 355.62kg/candy	50000

Example- Conversion charges for 40s Ne: 40*1.15= Rs 46/Kg

(PV Yarn +Cotton yarn)

(June- March) – Avg. 29.17 Day/ Month

YARN SALES DIVISION

A. RAW MATERIAL COSTING

PARTICULARS	Cotton	Poly. Grey	Viscose Grey	PV 65:35
Rs./Candy (355.62 kg/candy)	50000			
Rs./ Kgs	140.60			
VAT 4 % / CST 2%				
Freight	0.80			
Insurance				
TOTAL	**141.40**			
Entry tax (1%)				
Mixing Cost	**141.40**			
Waste credit	2.00			
NET RATES	**139.40**	**130.00**	**170.00**	
Recovery %	85.00	97.00	94.00	
Net raw material rate Rs /Kg	**164.00**	**134.02**	**180.85**	**150.41**

B. YARN SALES DIVISION

Count	15s PV	15s PV	2/30s PV	2/30s PV	27s K	34s K	2/40s K	Total
	GC	Mkt.	GC	Mkt.	Mkt.	Mkt.	Mkt.	
Production / Month MT	**40.00**	**29.90**	**60.00**	**33.90**	**48.00**	**14.40**	**36.60**	**262.80**
Total Mixing Cost Rs./ Kgs.	150.41	150.41	150.41	150.41	164.00	164.00	164.00	408.73
Spinning. Charges Rs. / Kgs.	16.50	16.50	48.00	48.00	29.70	37.40	58.00	97.48
Ex Mill Cost Rs./ Kgs.	166.91	166.91	198.41	198.41	193.70	201.40	222.00	
CST 2% / VAT 4 % (as applicable)		3.34		3.97	7.75	8.06	4.44	8.85
Brokerage & CD 2%		3.34		3.97	3.87	4.03	4.44	6.41

Transport	0.60	1.50	0.60	1.50	0.60	0.60	1.50	2.48
Total Cost Rs./ Kg A	167.51	175.09	199.01	207.85	205.92	214.08	232.38	523.94
Sale Rate Rs./ Kg B	183.00	183.00	220.00	220.00	212.00	222.00	245.00	557.90
Contribution Rs./ Kg (B-A)	15.49	7.91	20.99	12.15	6.08	7.92	12.62	12.92
Total Contribution Rs.in Lacs	6.20	2.37	12.59	4.12	2.92	1.14	4.62	33.95

Remark: GC-Group Company

Raw material & yarn rates are considered based on market rate.

Sale rates of PV Yarn is considered Rs. 4/- to Rs.5 /- less than the market rate.

(PV yarn + Cotton yarn)

(June- March) - 29.17 Days Each Month

MFG. DIVISION - Spinning (Expenses)

PARTICULARS	Total for Month MT	Per day MT	Conversion Rs. / Kg	AMT. lacs/ Day	Total Amount / month Rs. in lacs
FOR MARKET					
15s pv Grey	69.90	2.33	16.50	0.38	11.53
2/30s pv Grey	93.90	3.13	48.00	1.50	45.07
27s k	48.00	1.60	29.70	0.48	14.26
34s k	14.40	0.48	37.40	0.18	5.39
2/40s k	36.60	1.22	58.00	0.71	21.23
Total	262.80	8.76		3.25	97.48

PARTICULARS	Total Spg.	PER DAY
Working Days	29.17	29.17
Ring Frames Spindle / Machine/shift	19442	44
Average Count		27.07
Avg Spindle. Speed		13578
Efficiency %		90
Utilisation %		96
Production - MT	262.80	8.76
No. of Workers. (P.)	8700	290
No. of Workers. (C.)	1050	35
Engg. Workers. (P.)	810	27
Engg. Workers. (C.)	60	2
General Workers. (P.)	450	15
General Workers. (Contrs.)	660	22
Total Manpower	11730	391
Power Consumed Kwh	645000	21500

*P-Permanent, C-Contract, Engg-Engineering

	Amt (Rs.in lacs)	Rs/Kg	ps./count	Remark
CONVERSION CHARGES	97.48	37.09	1.37	
Wages	24.23	9.22		Perm. @ Rs. 255/- & Cont. @ Rs. 195/- per day.
Wages - Engineering	2.13	0.81		

Winding & TFO Charges	2.07	0.79	
Garden / Trainee (25*138*30) etc.	1.14	0.43	
Packing / Tipping	0.71	0.27	
Wages - Total	**30.28**	**11.52**	**0.43**
Salaries Direct	3.08	1.17	
Salaries - Ancillary & Supporting	3.61	1.38	
Stores/Spares	3.50	1.33	20p/Spdl considered less Last one year Avg - 9 p.
Stores/Spares (Winding)	0.69	0.26	
Stores/Spares (Generators./Engineering.)	0.30	0.11	
Power Expenses	31.28	11.90	Unit Rate - Rs. 4.85 per Unit
Packing Material	5.26	2.00	
Year End	3.50	1.33	
Total A (Direct Spg. Expenses)	**51.23**	**19.49**	**0.72**
Wages - General	2.37	0.90	
Salaries - Admin & Account	4.20	1.60	
Admin. Factory Overheads	4.20	1.60	
Total B (Common Exp. - 70%)	**10.77**	**4.10**	**0.15**
Total Expenses	**92.28**	**35.12**	**1.30**
Net Cash Generation	**5.19**	**1.98**	**0.07**

SPINNING PLAN SUMMARY (PV & Cotton yarn) (June to March)

Working Days: 30   29.17

Count	Single/Double	Blend	Kgs	Conversion Rs./Kg.	Amount Rs.	R/F TYPE	Spindle Speed	T. M.	T. P. I.	Gms / spdl per Shift	Spindles Required	M/Cs Reqd.
15	S	PV Grey	2330	16.50	38445	476 ss	13500	3.50	13.56	430.23	1805	4
30	D	PV Grey	3130	48.00	150240	476 ss	13500	3.50	19.17	152.11	6859	15
27	S	K	640	29.70	19008	400 ss	13500	4.50	23.38	138.56	1540	4
34	S	K	480	37.40	17952	LR	14500	4.50	26.24	105.32	1519	2
27	S	K	960	29.70	28512	476 ss	13500	4.50	23.38	138.56	2309	5
40	D	K	1220	58.00	70760	400 ss	13500	4.60	29.09	75.17	5410	14
	TOTAL ⇒		8760		324917						19442	44
Avg. Count ⇒		27.07	Avg. Speed ⇒		13578	Avg. TPI ⇒		22.80		Utilisation =		96%

PRODUCTION DURING THE PERIOD

Count	MT / Day	MT / Month	
15s PV Grey	2.33	69.90	For Group Co
2/30s PV Grey	3.13	93.90	"
27s K	1.6	48.00	Market
34s K	0.48	14.40	"
2/40s K	1.22	36.60	"
Total Mkt.	**8.76**	**262.80**	

Doubling / TFO:

Count
2/30s PV Grey
2/40s k

Prod.Summary
PV Yarn
Cotton Yarn
Total / Month

RAW-MATERIAL CONSUMPTION / MONTH

Particulars	Yarn prod. MT	Recovery %	Raw - Material MT	Rate / Kg	Amt. Rs. in lacs
Cotton	99.00	85.0	116.47	141.40	164.69
Polyester.	106.47	97.0	109.76	130.00	142.69
Viscose	57.33	94.0	60.99	170.00	103.68
Total	262.80		287.22		411.06

SPIN PLAN AND MACHINE BALANCING

RING FRAME									TOTAL
COUNT Ne	15s PV	2/30s PV	Total PV	27s K	27s K	34s K	2/40s K	Total Cotton	27.07
NO. OF M/C	4	15		4	5	2	14		44.0
SPINDLES/MACHINE	476	476		400	476	768	400		
SPINDLES ALLOTTED	1904	7140		1600	2380	1536	5600		20160
LESS 4 % MAINT	76.2	285.6		64.0	95.2	61.4	224.0		
ACTUAL.SPINDLES WORKED	1828	6854		1536	2285	1475	5376		19354
SPINDLE SPEED (AVG)	13500	13500		13500	13500	14500	13500		
TM	3.50	3.50		4.50	4.50	4.50	4.60		
TPI	13.56	19.17		23.38	23.38	26.24	29.09		
100% PROD / SPDL/SHIFT (Gms)	478.04	169.01		153.96	153.96	117.02	83.53		
90% PROD / SPDL/SHIFT(Gms)	430.23	152.11		138.56	138.56	105.32	75.17		
PROD / MC / SHIFTS (Kg)	205	72		55	66	81	30		
PRODN/MC/DAY (Kg)	614	217		166	198	243	90		
PROD / DAY (Kg)	2330	3130	5460	640	960	480	1220	3300	8760
WASTE IN R/F 2% PV & 5% COT	47	63						165	
SPEED FRAME (SF)	LR	LR						MMC	
SPINDLES/MACHINE	120	120						132	
PROD REQUIRED IN S/F	2377	3193	5569					3465	9034
S/F HANK	0.8	1.00						1.2	
TM	0.9	0.9						1.35	
TPI	0.80	0.90						1.48	
SPINDLE SPEED	600	600						600	
100%PRODN/SPL/SHIFT(Kg)	6.71	4.80						2.43	
80%PROD / SPL/ SHIFT	5.37	3.84						1.95	
SPDL REQD / SHIFT	148	277						593	
MC REQD / SHIFT	1.23	2.31	3.5					4.49	8.0
WASTE IN S/F 0.1% PV & 0.5% COT	2	3						17	
RSB DRAW FRAME									
PROD REQD IN RSB	2379	3196	5575					3482	9057
DEL. SPEED MTR / MIN	500	500						500	
HANK DELIVERED	0.100	0.115						0.115	
100%PRODN DEL / SHIFT	1417	1232						1232	
82% PROD / DEL / SHIFT	1162	1011						1011	
PROD / MC / DAY (Kg)	3486	3032						3032	

	C1	C2	C3				C7	C8
M/C REQD / DAY	0.68	1.05	1.7				1.15	2.9
WASTE IN RSB 0.5%	12	16					17	
PRE-DRAW FRAME								
PROD REQD IN DRAW FRAME	2391	3212	5603				3500	9102
DEL SPEED MTR / MIN	300	300					300	
HANK DELIVERED	0.100	0.115					0.120	
100% PROD / DEL / SHIFT	850	739					709	
82% PROD / DE L/ SHIFT	697	606					581	
PROD /MC / DAY (KG)	4184	3638					3486	
MACHINES REQD / DAY	0.57	0.88	1.5				1.00	2.5
WASTE IN DO/6 DRAWING 0.5%	12	16					17	
CARD DK 780								
REQD CARD PROD	2403	3228	5631				3517	9148
DEL SPEED MTR / MIN	150	150					150	
HANK DELIVERED	0.100	0.115					0.120	
100%PRODN/M/C/SHIFT	425	370					354	
88%PRODN/M/C/SHIFT	374	325					312	
PROD / M/C/ DAY (Kg)	1122	976					935	
M/C REQD/ DAY	2.14	3.31	5.4				3.76	9.2
WASTE IN CARDING PV 2.5% & COT 5%			141				176	457
BLOW ROOM								
REQD B/R PRODN			5771				3693	9465
LAP WEIGHT IN Kg			23				23	
LAP PROD / MACHINE			52				52	
LAP PRODUCTION			251				161	
WASTE IN B/R PV 1.0 % & COT 4.0%			58				148	
M/C REQUIRED			4.83				3.09	7.9
MIXING REQUIRED / DAY			5829				3041	9465

*Spdl-Spindles, Act- Actual, PROD-Production, M/C-Machine, Del-Delivery, B/R-Blow Room, SF-Speed Frame

	WINDING (M/C)				
		27	40		
		Winding	Winding		Total
	WINDING (MACHINES)	**27s K**	**40s K**		
RJK	DRUM / M/C	60	60		
	REQD WINDING PROD	1600	1220		2820
	DEL SPEED MTR / MIN	600	600		
	100% PRODN/DRUM / SHIFT	6.30	4.25		
	70% PROD /DRUM / SHIFT	4.41	2.98		
	PROD / M/C / SHIFT	265	179		
	M/C REQUIRED /SHIFT	2.02	2.28		4.29
		AC	AC	AC	PSM
		30s PV	34s K	15s PV	2/40s K
		30	34	15	40
	DRUM / M/C	60	60	60	120

	REQUIRED WINDING PROD	3130	480	2330	1220
	DEL SPEED MTR / MIN	1000	1000	900	500
	100% PROD / DRUM / SHIFT	9.45	8.34	17.01	3.54
	70% PROD / DRUM / SHIFT	6.61	5.84	11.90	2.48
	PROD / M/C / SHIFT	397	350	714	298
	M/C REQD / SHIFT	**2.63**	**0.46**	**1.09**	**1.37**
	CHEESE WINDING				**40**
	DRUM / M/C				120
	REQD WINDING PROD				1220
	DEL SPEED MTR / MIN				400
	100% PROD /DRUM / SHIFT				5.67
	70% PROD / DRUM / SHIFT				3.97
	PROD /M/C / DAY				1429
	M/C REQD / SHIFT				**0.85**
	DOUBLING				
	REQUIRED TFO PROD				**1220**
	SPINDLES / M/C				436
	SPINDLE SPEED				8000
	TPI				20.36
	100% PROD SPDL./SHIFT(Gm)				141.43
	90% PRODN/SPDL./SHIFT(Gm)				127.29
	PRODN/MC/SHIFTS (Kg)				54
	PROD / MC / DAY (Kg)				161
	M/C REQUIRED /DAY				**7.6**
	Remark : ⇒ 2/30s pv 3130 kg TFO out side				

PROFIT & LOSS ACCOUNT – Weaving Division (Manufacturing + Sales)

(April - March) - 29.17 Days Each Month

PARTICULARS	Grey Cloth Div.for Group Co		Grey Cloth Div. Market		Process Div.		Over All
Avg Working days / Month	29.17						
No. Looms (Air Jet - 48 + Sulzer Projectile 32)	80						
Avg. Loom Speed - Airjet / Sulzer	675 / 250						
Utilisation %	96						
Efficiency % - Air Jet / Sulzer	85 / 75						
Avg. Pick - Air Jet / Sulzer	52.08 / 61.18						
Avg. Pick - Over All	54.76						
Production Cloth - Mtrs lacs /day	0.23		0.07		0.07		0.36
Cloth Production for – Group Co / Month	6.83						6.83
Cloth Production for - Market / Month			1.98		1.98		1.98
Cloth Production - Mtrs / Month (in lacs)	**6.83**		**1.98**		**1.98**		**8.81**
	Amt. In lacs	Rs. / Mtrs	Amt. In lacs	Rs. / Mtrs	Amt. In lacs	Rs. / Mtrs	Rs.in lacs
CONVERSION CHARGES / SALE	**63.02**	**9.23**	**229.54**	**115.93**	**379.23**	**191.53**	**671.79**
Yarn Cost / Grey Cost			192.60	97.27	229.54	115.93	422.14

Hard Waste Impact			6.74	3.40			6.74
Weaving Charges	63.02	9.23	30.20	15.25			93.22
Processing Charges					69.55	35.13	69.55
VAT							
Transport					1.98	1.00	1.98
Sales Over Heads - Commission					7.58	3.83	7.58
Value Loss /Shrinkage					14.95	7.55	14.95
TOTAL EXPENSES	**63.02**	**9.23**	**229.54**	**115.93**	**323.61**	**163.44**	**616.17**
CASH GENERATION					55.62	28.09	**55.62**
Weaving Division	20.33	2.98					20.33
Processed Cloth Sales Division					55.62	28.09	55.62
TOTAL	**20.33**				**55.62**		**75.95**
EBITDA	**20.33**	**2.98**			**55.62**	**28.09**	**75.95**

Avg. GLM (Gram/l inear mtr) Sulzer	127
Avg. GLM Air Jet	435
Avg Yarn Rate / Kg	243.84
Avg. Wvg Charges Paisa/Pick	19.32
Avg. Wvg Cost Paisa/Pick	15.11
Avg. Processing Charges. Rs./Mtrs	35.13
Grey Fabric Sale rate	nil
Profit / Mtr	nil
% of Profi	nil
Processed Fabric Sale Rate Rs./ Mtrs	191.53
Profit Rs./ Mtrs	28.09
% of Profi	17.19
Conversion charges – Group Co Rs in lacs	63.02
Sales - Processed Fabric Rs in lacs	379.23
TOTAL	**442.25**

CONVERSION CHARGES		
Sulzer	**Paisa /pick**	
QN —	14.0	3 Width with sizing
QN —	12.0	3 Width with sizing
QN —	12.0	3 Width without sizing
Air-Jet		
QN —	24.0	without sizing with Dobby name writer
QN —	22.0	without sizing with Dobby name writer
For Market	18.0	Sizing Rs.15/- per kg

*Avg-Average, GLM-Grams per linear meter

WORKING CAPITAL REQUIREMENT

Stock	Days	Amt. – Rs lacs
Yarn Godown	10	66.77
Work in Progress (WIP)	15	100.16
Grey Fabric Godown / Packing	10	66.77
Process House	45	300.48
Finished Godown Including Grey for Sale	10	66.77
Debtors	21	140.22
Total	**111**	**741.18**

WEAVING PLAN - (April to March)

(48 AirJet + 32 Sulzer Projectile Looms)

Working Days ⇒ 29.17 Each Month

Quality No.	Construction	Width	Reed	Picks/ Inch	Looms / Day	R.P.M.	100% Mts /Day	Effi	Mtrs/ loom day	Mtrs. lacs/ Day	Mtrs. lacs/ Month	Con. Rs / mtrs
Sulzer For Group Co												
	28 PC x 28 PC 3 w	39.50"	72	60	22	250	457.2	75	342.9	0.07	2.20	8.40
	40 PC x 40 PC 3 w	38.50"	92	76	3	250	360.95	75	270.7	0.01	0.23	9.12
	2/80 PC x 2/80 PC 3 w	38.50"	96	76	3	250	360.95	75	270.7	0.01	0.23	9.12
	2/40 PV x 2/40 PV 3 w	38.50"	52	52	4	250	527.5	75	395.7	0.02	0.46	6.24
Total				61.18	32					0.10	3.12	
Air Jet												
For Group Co												
	2/15 PV x 2/15 PV	61"	66	42	12	675	587.8	85	499.7	0.06	1.73	10.08
	2/30 PV x 2/30 PV	63"	58	46	15	675	536.7	85	456.2	0.07	1.98	10.12
Total for Group Co				44.13	27					0.12	3.71	
Own Marketing												
	2/40 C x 2/40 C	63" 2/1	128	68	5	675	363.1	85	308.6	0.01	0.44	16.26
	2/40 C x 20 C	63" 2/1	114	64	5	675	385.8	85	327.9	0.02	0.48	15.10
	2/40 C x 300 D	63" 2/1	128	60	6	675	411.5	85	349.8	0.02	0.61	14.82
	30 C x 10 OE	63" 3/1	132	68	5	675	363.1	85	308.6	0.02	0.45	15.01
Total for Marketing				64.57	21					0.07	1.98	
AirJet Total				51.24	48					0.19	5.69	
Grand Total					80					0.29	8.81	

Average Picks ⇒ 54.76

B – Quality-wise Yarn Requirement & Usage   **C – Count-wise Yarn Requirement**

Quality No.	Usage	Count		Weight / Mtrs (Gms)		Yarn Reqd./ Day		Count	Kgs.	MT	Yarn	Amt.
		wp	wt	wp	Wt	wp	wt		day	Month	Rs./kg	in lacs
2828	Job work Group Co	28 PC	28 PC	0.070	0.052	534	397	28s pc	930	27.91		
4049	"	40 PC	40 PC	0.062	0.046	49	37	40s pc	86	2.58		
280292	"	2/80 PC	2/80 PC	0.065	0.046	52	37	2/80s pv	89	2.66		
240241	"	2/40 PV	2/40 PV	0.071	0.063	113	100	2/15s pv	3556	106.69		
1005	"	2/15 PV	2/15 PV	0.378	0.215	2267	1289	2/30s pv	1949	58.48		
230232	"	2/30 PV	2/30 PV	0.166	0.118	1139	810	2/40s pv	214	6.41		
Total for Group Co- A						4155	2670		6825	204.74		
AJ 2417	Dyed Fabric sale	2/40 C	2/40 C	0.268	0.130	409	198	10s OE	404	12.12	195.0	23.64
AJ 2412	"	2/40 C	20 C	0.239	0.122	398	203	20s c	203	6.09	240.0	14.62
AJ 2429	"	2/40 C	300 D	0.268	0.129	567	273	30s c	287	8.61	220.0	18.94
AJ 3008	"	30 C	10 OE	0.184	0.259	287	404	2/40s c	1572	47.15	285.0	134.36
								300d	273	8.18	107.0	8.76
Total for Market- B						1660	1078		2738	82.15		
G.Total (A + B)							9563		9563	286.89		200.32

CONVERSION CHARGES ⇒

	Q,NO	Paisa / Pick	
Sulzer	2828	14.0	3 Width with sizing
	4049,280292	12.0	3 Width with sizing
	240241	12.0	3 Width without sizing
AirJet	1005	24.0	without sizing with name write
	230232	22.0	without sizing with name write
	For Market	18.0	Sizing Rs.15/- per kg

Remark: Required one Sizing m/c to achieve above Quantity

PROJECTIONS (April - March) - 29.17 Days Each Month

A. GREY CLOTH - JOB TRANSFER (Group Company)

Quality No.	2828	4049	280292	240241	1005	230232	Avg/mtrs	Total
CONSTRUCTION	28s x 28s PC	40s x 40s PC	(2/80s x 2/80s) PC	(2/40s x 2/40s) PV	(2/15s x 2/15s) PV	(2/30s x 2/30s) PV		
Looms	Sulzer	Sulzer	Sulzer	Sulzer	Air-Jet	Air-Jet		
Width	39.50"	38.50"	38.50"	38.50"	61"	63"		
Ends/Inch	72	92	96	52	66	58		
Picks/Inch	60	76	76	52	42	46		
Weight of warp Kg/Mtr	0.070	0.062	0.065	0.071	0.378	0.166		
Weight of weft Kg/Mtr	0.052	0.046	0.046	0.063	0.215	0.118		
Sizing weight @ 10%	0.007	0.006	0.007					
Total Weight Kg / Mtrs	0.129	0.115	0.118	0.133	0.593	0.284		
Mtrs/Kg	7.78	8.72	8.50	7.49	1.69	3.52		
Prod. Mtrs in Lacs	2.20	0.23	0.23	0.46	1.73	1.98		6.83
Hard waste . Sulzer .2%, Air Jet.4%								
Weaving.Charges. Rs./Mtrs	8.40	9.12	9.12	6.24	10.08	10.12		
Weaving.Charges. Amt. Rs.in lacs	18.48	2.10	2.10	2.87	17.44	20.04		
Total Cost Amt. Rs.in lacs	18.48	2.10	2.10	2.87	17.44	20.04		63.02
Billing Rate / Mtrs	8.40	9.12	9.12	6.24	10.08	10.12	9.23	
Brokerage & Commission								
Total Rs. in lacs	18.48	2.10	2.10	2.87	17.44	20.04		63.02

B. TRANSFER FOR PROCESSING

Sort No.	AJ 2417	AJ 2412	AJ 2429	AJ 3008		Total
CONSTRUCTION	2/40 x 2/40 C	2/40 x 20 C	2/40 C x 300 D Poly	30 C x 10 C OE		
Looms	Air-Jet	Air-Jet	Air-Jet	Air-Jet		
Weave	Twill 2/1	Twill 2/1	Twill 2/1	Twill 3/1		
Width	63"	63"	63"	63"		
Yarn Cost / kg Warp	285	285	285	220		
Weft	285	240	107	195		
Ends/Inch	128	114	128	132		
Picks/Inch	68	68	60	68		
Weight of warp Kg / Mtr	0.268	0.239	0.268	0.184		
Weight of weft Kg / Mtr	0.130	0.122	0.129	0.259		
Sizing weight @ 10%	0.027	0.024	0.027	0.018		

Total Weight Kg / Mtrs	0.424	0.385	0.424	0.462		
For Processed Mtrs in Lacs	0.44	0.48	0.61	0.45		1.98
ESTIMATED COST OF GREY						Total Amt. in lacs
Yarn Cost / Mtr - Warp	76.43	68.07	76.43	40.56		
Weft	36.91	29.25	13.80	50.51		
YARN COST -Total (RS/MTR)	113.34	97.32	90.23	91.07	97.27	192.60
Hard waste @ -3.5% (Rs / Mtr)	3.97	3.41	3.16	3.19	3.40	6.74
Wvg charges Rs./mtrs	16.26	15.10	14.82	15.01	15.25	30.20
TOTAL GREY COST (RS/MTR)	133.56	115.83	108.21	109.26	115.93	229.54

April to-March - 29.17 Days Each Month

C.PROCESSED CLOTH SALES DIVISION

Sort No.	AJ 2417	AJ 2412	AJ 2429	AJ 3008	Avg / Mtrs	Total
CONSTRUCTION	2/40 x 2/40 C	2/40 x 20 C	2/40 x 300D	30 C x 10 OE		
Looms	Air-Jet	Air-Jet	Air-Jet	Air-Jet		
Width	63"	63"	63"	63"		
Yarn Cost Rs./Kg Warp	285	285	285	220		
Weft	285	240	107	195		
Ends / Inch	128	114	128	132		
Picks / Inch	68	68	60	68		
Total Weight / Mtrs	0.418	0.381	0.478	0.470		
For Process Mtrs in lacs	0.44	0.48	0.61	0.45		1.98
						Total Amt. in lacs
TOTAL GREY COST (RS/MTR)	133.56	115.83	108.21	109.26	115.93	229.54
Dyeing Cost @ Rs 80/-/Kg	33.42	30.46	38.22	37.57	35.13	69.55
Pigment Blotch / Mtr						
Shrinkage 2.5%	4.17	3.66	3.66	3.67	3.78	7.48
Value Loss 2.5%	4.17	3.66	3.66	3.67	3.78	7.48
Transport	1.00	1.00	1.00	1.00	1.00	1.98
Total Cost of Processed fabric	176.33	154.61	154.75	155.18	159.61	316.03
Brokerage - 2%	4.23	3.71	3.71	3.72	3.83	7.58
Sale Rate Rs./ Mtr (Cost + 20%)	211.60	185.53	185.70	186.21	191.53	379.23
Contribution / Mtrs Rs	31.03	27.21	27.24	27.31	28.09	55.62

MANUFACTURING EXPENSES (Weaving) April to March Ave working days -29.17

PARTICULARS		PER-DAY
Avg Working days / Month	29.17	
Looms	80.0	
Avg. Picks/Inch	54.76	
Avg. Loom Speed - Airjet / Sulzer PPM(Picks / minute)	675 / 250	
Efficiency % - AirJet / Sulzer	85 / 75	
PRODUCTION Mtrs in lacs	8.81	0.29
No. of Workers (P.)	2610	87
No. of Workers. (C.)	1950	65

REMARK	
(48 AirJet + 32 Sulzer Looms)	
Summary of prod.	Mtrs lacs
Group Co - Airjet	3.71
- Sulzer	3.12
Total Group Co	6.83
Market sale - Air-jet	1.98
- Sulzer	nil
Total market sale	1.98
G.Total	8.81

Engg. Workers. (P.)	330	11	
Engg. Workers. (C.)	30	1	
General Workers. (P.)	180	6	
General Workers. (C.)	420	14	
Total Manpower	**5520**	**184**	
Power (Govt Supply) Kwh.	600000	20000	
Wood / Coal Tons.	120.0	4.00	Wood - 2.5 tons / Coal - 1.5 tons

	Rs. in lcs	Rs./ Mtrs	Ps./ Pick	
CONVERSION CHARGES	**93.22**	**10.58**	19.32	
Wages - Direct Weaving	10.46	1.19	2.17	Perm. @ Rs.255/- & Cont @ Rs.195/- per Day
Wages - Engineering	0.90	0.10	0.19	
Total - (A) Wages	**11.36**	**1.29**	2.35	
Salaries - Direct & Mktg.	3.82	0.43	0.79	
Year End	1.62	0.18	0.34	
Stores and Spares	4.32	0.49	0.90	Rs.60 per loom shift
Store/Spares (Wdg / Wvg.Prep)	0.31	0.03	0.06	
Stores/Spares for Compressor/Engg	0.70	0.08	0.15	
Sizing/ Material	5.14	0.58	1.07	
Wood / Coal	4.35	0.49	0.90	Rs. 2500 / Rs. 5500 per MT
Power Expenses	29.10	3.30	6.03	Unit Rate Rs. 4.85 per Unit.
Grey & Mended Inspection / Fabric Handling	1.23	0.14	0.26	
Mending Charges	3.08	0.35	0.64	
Packing Materials	2.20	0.25	0.46	
Total B (Direct Wvg. Expenses)	**55.88**	**6.34**	**11.58**	
Wages - General	1.28	0.15	0.26	
Salaries - Common (30% Loading)	2.58	0.29	0.53	
Admn. & Factory OH.	1.80	0.20	0.37	
Total C (Common Expenses - 30%)	**5.66**	**0.64**	**1.17**	
TOTAL EXPENSES (A +B + C)	**72.89**	**8.27**	**15.11**	
NET CASH GENERATION	**20.33**	**2.31**	**4.21**	

* P-Permanent, C-Contracts, Wdg-winding, Wvg-weaving, Prep-Preparatory, Engg-Engineering, Admin-Administration, OH-Overheads

BREAK-UP OF MANUFACTURING EXPENSES / P & L - April to March - 29.17 days/month

PARTICULARS	Sulzer			Air-Jet			Total		
	For month	Per Mtrs.	Per Pick	For month	Per Mtrs.	Per / Pick	For month	Per Mtrs.	Per / Pick
Working Days	29.17			29.17			29.17		
Installed looms	64			48			112		
Working Looms	32			48			80		
Looms Speed	250			675					
Avg. Picks	61.18			51.24			54.76		
Production Mtrs in lacs	**3.12**			**5.69**			**8.81**		

A. Billing Rs. in lacs	25.56			67.85			93.41		
Rate per Mtrs.		8.19	13.39		11.92	23.27		10.60	19.36
B. EXPENSES									
Wages -	4.17	1.34	2.18	8.47	1.49	2.91	12.64	1.43	2.62
Salaries -	1.60	0.51	0.84	4.80	0.84	1.65	6.40	0.73	1.33
Mending	1.60	0.51	0.84	2.71	0.48	0.93	4.31	0.49	0.89
Power	7.27	2.33	3.81	21.82	3.84	7.49	29.10	3.30	6.03
Stores & Spares cons.	1.99	0.64	1.04	3.34	0.59	1.15	5.33	0.60	1.10
Sizing Materials	1.42	0.46	0.74	3.72	0.65	1.28	5.14	0.58	1.07
Wood / Coal	1.56	0.50	0.82	2.79	0.49	0.96	4.35	0.49	0.90
Packing Materials	0.79	0.25	0.41	1.41	0.25	0.48	2.20	0.25	0.46
Common Expenses	0.86	0.28	0.45	2.56	0.45	0.88	3.42	0.39	0.71
B. Expenses Rs. in lacs	21.26	6.82	11.14	51.62	9.07	17.71	72.89	8.27	15.11
Net Cash Generation (A-B) in lacs	4.30	1.38		16.22	2.85		20.52	2.33	
Expenses Per Pick Paisa			11.14			17.71			15.11
Avg.Picks/Inch	61.18			51.24			54.76		
Billing Per Pick Paisa			13.39			23.27			19.36

CURRENT ASSETS & CURRENT LIABILITIES (CA/CL)

Rs. In Lacs

PARTICULARS	Provisional as on 31.03.15	Budgeted as on 31.03.16	Increase (+)/Decrease (-)
Yarn Production		262800 Kgs	
Cloth Production		873000 Mtrs	
CURRENT ASSETS			
Cash & Bank balance	4.6	5.0	0.4
Loans & Advances	2.2	5.0	2.8
Prepaid Expenses	4.2	6.0	1.8
Security Deposits	61.3	61.0	-0.3
Receivables (TDS)	2.7	3.0	0.3
Stocks			
Raw Material (Godown / Transit)	77.8	195.2	117.4
Purchase Yarn (Godown)	4.7	68.3	63.6
Stock in Process			
Spinning Deptt.	49.9	99.2	49.3
Doubling / Winding / Loose	19.4	24.6	5.2
Weaving Deptt.	1.0	80.5	79.5
Grey Fabric under mending		23.2	23.2
Grey Fabrics in Godown		76.6	76.6
Grey Fabrics with Processor		344.5	344.5
Finished stock			
Yarn	24.0	88.9	64.9

Grey Fabrics			
Finished Fabrics		126.7	126.7
Stores & Consumables	8.9	30.0	21.1
Waste Stock	3.7	5.0	1.3
Sundry Debtors			
Yarn	8.6	125.5	116.9
Grey Fabrics	0.1		−0.1
Finished Fabrics	1.3	266.9	265.6
Others − Weaving Job Work	26.6	63.0	36.4
Advance to Processors			
Advance to Store Suppliers	18.8		−18.8
(Capital Commitment & Others)			
Total Current Assets	**319.7**	**1698.1**	**1378.4**

CURRENT LIABILITIES			
Sundry Creditors			
Raw Material - Cotton & Yarn	289.7		−289.7
Power & Fuel	48.2	60.0	11.8
Stores & Consumables	86.1	25.0	−61.1
Statutory / Salary & Wages Prov.	335.6	85.5	−250.1
Other Creditors	63.2	18.5	−44.7
Advance against Grey Sales	1.6		−1.6
Advance against Waste Sales	1.8		−1.8
Total Current Liabilities	**826.3**	**189.0**	**−637.3**
Net Current Assets	**−506.6**	**1509.1**	**2015.7**
Expected PBDIT Apr 10 to Mar 11			**1079.4**
Net Induction Required			**936.3**

DETAILS OF SHOULD BE STOCK (Yarn - 262800 Kgs)

Particulars	Remark	Quantity	Amount Rs.in Lac	Avg.Rate / Kg /Mtr
Raw Material Stock (Godown)				
Cotton	15 Days in RM Godown	58200 Kgs	79.02	135.78
Polyester	15 Days in RM Godown	54900 Kgs	65.88	120.00
Viscose	15 Days in RM Godown	30500 Kgs	50.33	165.00
Total			**195.23**	
Process Stock				
Spinning Section	7 Days in Process	67000 Kgs	99.16	148.00
Doubling / Winding /Loose	3 Days in Process	12500 Kgs	24.63	197.00
Total			123.79	
Finished Stock				
Yarn For Sale	5 Days	43600 Kgs	88.94	204.00
Total			88.94	
A: Grand Total			**407.96**	

DETAILS OF SHOULD BE DEBTORS

B: Yarn Sales Market	7 Days	61500 Kgs	125.46	204.00
Grand Total (A+B)			**533.42**	
			533.42	

DETAILS OF SHOULD BE STOCK (Fabric - 881000 Mtrs.)

(Fabric - 198000 Mtrs. & Group Co Job work - 683000 Mtrs.)

Particulars	No of Days	Quantity Kg	Amount Rs.in Lac	Avg.Rate / Kg /Mtr
Raw Material Stock (Godown)				
Yarn	10	28000	68.32	244.00
C: Total			68.32	
Process Stock				
Weaving Dept (Warping, Sizing, Weaving)	12	33000	80.52	244.00
Grey Fabric Inspection / Mending	3	20000	23.20	116.00
Grey Fabric - In Godown / Packing	10	66000	76.56	116.00
(For Lot Making)				
Grey Fabric with Processor	45	297000	344.52	116.00
D: Total			524.80	
Finished stock				
Finished Fabric for Sale	10	66000 Mtrs	126.72	192.00
E: Total			126.72	
F: Grand Total (C+D+E)			719.84	
SHOULD BE DEBTORS				
Finished Fabric Sales	21	139000 Mtrs	266.88	192.00
G: Total			266.88	
Grey Cloth Job Charges	30	682600 Mtrs	63.00	9.23
H: Total			63.00	
I: Grand Total (F+G+H)			1049.72	

OVERALL CASH FLOW PLANNING SUMMARY

Particulars / (Amt Rs in lacs)	APR	MAY	JUN	JUL	AUG	SEP	OCT	NOV	DEC	JAN	FEB	MAR	TOTAL
Opening Balance													
INFLOWS													
Yarn Sales - Market	257.00	257.00	300.00	558.00	558.00	558.00	558.00	558.00	558.00	558.00	558.00	558.00	5,836.00
Grey Sales – Market													
Grey Job Work	40.00	40.00	40.00	30.00	63.00	63.00	63.00	63.00	63.00	63.00	63.00	63.00	654.00
Finished Fabric Sales				368.00	368.00	368.00	368.00	368.00	368.00	368.00	368.00	368.00	3,312.00
Waste Sales	3.00	3.00	2.00	2.00	2.00	2.00	2.00	2.00	2.00	2.00	2.00	2.00	26.00
Corporate - Working Fund	481.00	350.00	868.00										1,699.00
Corporate – Capex	696.00												696.00
Corporate - Capex (Civil, General, Computers)	347.00												347.00
Corporate – Backlog	41.00	41.00	41.00	40.00	41.00								204.00
TOTAL INFLOWS	1,865.00	691.00	1,251.00	998.00	1,032.00	991.00	991.00	991.00	991.00	991.00	991.00	991.00	12,774.00
OUTFLOW													
RAW MATERIAL													
Cotton	231.00	214.00	300.00	164.00	164.00	164.00	164.00	164.00	164.00	164.00	164.00	164.00	2,221.00
Polyester			262.00	143.00	143.00	143.00	143.00	143.00	143.00	143.00	143.00	143.00	1,549.00
Viscose			190.00	104.00	104.00	104.00	104.00	104.00	104.00	104.00	104.00	104.00	1,126.00
Purchase Yarn	375.00	204.00	204.00	204.00	204.00	204.00	204.00	204.00	204.00	204.00	204.00	204.00	2,619.00
Processing Charges		70.00	70.00	70.00	70.00	70.00	70.00	70.00	70.00	70.00	70.00	70.00	770.00
Entry Tax/Sales Tax, etc.	3.00	3.00	8.50	8.50	8.50	8.50	8.50	8.50	8.50	8.50	8.50	8.50	91.00

Transportation, etc.	4.00	4.00	4.50	6.50	4.50	3.50	3.50	3.50	3.50	3.50	3.50	3.50	48.00
Comm./Selling Overhead Expenses	3.00	3.00	7.00	14.00	14.00	14.00	14.00	14.00	14.00	14.00	14.00	14.00	139.00
TOTAL	**616.00**	**498.00**	**1,046.00**	**714.00**	**712.00**	**711.00**	**711.00**	**711.00**	**711.00**	**711.00**	**711.00**	**711.00**	**8,563.00**
MANUFACTURING EXPENSES													
Wages/Salary/Cont. Wages	48.00	48.00	53.00	54.00	54.00	52.00	55.00	55.00	55.00	57.00	57.00	57.00	645.00
Provident Fund	10.00	10.00	10.00	10.00	10.00	10.00	10.00	10.00	10.00	10.00	10.00	10.00	120.00
E.S.I.C.	3.00	3.00	2.00	2.00	2.00	2.00	2.00	2.00	2.00	2.00	2.00	2.00	26.00
L.I.C././Society	3.00	3.00	2.00	2.00	2.00	2.00	2.00	2.00	2.00	2.00	2.00	2.00	26.00
Power Govt Supply.	45.00	45.00	60.00	60.00	60.00	60.00	60.00	60.00	60.00	60.00	60.00	60.00	690.00
Stores Spares & Packing Materials	19.00	19.00	28.00	27.00	26.00	25.00	25.00	25.00	25.00	25.00	25.00	25.00	294.00
Administrative Overhead Expenses	6.00	6.00	6.00	6.00	6.00	6.00	6.00	6.00	6.00	6.00	6.00	6.00	72.00
TOTAL	**134.00**	**134.00**	**161.00**	**161.00**	**160.00**	**157.00**	**160.00**	**160.00**	**160.00**	**162.00**	**162.00**	**162.00**	**1,873.00**
Others													
Vehicle EMI	2.00	2.00	2.00	2.00	2.00	2.00	2.00	2.00	2.00	2.00	2.00	2.00	24.00
TDS\Service Tax\FBT Tax	2.00	2.00	2.00	2.00	2.00	2.00	2.00	2.00	2.00	2.00	2.00	2.00	24.00
Gratuity Fund (LIC)	45.00	35.00	27.00	27.00	27.00	2.00	2.00	2.00	2.00	2.00	2.00	2.00	175.00
Hank Yarn Obligation	3.00	3.00	1.00	1.00	1.00	1.00	1.00	1.00	1.00	1.00	1.00	1.00	16.00
Property Tax	20.00	17.00	12.00	12.00	14.00	2.00	2.00	2.00	2.00	2.00	2.00	2.00	89.00
Capex	696.00												696.00
Capex (General, Civil & Computers)	347.00												347.00
Corporate				79.00	114.00	114.00	111.00	111.00	111.00	109.00	109.00	109.00	967.00
TOTAL	**1,115.00**	**59.00**	**44.00**	**123.00**	**160.00**	**123.00**	**120.00**	**120.00**	**120.00**	**118.00**	**118.00**	**118.00**	**2,338.00**
TOTAL OUTFLOWS	**1,865.00**	**691.00**	**1,251.00**	**998.00**	**1,032.00**	**991.00**	**991.00**	**991.00**	**991.00**	**991.00**	**991.00**	**991.00**	**12,774.00**
SHORT FALL/ SURPLUS													

* FBT-Fringe benefits tax, TDS-Tax deducted at source

CASH FLOW PLANNING SPINNING DIVISION

Particulars/ (Amt Rs in lacs)	APR	MAY	JUN	JUL	AUG	SEP	OCT	NOV	DEC	JAN	FEB	MAR	TOTAL
Opening Balance													
INFLOWS													
Yarn Sales - Market	257.00	257.00	300.00	558.00	558.00	558.00	558.00	558.00	558.00	558.00	558.00	558.00	5,836.00
Grey Sales – Market													
Weaving Job Work													
Finished Fabric Sales													
Waste Sales	3.00	3.00	2.00	2.00	2.00	2.00	2.00	2.00	2.00	2.00	2.00	2.00	26.00
Corporate - Working Fund	80.00	50.00	560.00										690.00
Corporate – Capex	366.00												366.00
Corporate - Capex (Civil, General, Computers)	**347.00**												**347.00**
Corporate – Backlog													
TOTAL INFLOWS	**1,053.00**	**310.00**	**862.00**	**560.00**	**560.00**	**560.00**	**560.00**	**560.00**	**560.00**	**560.00**	**560.00**	**560.00**	**7,265.00**

OUTFLOW													
RAW MATERIAL													
Cotton	231.00	214.00	300.00	164.00	164.00	164.00	164.00	164.00	164.00	164.00	164.00	164.00	2,221.00
Polyester			262.00	143.00	143.00	143.00	143.00	143.00	143.00	143.00	143.00	143.00	1,549.00
Viscose			190.00	104.00	104.00	104.00	104.00	104.00	104.00	104.00	104.00	104.00	1,126.00
Purchase Yarn													
Processing Charges													
Entry Tax/Sales Tax, etc.	3.00	3.00	8.50	8.50	8.50	8.50	8.50	8.50	8.50	8.50	8.50	8.50	91.00
Transportation, etc.	2.00	2.00	2.50	2.50	2.50	2.50	2.50	2.50	2.50	2.50	2.50	2.50	29.00
Comm./Selling Overhead Expenses	2.00	2.00	6.00	6.00	6.00	6.00	6.00	6.00	6.00	6.00	6.00	6.00	64.00
TOTAL	**238.00**	**221.00**	**769.00**	**428.00**	**428.00**	**428.00**	**428.00**	**428.00**	**428.00**	**428.00**	**428.00**	**428.00**	**5,080.00**
MANUFACTURING EXPENSES													
Wages/Salary/Cont. Wages	32.00	32.00	35.00	36.00	36.00	36.00	37.00	37.00	37.00	38.00	38.00	38.00	432.00
Provident Fund	6.00	6.00	7.00	7.00	7.00	7.00	7.00	7.00	7.00	7.00	7.00	7.00	82.00
E.S.I.C.	2.00	2.00	1.00	1.00	1.00	1.00	1.00	1.00	1.00	1.00	1.00	1.00	14.00
L.I.C././Society	2.00	2.00	1.00	1.00	1.00	1.00	1.00	1.00	1.00	1.00	1.00	1.00	14.00
Power Govt supply	27.00	27.00	31.00	31.00	31.00	31.00	31.00	31.00	31.00	31.00	31.00	31.00	364.00
Stores Spares & Packing Materials	7.00	7.00	9.00	9.00	9.00	9.00	9.00	9.00	9.00	9.00	9.00	9.00	104.00
Administrative Overhead Expenses	4.00	4.00	4.00	4.00	4.00	4.00	4.00	4.00	4.00	4.00	4.00	4.00	48.00
TOTAL	**80.00**	**80.00**	**88.00**	**89.00**	**89.00**	**89.00**	**90.00**	**90.00**	**90.00**	**91.00**	**91.00**	**91.00**	**1,058.00**
Others													
Vehicle EMI	1.00	1.00	1.00	1.00	1.00	1.00	1.00	1.00	1.00	1.00	1.00	1.00	12.00
TDS\Service Tax\FBT Tax	1.00	1.00	1.00	1.00	1.00	1.00	1.00	1.00	1.00	1.00	1.00	1.00	12.00
Gratuity Fund (LIC)	12.00	2.00	1.00	1.00	1.00	1.00	1.00	1.00	1.00	1.00	1.00	1.00	24.00
Hank Yarn Obligation	3.00	3.00	1.00	1.00	1.00	1.00	1.00	1.00	1.00	1.00	1.00	1.00	16.00
Property Tax	5.00	2.00	1.00	1.00	1.00	1.00	1.00	1.00	1.00	1.00	1.00	1.00	17.00
Capex	366.00												366.00
Capex (General, Civil & Computers)	347.00												347.00
Corporate				38.00	38.00	38.00	37.00	37.00	37.00	36.00	36.00	36.00	333.00
TOTAL	**735.00**	**9.00**	**5.00**	**43.00**	**43.00**	**43.00**	**42.00**	**42.00**	**42.00**	**41.00**	**41.00**	**41.00**	**1,127.00**
TOTAL OUTFLOWS	**1,053.00**	**310.00**	**862.00**	**560.00**	**560.00**	**560.00**	**560.00**	**560.00**	**560.00**	**560.00**	**560.00**	**560.00**	**7,265.00**
SHORT FALL/ SURPLUS													

CASH FLOW PLANNING WEAVING DIVISION

Particulars/ (Amt Rs in lacs)	APR	MAY	JUN	JUL	AUG	SEP	OCT	NOV	DEC	JAN	FEB	MAR	TOTAL
Opening Balance													
INFLOWS													
Yarn Sales - Market													
Grey Sales – Market													
Grey Job Work	40.00	40.00	40.00	30.00	63.00	63.00	63.00	63.00	63.00	63.00	63.00	63.00	654.00

												Total
Finished Fabric Sales				368.00	368.00	368.00	368.00	368.00	368.00	368.00	368.00	368.00	3,312.00
Waste Sales													
Corporate - Working Fund	401.00	300.00	308.00										1,009.00
Corporate – Capex	330.00												330.00
Corporate - Capex (Civil, General, Computers)													
Corporate – Backlog													
TOTAL INFLOWS	**771.00**	**340.00**	**348.00**	**398.00**	**431.00**	**431.00**	**431.00**	**431.00**	**431.00**	**431.00**	**431.00**	**431.00**	**5,305.00**
OUTFLOW													
RAW MATERIAL													
Cotton													
Polyester													
Viscose													
Purchase Yarn	375.00	204.00	204.00	204.00	204.00	204.00	204.00	204.00	204.00	204.00	204.00	204.00	2,619.00
Processing Charges		70.00	70.00	70.00	70.00	70.00	70.00	70.00	70.00	70.00	70.00	70.00	770.00
Entry Tax/Sales Tax, etc.													
Transportation, etc.	1.00	1.00	1.00	3.00	1.00	1.00	1.00	1.00	1.00	1.00	1.00	1.00	14.00
Comm./Selling Overhead Expenses	1.00	1.00	1.00	8.00	8.00	8.00	8.00	8.00	8.00	8.00	8.00	8.00	75.00
TOTAL	**377.00**	**276.00**	**276.00**	**285.00**	**283.00**	**283.00**	**283.00**	**283.00**	**283.00**	**283.00**	**283.00**	**283.00**	**3,478.00**
MANUFACTURING EXPENSES													
Wages/Salary/Cont. Wages	14.00	14.00	16.00	16.00	16.00	16.00	18.00	18.00	18.00	19.00	19.00	19.00	203.00
Provident Fund	4.00	4.00	3.00	3.00	3.00	3.00	3.00	3.00	3.00	3.00	3.00	3.00	38.00
E.S.I.C.	1.00	1.00	1.00	1.00	1.00	1.00	1.00	1.00	1.00	1.00	1.00	1.00	12.00
L.I.C./Society	1.00	1.00	1.00	1.00	1.00	1.00	1.00	1.00	1.00	1.00	1.00	1.00	12.00
Power Govt Supply	18.00	18.00	29.00	29.00	29.00	29.00	29.00	29.00	29.00	29.00	29.00	29.00	326.00
Stores Spares & Packing Materials	9.00	9.00	16.00	16.00	16.00	16.00	16.00	16.00	16.00	16.00	16.00	16.00	178.00
Administrative Overhead Expenses	2.00	2.00	2.00	2.00	2.00	2.00	2.00	2.00	2.00	2.00	2.00	2.00	24.00
TOTAL	**49.00**	**49.00**	**68.00**	**68.00**	**68.00**	**68.00**	**70.00**	**70.00**	**70.00**	**71.00**	**71.00**	**71.00**	**793.00**
Others													
Vehicle EMI	1.00	1.00	1.00	1.00	1.00	1.00	1.00	1.00	1.00	1.00	1.00	1.00	12.00
TD&\Service Tax\FBT Tax	1.00	1.00	1.00	1.00	1.00	1.00	1.00	1.00	1.00	1.00	1.00	1.00	12.00
Gratuity Fund (LIC)	8.00	8.00	1.00	1.00	1.00	1.00	1.00	1.00	1.00	1.00	1.00	1.00	26.00
Hank Yarn Obligation													
Property Tax	5.00	5.00	1.00	1.00	1.00	1.00	1.00	1.00	1.00	1.00	1.00	1.00	20.00
Capex	330.00												330.00
Capex (General, Civil & Computers)													
Corporate				41.00	76.00	76.00	74.00	74.00	74.00	73.00	73.00	73.00	634.00
TOTAL	**345.00**	**15.00**	**4.00**	**45.00**	**80.00**	**80.00**	**78.00**	**78.00**	**78.00**	**77.00**	**77.00**	**77.00**	**1,034.00**
TOTAL OUTFLOWS	**771.00**	**340.00**	**348.00**	**398.00**	**431.00**	**431.00**	**431.00**	**431.00**	**431.00**	**431.00**	**431.00**	**431.00**	**5,305.00**
SHORT FALL/ SURPLUS													

Particulars/ (Amt Rs in lacs)	CORPORATE BACK LOG CASH FLOW PLAN					TOTAL
	APR	MAY	JUN	JUL	AUG	
Opening Balance						
INFLOWS						
Yarn Sales - Market						
Grey Sales - Market						
Weaving Job Work						
Finished Fabric Sales						
Waste Sales						
Corporate - Working Fund						
Corporate - Capex						
Corporate - Capex (Civil, General, Computers)						
Corporate - Backlog	41.00	41.00	41.00	40.00	41.00	204
TOTAL INFLOWS	41.00	41.00	41.00	40.00	41.00	204.00
OUTFLOW						
RAW MATERIAL						
Cotton						
Polyester						
Viscose						
Purchase Yarn						
Processing Charges						
Entry Tax/Sales Tax, etc.						
Transportation, etc.	1.00	1.00	1.00	1.00	1.00	5
Comm./Selling Overhead Expenses						
TOTAL	1.00	1.00	1.00	1.00	1.00	5.00
MANUFACTURING EXPENSES						
Wages/Salary/Cont. Wages	2.00	2.00	2.00	2.00	2.00	10
Provident Fund						
E.S.I.C.						
L.I.C./.Society						
Power M.P.E.B.						
Stores Spares & Packing Materials	3.00	3.00	3.00	2.00	1.00	12
Administrative Overhead Expenses						
TOTAL	5.00	5.00	5.00	4.00	3.00	22.00
Others						
Vehicle EMI						
TDS\Service Tax\FBT Tax						
Gratuity Fund (LIC)	25.00	25.00	25.00	25.00	25.00	125
Hank Yarn Obligation						
Property Tax	10.00	10.00	10.00	10.00	12.00	52
Capex						
Capex (General, Civil & Computers)						
Corporate						
TOTAL	35.00	35.00	35.00	35.00	37.00	177.00
TOTAL OUTFLOWS	41.00	41.00	41.00	40.00	41.00	204.00
SHORT FALL/SURPLUS						

ADMIN & FACTORY OVERHEAD EXPENSES FOR THE YEAR 2011 -12

PARTICULARS	BUDGET
ADMINISTRATION EXPENSES	
ADVERTISEMENT EXPENSES	2000
AUDITORS' EXPENSES	1000
BANK CHARGES	25000
BOOKS & PERIODICALS	1000
COMPUTER EXPENSES	6000
CONVEYANCE CHARGES	12000
CHANDALA & DONATION EXPENSES	3000
LEGAL & PROFESSIONAL CHARGS	20000
POSTAGE & COURIERS EXPENSES.	6000
RECRUITMENT EXPENSES	2000
STAFF WELFARE EXPENSES	6000
STATIONERY & PRINTING EXPENSES	20000
SUBSCRIPTION CHARGES	5000
TRAVELLING EXPENSES	50000
UNIFORM EXPENSES	3000
VEHICLE EXPENSES	50000
GUEST HOUSE/STAFF MESS EXPENSES	15000
REFRESHMENT TO STAFF	10000
QUARTER ELECTRICITY EXPENSES	20000
GARDEN EXPENSES	4000
GENERAL & OFFICE EXPENSES	6000
DIRECTOR FEES	20000
WASHING & PARADE EXPENSES	2000
TEMPLE EXPENSES	8000
TELEPHONE EXPENSES	20000
MOBILE PHONE EXPENSES	10000
AGRICULTURAL EXPENSES	5000
DEEPAWALI EXPENSES	4000
VEHICLE HIRE CHARGES	2000
ENTERTAINMENT EXPENSES	1000
INTERNET EXPENSES	1000
REPAIRS TO OTHERS	12000
RENT EXPENSES	12000
TRAINING EXPENSES	1000
TOTAL ADMIN. EXPENSES	**365000**
FACTORY EXPENSES	
INSURANCE CHARGES	82000
WORKERS CANTEEN EXPENSES	30000
FACTORY CLEANING EXPENSES	10000
FACTORY EXPENSES	5000
REPAIRS TO BUILDING EXPENSES	25000
PROPERTY TAX	57000
RATES & TAXES	16000
WATER CHARGES	10000
TOTAL FACTORY EXPENSES	**235000**
GRAND TOTAL	**600000**
Weaving Division 30%	180000
Spinning Division 70%	420000

SUMMARY OF CAPEX (WEAVING DIVISION)

	PARTICULARS	QTY	Rs in Lacs	REMARK
	Weaving Section Upgradation of old machines			
a.	**Airjet Looms**			
	- Shaft Tappet Gear (Staubli) IST	17	2.55	
	- Cutter Grinding M/c - A 9300	1	3.42	
	- Oil Change Unit (Mengini)	1	3.50	
	- Test & Control Unit (Air Buggi)	1	5.23	
	- Main Nozzle Tools (Various Type)	4	5.61	
	- Relay Nozzle Tools (Various Type)	3	6.22	
	- Ultra Sonic Machine	1	2.00	Essential spares & gauges
	- Weft Cutter Setting Machine	1	1.28	
	- Weft Stop Motion BE - 153278		5.12	
	- Heald Wire & Drop Wire		3.50	
	- Heald Frame		10.00	
	- Armature Flange B - 163593		9.66	
	- Pre-winder Magnet (IROI) 310025-102		2.60	
	- Mixing Return Pump		0.30	
	- Air Compressor	1	11.17	
	TOTAL		**72.16**	
b.	**Sulzer Looms**			
	- Sulzer Looms Spares	20	10.00	Replacement of worn out parts
	(for existing 20 Looms) @ Rs. 50000/- Each			
	- Sulzer Looms Spares	12	30.00	
	(for existing 12 Looms) @ Rs.250000/- Each			
	- Name Writer for Sulzer	32	64.00	
	(for 32 Looms) @ Rs.200000/- Each			
	- Heald Wire & Drop Wire Pins, etc.		4.00	
	TOTAL		**108.00**	
c.	**Sizing and Warping Machine**			
	- New Sizing Machine		105.00	
	- Sizing Machine Maintenance		1.00	Gear box
	- Warping Machine Maintenance		1.00	Stop motion
	- Add 15% for Taxes, Transportation & Miscellaneous		43.05	
	Expenses for Installation, etc.			
	TOTAL		150.05	
	GRAND TOTAL		330.21	

SUMMARY OF CAPEX (GENERAL & CIVIL)

		PARTICULARS	QTY	AMOUNT
A		CIVIL WORK		
	a	Mill Premises Compound Wall		95
		(2000 Mtrs x 9 Ft.)		
	b	Spinning & Weaving shed Flooring Work		11
	c	Spinning & Weaving shed Roof & Rainwater Gutter Repairing		55
	d	Spinning Preparatory Humidification Tower Plant New Construction		55
	e	Staff Houses - Repairing Works	50 Qtrs	25

	f	Factory Premises Building - Repairing & Painting Work		11
	g	Telephone Line Cable - for Intercom	400 Mtrs	2
	h	Fire Fighting Instruments (Extinguisher, Buckets, Hydrant Post		15
	i	Horse Pipe, Nozzle & Sprinkler Points, etc.		
	j	Factory Temple -		16.5
		Add 15% for Taxes Transportation & Misc. Expenses for Installation, etc.		43
		TOTAL		**328.5**
B		**COMPUTERS & PRINTERS**		
	a	Computers Core 2 Duel / I 3	20 Nos	7
	b	Dot Matrix Printers	4 Nos	
	c	Laser printers	6 Nos	
	d	UPS, Attendance Machine with Software	20 Nos	
C		**GENERAL**		
	a	Cloth Handling Trolly	1 No	1
	b	Forklift - 4 Ton Capacity	1 No	9
		Add 15% for Taxes Transportation & Misc. Expenses for Installation, etc.		1.5
		GRAND TOTAL		**347**

LABOUR COMPLIMENTS – Spinning Division

For 44 Ring Frames

	Ist	IInd	IIIrd	Today
No of Ring Frames	44	44	44	44.0
Production MT /day	2920	2920	2920	8760
Average Count Ne				27.07
BLOW ROOM				
Mixing (contract worker)	5			5
Bale Breaker (Cotton feeder)	1	1	1	3
Roving end opener				0
Sweeper	1	1	1	3
NSE Tenter	2	2	2	6
Fitter	1			1
Total A	**10**	**4**	**4**	**18**
CARDING				
Jobber	1	1	1	3
Card Tenter 5 cards/tenter	2	2	2	6
Lap Man / Sweeper	2	2	2	6
Fitter	4			4
Total B	**9**	**5**	**5**	**19**
FRAME				
Jobber	1	1	1	3
Helper	1	1	1	3
Inter Tenter 2 m/c/tenter	3	3	3	9
Dr. Tenter 4 m/cs/tenter	2	2	2	6
Doffer 1200 Bobbin/doffer	5	5	5	16
Can boy	2	2	2	6
Sweeper	1	1	1	3
M/c Cleaner / Fitter Mazdoor	7			7

Total C	22	15	15	53
RING FRAME- 19442 Spindles				
Jobber 24 m/c / jobber	2	2	2	6
Helper 24 m/c / jobber	2	2	2	6
Sider	11	11	11	33
Doffer 4000 bobbin/doffer	15	15	15	44
Bobbin carrier 24 m/c /carrier	2	2	2	6
Beldar 24 m/c / beldar	2	2	2	6
Sweeper	1	1	1	3
Bottom (con. worker)/reliver	4	4	4	12
Maintenance	22	2	1	25
Humidity	1	1	1	3
Wrapping boy				0
Semi clerk	2			2
Roof cleaning (Extra sanction)	2			2
Total	66	42	41	148
G. Total Up to Ring Frame	107	66	65	238

Doubling / Ply winding

	Ist	IInd	IIIrd	Today	
M/C Worked				0.0	
Production	407	407	407	1220	
Jobber	1	1	1	3	
Sider	2	2	2	6	
Doffer 1700/doffer	3	3	3	9	
Maintenance	5			5	
Winder 150 kg/winder	3	3	3	9	
Extra Winder for TFO					
Cone Carrier				0	
Total	14	9	9	32	
G. Total				270	
Shift	I	II	III	Today	Workload
Production Single Yarn	2920	2920	2920	8760	
Production Double Yarn	407	407	407	1220	
Winder/Cleaner	5	5	5	15	M/c (3+3+3)
Fitter /Maint.	1	1		2	3130 Kg
Lable boy	1	1	1	3	
Bottom Cleaner	1	1	1	3	
Jobber	1			1	
Total	9	8	7	24	
Winder/Cleaner	4	4	4	12	M/c (2+2+2)
Fitter /Maint.					1100 Kg
Total	4	4	4	12	
Winder/Cleaner	5	5	5	15	M/c (4+4+4)
Fitter /Maint.	3	2	2	7	3100 Kg
Total	8	7	7	22	

Total (A)	21	19	18	58	
Total				323	
Conventional winding price Rate	Kg/day	Rate	Amt.Rs.		
44sk Rt		2.39	0		
10s k sl		1.4	0		
27s k	1600	1.84	2944		
34sk		1.84	0		
40sk	1220	2.02	2464		
30s k			0		
Total / day	2820		5408		
For the Month Rs.			162252		
TFO outside	3130	0.5	46950		
Packing	4930	11.00	54230		
Reeling			0		
Tipping	175200	0.10	17520		
Total			71750	2392	

ORGANISATION CHART FOR MANAGEMENT STAFF

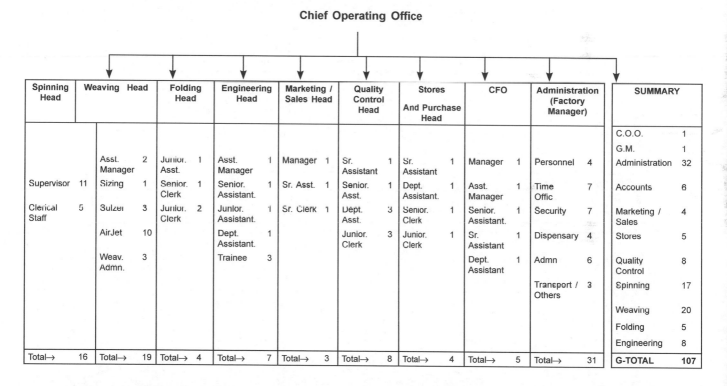

Chief Operating Office

Spinning Head	Weaving Head		Folding Head		Engineering Head		Marketing / Sales Head		Quality Control Head		Stores And Purchase Head		CFO		Administration (Factory Manager)		SUMMARY	
																	C.O.O.	1
																	G.M.	1
	Asst. Manager	2	Junior. Asst.	1	Asst. Manager	1	Manager	1	Sr. Assistant	1	Sr. Assistant	1	Manager	1	Personnel	4	Administration	32
Supervisor 11	Sizing	1	Senior. Clerk	1	Senior. Assistant.	1	Sr. Asst.	1	Senior. Asst.	1	Dept. Assistant.	1	Asst. Manager	1	Time Offic	7	Accounts	6
Clerical 5 Staff	Sulzer	3	Junior. Clerk	2	Junior. Assistant.	1	Sr. Clerk	1	Dept. Asst.	3	Senior. Clerk	1	Senior. Assistant.	1	Security	7	Marketing / Sales	4
	AirJet	10			Dept. Assistant.	1			Junior. Clerk	3	Junior. Clerk	1	Sr. Assistant	1	Dispensary	4	Stores	5
	Weav. Admn.	3			Trainee	3							Dept. Assistant	1	Admn	6	Quality Control	8
															Transport / Others	3	Spinning	17
																	Weaving	20
																	Folding	5
																	Engineering	8
Total→ 16	Total→	19	Total→	4	Total→	7	Total→	3	Total→	8	Total→	4	Total→	5	Total→	31	G-TOTAL	107

B. BUDGET ASSUMPTIONS: MANUFACTURING EXPENSES PROJECTIONS FOR ANNUAL BUDGET

Manufacturing and other departments are required to submit their manpower budget and other expenses directly related to their respective departments on the basis of the targeted goals with proper justification to incorporate in the annual budget. The MIS section checks the authenticity of this data, verifies on logical basis, compared with the past results and scope for improvement and then incorporates in annual budget. Once these figures are incorporated in budget, it is the responsibility of each department head to achieve these as a target and explain reasons for variations on daily or monthly basis, if any.

Example (Worsted Unit):

S. No.	Particulars	Unit	Dyeing	Recombing	Spinning	Weaving	Finishing	Folding
1	Working Days	Nos	312	312	312	312	312	312
2	Installed Capacity (Mtr)	MTR	2252016			3017244	3600000	
3	Installed Capacity (Top/Yarn)	KG	1310400	1044000				
4	Production (Mtr) (3.0Mtr/Kg)	MTR	600000			3228797	3228797	3075045
5	Production (Top/Yarn)	KG	1063920	1042644	969659			
6	Avg Count Nm				43			
7	Avg Picks/Inch					55		
8	Waste%	%	0.09	2.5	5.5	2	0.31	
9	Regular+Temporary Workers	Nos	432	18096	3060	2736	936	9984
10	Other Basket Nos	Nos			420			
11	Power-Lacs Units – M/C	Units	909600	1080000	3900000	1812000	1420668	
		Kwh/Mtr	0.72	1.04	4.02	0.56	0.44	
12	Power-Lacs Units - Fixed General	Units						
13	Steam MT	MT	10800	2880	2040		16176	
		MT/Mtr	0.01	0	0		0.01	
14	Water KL	KL	82800	22800			92976	
		KL/Mtr	0.07	0.002			0.03	
Sr No	Particulars	UOM	Rate	Rate	Rate	Rate	Rate	Rate
1	Power Rate	Rs/kwh	8.04	8.04	8.04	8.04	8.04	8.04
2	Steam Rate	Rs/Mt	1247.23	1247.23	1247.23	1247.23	1247.23	1247.23
3	Water Rate	Rs/KL	57.3	57.3	57.3	57.3	57.3	57.3
4	CNT Rate	Rs/Head						
	Total No of Staff	Nos	12	6	18	33	9	11
	Total No of Workers	Nos	36	58	290	228	78	32
	Ave CTC Per Staff /Month	Rs	27863	39096	34277	27011	31800	22400
	Ave CTC Per Worker/Month	Rs	16510	16554	16395	16246	16601	18750
Total Cost/Year								
1	Salaries/Wages Cost	INR	11144520	14336460	64458456	55145376	18972958	10087704
2	Other Basket Cost	INR			100000	S		
3	Outside Processing Cost	INR			50000			
4	Dyes/Chemical Cost	INR	30334080	2220000	60000		7800000	
5	Packing Materials Cost	INR		180000	40000			4963092
6	Power Cost	INR	7313184	8683200	31356000	14568480	11422171	
7	Steam Cost	INR	13470084	3592022	3525120		20175192	
8	Water Cost	INR	4744440	1306440			5327525	
9	Stores/Spares Cost	INR	3791760	2640000	7200000	6600000	1.4E+07	
10	Repair Cost	INR	420000	360000	300000	300000	360000	
11	**Total**	**INR**	**71218068**	**33318122**	**107089576**	**76613856**	**78457846**	**15050796**
Rs/kg	Rs/kg	Rs/kg	Rs/Mtr	Rs/Mtr	Rs/Mtr			
1	Salaries/Wages Cost		8.82	13.75	66.48	17.08	5.88	3.28
2	Other Basket Cost				0.1			

3	Outside Processing Cost				0.05			
4	Dyes/Chemical Cost		24	2.13	0.06		2.42	
5	Packing Materials Cost			0.17	0.04			1.61
6	Power Cost		5.79	8.33	32.34	4.51	3.54	
7	Steam Cost		10.66	3.45	3.64		6.25	
8	Water Cost		3.75	1.25			1.65	
9	Stores/Spares Cost		3	2.53	7.43	2.04	4.46	
10	Repair Cost		0.33	0.35	0.31	0.09	0.11	
11	**Total**		**56.35**	**31.96**	**110.44**	**23.73**	**24.3**	**4.89**

C. ANNUAL BUDGET OF A COMPOSITE TEXTILE MILL

UNIVERSAL TEXTILE LTD. (Manufacturer of Worsted Suitings)

Financial Year- 20XX-20XY

ASSUMPTIONS

Exchange Rates	Previous Year	Current
Conversion Rate US$ to INR (Rs)	44.00	54.00
AS$ (Australian Dollar) to INR (Rs)	45.00	55.00
Working Day	309	348
Production (000' Mtrs)		
Domestic market	2,240	2,912
Government /Institutional Supply	707	790
Export	980	1,700
Inhouse Total	3927	5402
Outsourced	540	435
Total	4467	5836
Spinning Job work (000' Kgs)	984	660
Avg Count Nm	45	45
Yarn Conversion Rate Rs/Count	2.25	2.30
Sales Quantity (000' Mtrs)		
Domestic Market	2,734	3,165
Government /Institutional Supply	710	889
Export	1005	1,650
In house Total	4,449	5,704
Outsourced Fabric	510	435
Total	4,959	6,139
Revenues Net (000'Rs)		
Domestic Market		11,39,422
Government /Institutional Supply		4,23,604
Export		5,50,540
In house Total		**21,13,566**
Outsourced Realisation		1,05,215
Grand Total		22,18,782
Avg Net Selling Price (Rs.)/Mtr		
Domestic market		360

Government /Institutional Supply		476
Export		334
Outsourced Realisation		242
TOI % (Turn over Incentives)		6.0%
Selling Expenses %		
Commission %		
Domestic	2.1%	3.6%
Group Company	6.7%	3.6%
Govt Supply (Agent)	4.8%	7.0%
Institutional Supply (Agent)	0.5%	0.0%
Export, Others	4.0%	3.5%
Export Europe & US	5.0%	5.0%
Sampling	1.0%	1.1%
Export Benefits - % of Export revenues	4.5%	5.0%
Flag & Defective % of Fabric		
Domestic		
Economy	2.5%	6.0%
Standard	3.0%	6.0%
Premium	4.0%	6.0%
Super Premium	6.0%	6.0%
Government /Institutional Supply	1.0%	1.0%
Export	6.0%	6.0%
Outsourced	2.5%	2.5%
Employee (Workers) Nos	1,000	1,214
Employee (Workers) Cost Rs 000	85,712	1,21,554
Add 6% Increment	5,143	-
Sub Total (For the year)	90,855	1,21,554
Per Head cost /Month	7,571	8,344
Employee (staff) Nos	295	307
Employee (staff) Cost		
Existing- INR'000		1,16,424
Add 10% Increment		11,642
Sub Total		1,28,066
Per Head cost/Month		34,763
Total Employee Nos		1,521
Total Employee Cost/Year		2,49,620
Raw Material Prices		
Greasy Wool Price / Kg. (ACK)		
17.5M	2200	1,700
18.5M	1700	1,600
19.5M	1580	1,400

20.5M	1400	1,410
21.5M	1350	1,410
22.5M	1050	1,385
22.5M NG	900	1000
24.5M	1000	1,150
28.0M	600	600
Polyester Tow Rs./ Kg.		
Polyester	111	109
Low Pill Poly	135	135
Trilobal Poly	135	119
Viscose	300	290
Flax Tops	525	650
Silk	2800	2,800
Cashmere	6000	7,000
Lycra	1380	1,200
Purchased Yarn	280	275
Purchased Cloth-Imp.	250	250
Suiting special	1325	-
Purchased Cloth-Local	137	130
Dyes & Chemicals Cost /kg		
Wool	35.5	26.0
Poly	30.0	24.0
Avg. Cost/kg	32.5	24.6
Stores and Consumables Cost/Mtr		
Stores & Spares		12.00
Repairs & Maintenance		1.20
Ave Power Rates Rs/Unit		4.50
Rate of Interest		
Interest Rate on Bank C/C (Cash Credit)		13.50%
Interest Rate on Existing Term Loan		15.00%
Interest Rate on New term Loan		14.00%
Interest on New Capex loan		14.50%
Depreciation rates: (SLM)		
Plant and machinery (Double shift)		7.42%
Building		3.34%
Furniture		6.33%
Interest Rate on TD (Term deposit)		12.00%

*Plant does not have wool scouring facility. Therefore, either the ready tops will be procured, or greasy wool will be imported. Repayment of instalment of capex loan will be made at the beginning of the month. Amount of capex loan will be received on 1st Month. *ACK- Australian Dollar Cents*

MONTHLY ASSUMPTION SHEET

All figures in Rupees, millions unless otherwise specified

	Apr	May	June	Jul	Aug	Sep	Oct	Nov	Dec	Jan	Fab	Mar	Total	PFY
Exchange Rates														
US$ / INR													54.00	44.50
AS$ / INR													55.00	47.00
Working days	27	29	30	31	29	30	29	28	30	28	27	30	348	309
Product Mix (Sales)														
Quantity (mn meters)														
Domestic	0.18	0.16	0.14	0.20	0.36	0.41	0.41	0.38	0.28	0.30	0.32	0.44	3.60	3.10
Govt./ Institutional	0.04	0.08	0.09	0.09	0.08	0.10	0.08	0.08	0.06	0.08	0.06	0.05	0.89	0.70
Export -US	0.05	0.06	0.05	0.06	0.08	0.05	0.05	0.06	0.07	0.04	0.04	0.04	0.66	0.30
Export -EU	-	0.01	-	0.02	0.03	0.01	0.02	0.01	0.04	0.02	0.03	0.03	0.22	0.14
Export -Others	0.04	0.06	0.09	0.05	0.08	0.07	0.09	0.09	0.10	0.03	0.06	0.07	0.83	0.51
Total	0.32	0.38	0.37	0.43	0.64	0.64	0.65	0.62	0.54	0.47	0.51	0.63	6.19	4.75
Own	0.10	0.11	0.14	0.15	0.16	0.17	0.17	0.16	0.16	0.15	0.13	0.12	1.73	1.40
Job Spg.	0.06	0.06	0.06	0.06	0.06	0.06	0.06	0.06	0.06	0.06	0.06	0.06	0.66	0.98
Percentage														
Domestic	57%	44%	39%	48%	56%	64%	63%	62%	51%	64%	64%	70%	58%	65.26%
Govt./ Institutional	14%	22%	23%	21%	13%	15%	12%	13%	12%	17%	11%	8%	14%	14.74%
Export -US	16%	15%	15%	14%	13%	8%	8%	9%	12%	8%	9%	7%	11%	6.32%
Export -EU	0%	3%	0%	5%	5%	2%	3%	2%	7%	3%	5%	5%	3%	2.95%
Export -Others	13%	17%	24%	12%	13%	11%	14%	14%	18%	7%	11%	11%	13%	10.74%
Total	100%	100%	100%	100%	100%	100%	100%	100%	100%	100%	100%	100%	100%	100.00%
Revenue per metre (Rs)														
Domestic	315.52	307.76	289.29	304.22	313.83	361.14	406.30	392.75	379.48	360.53	314.90	297.32	343.56	397.05
Govt./ Institutional	516.00	479.58	466.32	459.43	459.26	460.34	483.41	517.37	461.48	482.30	474.75	475.78	476.49	356.29
Export -US	391.04	391.04	391.04	391.04	391.04	391.04	391.04	391.04	391.04	391.04	391.04	391.04	391.04	364.19
Export -EU	-	263.95	-	335.78	368.04	391.04	327.50	391.04	337.01	391.04	340.20	348.68	348.30	310.59
Export -Others	285.10	288.70	289.34	288.97	285.95	292.93	289.77	285.41	286.08	269.60	287.78	289.45	287.42	238.53
Revenues (Rs mn)														
Domestic	57.11	50.47	41.66	62.06	112.67	148.43	168.21	150.03	105.50	108.34	101.40	131.12	1236.98	1230.85
Govt./ Institutional	22.95	39.80	39.64	40.89	36.74	44.65	37.69	43.20	29.53	38.10	27.54	22.84	423.56	249.40
Export -US	19.94	21.51	21.12	23.46	32.77	20.73	21.12	22.68	25.42	15.41	17.21	16.81	258.16	109.26
Export -EU	-	2.64	0.00	7.72	12.51	3.91	6.55	3.91	12.81	5.87	8.51	10.46	74.88	43.48
Export -Others	12.26	18.48	25.46	14.74	22.88	20.51	25.79	24.55	27.18	8.63	16.12	20.55	237.12	121.65
Surplus spinning income	5.69	5.69	5.69	5.69	5.69	5.69	5.69	5.69	5.69	5.69	5.69	5.69	68.31	99.63
Total	117.95	138.58	133.56	154.56	223.26	243.91	265.04	250.06	206.12	182.03	176.45	207.47	2299.03	1854.27
Other Income														
Export benefits - % of export revenues	5.00%	5.00%	5.00%	5.00%	5.00%	5.00%	5.00%	5.00%	5.00%	5.00%	5.00%	5.00%	5.00%	4.50%
Total Export Revenues	32.20	42.62	46.58	45.92	68.16	45.14	53.46	51.14	65.40	29.90	41.83	47.83	570.17	274.39
Export benefit	1.61	2.13	2.33	2.30	3.41	2.26	2.67	2.56	3.27	1.49	2.09	2.39	28.51	12.35
Waste	0.75	0.79	1.01	1.09	1.14	1.21	1.18	1.12	1.13	1.07	0.95	0.90	12.34	6.29
Interest Received	0.83	0.83	0.83	1.09	1.29	1.29	1.29	1.09	1.16	1.36	1.50	1.26	13.78	9.00
Costs														
Raw Material														
Greasy Wool Price / Kg. (ACK)														
17.5M													1700.00	2200.00
18.5M													1600.00	1700.00
19.5M													1400.00	1580.00
20.5M													1410.00	1400.00

21.5M													1410.00	1350.00
22.5M													1350.00	1050.00
22.5M NG													1000.00	900.00
24.5M													1150.00	1000.00
28.0M													600.00	600.00
Polyester Tow Rs./Kg.														
Polyester													108.75	111.00
Low Pill													135.00	135.00
Trilobal													118.75	135.00
Viscose													290.00	300.00
Flax Tops													650.00	525.00
Silk													2800.00	2800.00
Cashmere(Rs./kg)													7000.00	6000.00
Lycra (Rs./kg)													1200.00	1380.00
Purchased Yarn (Rs./kg)													275.00	280.00
Purchased Cloth-Imp.													250.00	250.00
Suiting (Rodina)													0.00	1325.00
Purchased Cloth-Local													130.00	136.50
Manufacturing Expenses														
Dyes & Chemicals														
Wool Cost / Kg. (Rs.)	26.00	26.00	26.00	26.00	26.00	26.00	26.00	26.00	26.00	26.00	26.00	26.00	26.00	
Poly Cost / Kg. (Rs.)	24.00	24.00	24.00	24.00	24.00	24.00	24.00	24.00	24.00	24.00	24.00	24.00	24.00	
Cost / Kg. (Rs.)	24.60	24.60	24.60	24.60	24.60	24.60	24.60	24.60	24.60	24.60	24.60	24.60	24.60	32.50
Stores and Consumables														
Stores & Spares	3.95	4.63	5.45	5.66	5.82	6.03	5.94	5.86	6.00	5.27	5.15	5.07	64.82	33.38
Repairs & Maint	0.51	0.51	0.51	0.51	0.51	0.51	0.51	0.51	0.51	0.51	0.51	0.51	6.16	10.20
Addnl. Spares & Repair	-	-	-		-	-	-	-	-	-	-	-	-	-
Packing Primary	0.53	0.57	0.53	0.70	0.93	1.12	1.61	1.51	1.18	1.20	0.97	0.91	11.76	-
Packing Secondary	0.87	0.92	0.87	1.15	1.52	1.82	2.63	2.46	1.92	1.95	1.57	1.48	19.18	31.23
Total cost of Stores & Consumables	5.86	6.64	7.37	8.03	8.79	9.49	10.69	10.34	9.62	8.93	8.20	7.97	101.92	74.81
Power and Fuel Expenses														
Power														
Power units required Units/mtr	4.39	4.39	4.39	4.39	4.39	4.39	4.39	4.39	4.39	4.39	4.39	4.39	4.39	4.61
Units for Fabric (Mn Kwh)	1.45	1.70	1.99	2.07	2.13	2.21	2.17	2.14	2.20	1.93	1.88	1.86	23.73	18.11
Power units required (Units/Kg)	7.00	7.00	7.00	7.00	7.00	7.00	7.00	7.00	7.00	7.00	7.00	7.00	7.00	4.05
Units for Job Spinning (Mn Kwh)	0.39	0.39	0.39	0.39	0.39	0.39	0.39	0.39	0.39	0.39	0.39	0.39	4.62	3.99
Total units (Mn Kwh)	1.83	2.08	2.38	2.46	2.51	2.59	2.56	2.53	2.58	2.31	2.27	2.24	28.35	22.10
Avg. Rate	5.04	5.20	5.35	5.38	5.40	5.43	5.42	5.41	5.43	5.32	5.30	5.28	5.34	5.72
Cost of Power	9.22	10.82	12.72	13.23	13.58	14.09	13.87	13.67	14.01	12.30	12.02	11.83	151.35	126.42
Coal														
Coal Consumption (Kgs/mtr.)	1.02	1.02	1.21	1.10	0.76	0.78	0.75	0.78	0.92	0.94	0.84	0.66	0.87	1.36
Quantity required (MT)	327	383	451	469	481	499	491	484	496	436	426	419	5362	6452
Average rate (Rs. MT)	11508	11508	11508	11508	11508	11508	11508	11508	11508	11508	11508	11508	11508	9526

Cost of Coal (Mn Rs.)	3.76	4.41	5.19	5.39	5.54	5.74	5.65	5.57	5.71	5.01	4.90	4.82	61.71	61.46
Furnace Oil														
Fuel Consumption/ mtr (Ltr/mtr.)	0.14	0.14	0.17	0.15	0.10	0.11	0.10	0.11	0.13	0.13	0.11	0.09	0.12	0.03
Quantity required (Mn Ltr)	0.04	0.05	0.06	0.06	0.07	0.07	0.07	0.07	0.07	0.06	0.06	0.06	0.73	0.14
Average rate (Rs./ Ltr)	43.00	43.00	43.00	43.00	43.00	43.00	43.00	43.00	43.00	43.00	43.00	43.00	43.00	47.00
Cost of Furnace Oil	1.91	2.25	2.64	2.75	2.82	2.93	2.88	2.84	2.91	2.55	2.50	2.46	31.43	6.44
Total Power & Fuel Expenses	14.89	17.47	20.55	21.36	21.94	22.76	22.41	22.09	22.63	19.87	19.41	19.11	244.49	194.32
Administrative Selling & Other Expenses														
Administrative Expenses	6.98	6.98	6.98	6.98	6.98	6.98	6.98	6.98	6.98	6.98	6.98	6.98	83.71	74.10
Selling Expenses														
Commission & Freight, etc.	4.13	4.99	4.68	5.60	9.16	10.60	11.11	10.39	8.24	7.14	7.20	9.06	92.31	81.64
Advertisement & Publicity	6.82	6.82	6.82	13.64	20.45	20.45	13.64	13.64	13.64	10.91	9.55	13.64	150.00	110.00
Royalty (8 %)	-	-	-	0.28	-	0.23	1.48	0.33	-	0.20	-	-	2.52	3.43
Total Selling Expenses	10.95	11.81	11.50	19.52	29.62	31.29	26.23	24.35	21.88	18.25	16.75	22.69	244.83	195.07
Total Administrative & Selling Expenses	17.93	18.78	18.48	26.50	36.59	38.26	33.20	31.33	28.85	25.22	23.72	29.67	328.54	269.16
Employee Cost														
Salaries and Wages														
Salary	10.84	10.84	10.84	10.84	10.84	10.84	10.84	10.84	10.84	10.84	10.84	10.84	130.14	135.68
Wages	9.84	10.61	11.05	11.42	10.74	11.11	10.75	10.39	11.10	10.33	9.97	11.01	128.31	102.22
Total Employee Cost	20.69	21.46	21.89	22.26	21.58	21.95	21.59	21.24	21.95	21.17	20.81	21.85	258.44	237.91
Corporate Expenses														
Fees (Incld. Service Tax	1.52	1.52	1.52	1.52	1.52	1.52	1.52	1.52	1.52	1.52	1.52	1.52	18.20	2.95
Commission on Export	1.00	1.08	1.06	1.17	1.64	1.04	1.06	1.13	1.27	0.77	0.86	0.84	12.91	6.56
Total	2.51	2.59	2.57	2.69	3.16	2.55	2.57	2.65	2.79	2.29	2.38	2.36	31.11	9.50
MAT Rate	20.39%	20.39%	20.39%	20.39%	20.39%	20.39%	20.39%	20.39%	20.39%	20.39%	20.39%	20.39%	20.39%	20.39%

Export benefit is on % age of export revenue, ACK-Australian Dollar Cents – (1800 ACK = 18.00 Australian Dollar), PFY-Previous financial year

BREAKEVEN ANALYSIS

	Actual	@	BEP
All figures in Rupees, 000	Previous Year		
Particulars	TOTAL	@	
Sales Quantity			
Civil - Domestic	3,274		4,107
Govt/Institution	708		888
Export -EU	135		169
Export -US	349		438
Export -Others	496		622
	4,962		6223.41
Job Spinniing. Kg	984		542
Sales			
Civil - Domestic	13,09,410		1642446

Less: TOI (Turn Over Disscount)	78,565	6.0%	98547
Net Sales	12,30,846	375.95	15,43,899
Govt./ Institutional	2,49,400	352.26	3,12,833
Export -EU	43,482	322.09	54,542
Export -US	1,09,257	313.06	1,37,046
Export -Others	1,21,653	245.51	1,52,594
	17,54,638	353.65	22,00,914
Comm. Spinning	99,630	101.25	54,911
Waste	6,244	1.26	7,833
Other Income (Interest)	9,000		9,000
Export Benefi	12,348	4.50%	15,488
Total Income	18,81,860		22,88,145
Manufacturing Expenses			
Increase (Incr.) Decrease (Decr). in Finished goods (FG)	1,32,250		1,32,250
(Incr.) Decr. in WIP	-		-
Raw Material	6,80,220	39.28%	8,64,517
Out side Purchases	1,11,700		1,11,700
Dyes & Chemicals	49,133	9.90	61,630
Packing	31,227	6.29	39,169
Stores & Consumables	43,582	8.78	54,666
Power & Fuel	1,94,318	39.17	2,43,741
Total	12,51,430		15,07,672
Admin and Other Expenses	74,097		74,097
Publicity	1,10,000		1,10,000
Selling Expenses	85,066	4.85%	1,06,701
Employees Cost Salary	1,02,221		1,02,221
Employees Cost Wages	1,35,685	27.35	1,70,195
Total Operating Exps.	17,58,198		20,70,887
EBITDA	1,23,362	6.6%	2,17,259
% of total income	6.6%		
Corporate Expenses			
Fees	2,945		2,945
Commission	6,555	6.0%	8,222
Expenses	-		-
Management Cost for Special cause	30,436		30,436
Net EBITDA	89,981		1,83,878
% of total income	4.8%		8.0%
Interest	63,620		63,620
PBDT	26,361		1,20,258
Depreciation	1,20,258		1,20,258
Net PBT	(93,897)		(0)
Income Tax	-		-
Net Income / (Loss)	(93,897)		(0)
Net Income / (Loss)	(1,00,452)		

*Break Even Point (BEP): The break-even point is the production level at which total revenues generated for a product equal total expense. When a business reaches its break-even point, the total sales equal the total expenses. At the break-even point, a business does not profit or generate a loss. Break-even analysis tells us no. of units of a product that must be sold to cover the fixed and variable costs of production. In accounting, the break-even point formula is determined by dividing the total fixed costs associated with production by the revenue per individual unit minus the variable costs per unit. It is important to analyse the break-even point figures of the business to make action plan to increase sales prices, reduce expenses, or both, to cross the break-even point and make profit.

Break Even point in units = Fixed Cost / (sales price per unit – Variable cost per unit)

Fixed Costs: Costs that do not change in relation to production volume. Depreciation, insurance, advertisement, etc.

Sales Price per Unit: This is the amount a company charges consumers for one unit of products.

Variable Costs per Unit: Variable costs are costs directly proportional to the volume of production. Variable costs often fluctuate and are typically a company's largest expense.

PBDT: Profit before depreciation and taxes, PBT: Profit before taxes

CALCULATION FOR BREAK EVEN SALES

Particulars	Units in lacs		Rs in Lacs	
	Fixed cost	Variable cost	Total	
Raw Material Consumption.	-	11969.65	11,969.65	
Stores & Spares Consumption.	486.14	162.05	648.19	
Packing		309.41	309.41	
Power, Water & Fuel	-	2444.92	2,444.92	
Salary & Wages	2584.44	-	2,584.44	
Processing Charges	-		-	
Repair/Maintenance	51.86	9.72	61.58	
Rent	42.66	-	42.66	
Rate & Taxes	26.95	-	26.95	
Insurance	43.75	-	43.75	
Commission on Sales		789.56	789.56	TOI
Travelling Exps	181.49			
Publicity	1500.00	-	1,500.00	
Lease Rent	0.00	-	-	
Corporate Expenses		-	-	
Other Expenses	149.16	-	149.16	
Freight & Forwarding			-	
Audit Fees	12.00	-	12.00	
Interest	788.62	-	788.62	
Professional legal fees	381.06	-	381.06	
Depreciation	1185.00	-	1,185.00	
	7,433.13	15685.31	23,118.44	
	7433.13	15685.31	23118.44	
Stock Adjustment	382.17	806.45	1188.61	
Total	7815.30	16491.76	24307.05	
Sales	Lac Mtr	Rate/mtr	Lac Rs	
Domestic	36.01	343.56	12369.84	Net sales

Institutional		8.89	476.49	4235.61
Export		17.00	335.35	5701.71
Job commn				683.10
Misc income				546.29
Sales		61.90	380.26	23536.54
% on sales			70.07	
Contribution				7044.79
PV Ratio (Profit Volume)				29.93
Break-even sales		68.67	380.26	26110.80
Tally		7815.30	18295.51	26110.80
		Sales	Variable	
Contribution		23536.54	16491.76	7044.79
PV Ratio		23536.54		
PV Ratio (Contribution/Sales)		29.93		
BES		26110.80		

BUDGET COMPARISON OF PREVIOUS YEAR VS CURRENT YEAR

Sales (net)	Budget PY Sale Mtr'000	Production '000 Mtr	INR'000	Rate/mtr		Actual PY Sale Mtr'000	Production '000 Mtr	INR'000	Rate/mtr	Budget Current Year Sale Mtr'000	Production '000 Mtr	INR'000	Rate/mtr
Domestic	2,744	2,240	11,01,849	402		2,837	2,911	11,63,545	410	3,166	2,912	11,32,803	358
Govt /Institutional	708	707	2,49,400	352		707	800	2,72,564	386	889	790	4,23,561	476
Export	980	980	2,74,392	280		1,300	1,313	4,89,243	376	1,700	1,700	5,70,171	335
Total	4,432	3,927	16,25,641	367	367	4,844	5,024	19,25,352	397	5,755	5,402	21,26,535	370
PV Outsourced	530	530	1,28,997	243		327	339	79,461	243	430	430	1,04,181	243
Total	4,962	4,457	17,54,638			5,171	5,363	20,04,813		6,184	5,831	22,30,716	
Job Spinning			99,630	22				70,617	15			68,310	12
Waste			6,288	1				20,178	4			12,339	2
Other Income (Interest)			9,000	2				48,051	10			13,781	2
Export Benefit			12,348	3				17,639	4			28,509	5
Total Income			18,81,904	425				21,61,298	446			23,53,654	409
			Cost/Mtr	425				Cost/Mtr					Cost/Mtr
Raw Material (Wool/Poly/yarn)			8,31,763	211.81	213			11,96,124	238.08			11,84,379	219.26
PV			98,450	185.75				47,460	140.00			84,625	197.03
Dyes and Chemical			49,484	12.60	200			65,901	13.13			47,471	8.79
Packing (Secondary)	*		31,227	7.95	192			-				19,183	3.55
Stores & Consumables incld Primary Packing			43,582	11.10	181			77,849	15.50			82,735	15.32
Power & Fuel			1,94,318	49.48	132			2,16,047	43.00			2,44,492	45.26
Admin and Other Exps			74,097	18.87	113			92,650	18.44			83,707	15.50
Employees Cost			2,37,906	60.58	52			2,26,050	44.99			2,58,444	47.85
Publicity			1,10,000	28.01	24			55,720	11.09			1,50,000	27.77
Selling Expenses (Including Corporate comm)			91,621	23.33	1			72,369	14.40			1,07,738	19.95
Total Operating Expenses			17,62,448	448.80	(24)			20,50,250	408.09			22,62,774	418.91
EBITDA			1,19,456	30.42				1,11,048	22.10			90,880	16.82
Corporate Fee			2,945	0.75	(25)			15,000	2.99			18,202	3.37
Corporate Expenses.			-	-	(25)			-	-			-	-
Management Special Charges			30,436	7.75	(33)			36,464	7.26			-	-

Net EBITDA			86,075	21.92					59,584	11.86			72,678	13.45
Interest			61,941	15.77	(48)				42,219	8.40			78,862	14.60
PBDT			24,134	6.15					17,365	3.46			(6,184)	(1.14)
Depreciation			1,21,136	30.85	(79)				1,14,034	22.70			1,18,500	21.94
Net Income / (Loss)			(97,002)	(24.70)					(96,669)	(19.24)			(1,24,684)	(23.08)
Cash profi			24,134	6.15					17,365	4.42			(6,184)	(1.57)

COMPARATIVE PERFORMANCE RATIO

(Current and three previous consecutive years)

All figures in Rupees millions unless otherwise specified	Projected	Actual	Actual	Actual		Projected	Actual	Actual	Actual
	Current	Previous	Previous	Previous		Current	Previous	Previous	Previous
	FY	FY 1	FY 2	FY 3		FY	FY 1	FY 2	FY 3
Domestic	52.6%	54.7%	57.5%	69.9%	(Increase) / Decrease in FG/WIP	5.1%	0.9%	(9.0%)	(2.0%)
Govt./ Institutional	18.0%	14.3%	14.9%	13.3%	Raw Material	48.9%	51.7%	47.5%	36.0%
Export	24.2%	24.3%	19.4%	11.9%	Dyes & Chemicals	2.0%	2.9%	3.9%	3.5%
Job Spinning	2.9%	3.5%	4.4%	2.5%	Stores, Spares & Consumables	4.3%	4.2%	6.1%	4.8%
Other Income	2.3%	3.2%	3.8%	2.4%	Power & Fuel	10.4%	11.1%	11.9%	12.0%
Total Income	2,353.7	1,902.3	1,547.4	1,328.4	Manufacturing Expenses	70.7%	70.7%	60.3%	54.4%
					Gross Margin	29.3%	29.3%	39.7%	45.6%
Break-up of Manufacturing Expenses									
(Increase) / Decrease in FG/WIP	7.1%	1.2%	(15.0%)	(3.6%)					
Raw Material	69.2%	73.1%	78.7%	66.3%	Administrative and Other Expenses	3.6%	4.1%	3.5%	3.7%
Dyes & Chemicals	2.9%	4.1%	6.4%	6.4%	Selling Expenses	10.4%	8.3%	9.9%	11.5%
Stores, Spares & Consumables	6.1%	5.9%	10.1%	8.8%	Employees Cost	11.0%	11.6%	13.8%	14.4%
Power & Fuel	14.7%	15.6%	19.8%	22.1%	Total SG&A Expenses (before Inv.)	25.0%	24.0%	27.2%	29.6%
Total Manufacturing Expenses	1,662.9	1,344.7	933.6	722.3					
						4.4%	5.3%	12.4%	16.0%
Break-up of SG & A Expenses						1.3%	3.1%	10.1%	1.1%
Administrative and Other	14.2%	17.0%	12.8%	12.5%		3.1%	2.2%	2.3%	15.0%
Selling Expenses	41.8%	34.7%	36.3%	39.0%					
Employees Cost	44.0%	48.3%	50.8%	48.5%		(5.3%)	(5.3%)	(7.2%)	3.0%
Total SG&A Expenses	588.0	456.0	421.5	393.1					

* S – Sales, G - General, A – Administration, FG - Finished goods, WIP - Work in progress

MONTHLY PROFIT AND LOSS STATEMENT

Amount in Rupees, 000													
Particulars	Apr	May	June	Jul	Aug	Sep	Oct	Nov	Dec	Jan	Fab	Mar	TOTAL
Production Qnty	329	386	454	472	485	503	495	488	500	439	429	422	5,400
O/S Purchases	32	20	32	32	62	47	27	7	4	31	63	73	430
	361	406	486	504	547	550	522	495	504	469	492	495	5,831
Sales Qnty													
Domestic	181	164	144	204	359	411	414	382	278	301	322	441	3,601
Govt/ Institutional	44	83	85	89	80	97	78	84	64	79	58	48	889
Export -EU	-	10	-	23	34	10	20	10	38	15	25	30	215
Export -US	51	55	54	60	84	53	54	58	65	39	44	43	660

Export -Others	43	64	88	51	80	70	89	86	95	32	56	71	825
	319	376	371	427	637	641	655	620	540	466	505	633	6,190
Job Spg.	55	55	55	55	55	55	55	55	55	55	55	55	660
Sales													
Domestic	60,754	53,695	44,317	66,022	1,19,858	1,57,902	1,78,947	1,59,606	1,12,229	1,15,254	1,07,871	1,39,486	13,15,940
Less: TOI	3,645	3,222	2,659	3,961	7,191	9,474	10,737	9,576	6,734	6,915	6,472	8,369	78,956
Net Sales	57,109	50,473	41,658	62,060	1,12,667	1,48,428	1,68,210	1,50,030	1,05,495	1,08,339	1,01,399	1,31,117	12,36,984
Govt./ Institutional	22,951	39,795	39,637	40,889	36,741	44,652	37,686	43,201	29,534	38,101	27,535	22,837	4,23,561
Export -EU	-	2,640	-	7,723	12,513	3,910	6,550	3,910	12,807	5,866	8,505	10,460	74,884
Export -US	19,943	21,507	21,116	23,462	32,769	20,725	21,116	22,680	25,418	15,407	17,206	16,815	2,58,165
Export -Others	12,259	18,477	25,462	14,737	22,876	20,505	25,789	24,546	27,177	8,627	16,116	20,551	2,37,122
	1,12,262	1,32,892	1,27,872	1,48,872	2,17,566	2,38,221	2,59,351	2,44,367	2,00,431	1,76,340	1,70,761	2,01,781	22,30,716
Comm. Spinning	5,693	5,693	5,693	5,693	5,693	5,693	5,693	5,693	5,693	5,693	5,693	5,693	68,310
Waste	749	787	1,007	1,091	1,136	1,214	1,181	1,118	1,130	1,072	953	899	12,339
Other Income (Interest)	826	826	826	1,086	1,286	1,206	1,286	1,086	1,156	1,356	1,496	1,261	13,781
Export Benefit	1,610	2,131	2,329	2,296	3,408	2,257	2,673	2,557	3,270	1,495	2,091	2,391	28,509
Total Income	1,21,140	1,42,329	1,37,728	1,59,038	2,29,089	2,48,670	2,70,184	2,54,820	2,11,680	1,85,957	1,80,994	2,12,025	23,53,654
Manufacturing Exps.													
(Incr.) Decr. in FG	(13,766)	(9,953)	(38,127)	(25,528)	29,838	30,236	44,080	41,276	11,935	(1,160)	4,376	45,653	1,18,861
(Incr.) Decr. in WIP	-	-	-	-	-	-	-	-	-	-	-	-	-
Raw Material	67,000	69,342	86,813	93,556	97,926	1,05,208	1,11,076	96,906	98,007	89,091	77,136	73,457	10,65,518
Outsourced Purchases	6,800	3,800	6,200	6,200	12,500	9,350	4,350	1,150	760	5,225	12,750	14,890	83,975
Dyes & Chemicals	2,822	2,937	3,845	4,156	4,435	4,780	4,671	4,317	4,350	4,151	3,635	3,372	47,471
Packing	670	924	872	1,147	1,525	1,823	2,630	2,462	1,924	1,952	1,575	1,479	19,183
Stores & Consumables	4,461	5,145	5,961	6,177	6,331	6,547	6,453	6,360	6,613	5,780	5,659	5,581	70,977
Power & Fuel	14,892	17,471	20,549	21,364	21,943	22,758	22,405	22,088	22,631	19,866	19,409	19,115	2,44,492
Employees Cost	20,687	21,457	21,891	22,263	21,580	21,952	21,593	21,235	21,949	21,174	20,812	21,852	2,58,444
Depreciation	9,424	9,424	9,694	9,694	9,694	9,965	9,965	9,965	9,965	10,236	10,236	10,236	1,18,500
Total	1,13,190	1,20,547	1,17,700	1,39,029	2,05,774	2,12,619	2,27,223	2,05,769	1,78,034	1,56,314	1,55,588	1,95,634	20,27,422
Gross Margin of Sales	7,950	21,782	20,027	20,009	23,315	36,052	42,960	49,051	33,646	29,642	25,406	16,391	3,26,232
Publicity	6,818	6,818	6,818	13,636	20,455	20,455	13,636	13,636	13,636	10,909	9,545	13,636	1,50,000
Selling Expenses	4,134	4,988	4,684	5,885	9,161	10,831	12,591	10,715	8,240	7,340	7,203	9,058	94,829
Corporate Fees / Commission	2,514	2,592	2,573	2,690	3,155	2,553	2,573	2,651	2,788	2,287	2,377	2,358	31,111
Corporate Expenses	-	-	-	-	-	-	-	-	-	-	-	-	-
Special Management Cost	-	-	-	-	-	-	-	-	-	-	-	-	-
Admin and Other Exps	6,976	6,976	6,976	6,976	6,976	6,976	6,976	6,976	6,976	6,976	6,976	6,976	83,707
Total SG and A Expenses	20,442	21,374	21,050	29,187	39,746	40,815	35,776	33,978	31,639	27,511	26,101	32,028	3,59,647

Total Operating Exps.	1,33,632	1,41,921	1,38,751	1,68,216	2,45,520	2,53,433	2,62,999	2,39,747	2,09,674	1,83,826	1,81,689	2,27,662	23,87,069
EBITDA	(12,491)	408	(1,023)	(9,177)	(16,431)	(4,763)	7,185	15,073	2,007	2,131	(695)	(15,637)	(33,415)
% of total income	-10.3%	0.3%	-0.7%	-5.8%	-7.2%	-1.9%	2.7%	5.9%	0.9%	1.1%	-0.4%	-7.4%	-1.4%
Interest	5,826	5,767	6,184	6,250	6,090	6,675	7,828	6,681	6,636	7,081	7,021	6,826	78,862
PBDT	(18,317)	(5,359)	(7,207)	(15,427)	(22,521)	(11,437)	(643)	8,393	(4,629)	(4,950)	(7,716)	(22,462)	(1,12,277)
Income Tax	-	-	-	-	-	-	-	-	-	-	-	-	-
Net Income / (Loss)	(18,317)	(5,359)	(7,207)	(15,427)	(22,521)	(11,437)	(643)	8,393	(4,629)	(4,950)	(7,716)	(22,462)	(1,12,277)

*Incr-Increase, Decr-Decrease, FG-Finished goods, S-Sales, G-General, A-Administrative

ESTIMATED BALANCE SHEET

Estimated Balance Sheet													
All figures in Rupees millions unless otherwise specified													Budget
	Apr	May	Jun	Jul	Aug	Sep	Oct	Nov	Dec	Jan	Feb	Mar	PY
Current Assets (C/A)													
Cash & Bank Balances	6.0	6.0	6.0	6.0	6.0	6.0	6.0	6.0	6.0	6.0	6.0	6.0	2.0
Sundry Debtors	382.5	370.3	358.4	367.7	448.6	558.3	650.7	692.6	653.5	567.3	493.1	521.4	460.2
Inventories	590.1	637.0	660.1	693.7	664.8	617.6	601.5	518.2	496.2	502.4	529.1	479.4	539.2
Advances & Deposits	100.0	90.0	100.0	80.0	60.0	60.0	60.0	60.0	60.0	80.0	80.0	50.0	40.0
a-Total Current Assets	1,078.5	1,103.3	1,124.5	1,147.4	1,179.4	1,241.9	1,318.2	1,276.8	1,215.7	1,155.7	1,108.2	1,056.8	1,041.3
Fixed Assets													
Gross	2,019.9	2,019.9	2,063.7	2,063.7	2,063.7	2,107.5	2,107.5	2,107.5	2,107.5	2,151.3	2,151.3	2,151.3	1,987.9
Less: Depreciation	849.4	858.8	868.5	878.2	887.9	897.8	907.8	917.8	927.7	938.0	948.2	958.4	854.0
b-Net Fixed Assets	1,170.6	1,161.1	1,195.3	1,185.6	1,175.9	1,209.7	1,199.7	1,189.8	1,179.8	1,213.4	1,203.1	1,192.9	1,133.9
Investments	0.00	0.00	0.00	0.00	0.00	0.00	0.00	0.00	0.00	0.00	0.00	0.00	0.0
c-Total Assets (a+b)	2,249.1	2,264.4	2,319.7	2,332.9	2,355.3	2,451.6	2,518.0	2,466.6	2,395.5	2,369.0	2,311.3	2,249.7	2,175.2
d-Current Liabilities (C/L) & Provisions	491.4	515.5	541.1	578.1	634.3	694.6	761.9	717.1	659.2	603.0	565.7	558.0	377.4
Debt													
Bank Term Loans	146.3	152.7	194.9	192.6	185.3	227.6	225.2	217.9	216.4	257.8	250.6	224.0	203.8
Bank Cash Credit	200.0	200.0	200.0	200.0	200.0	200.0	200.0	200.0	200.0	200.0	200.0	200.0	170.0
e-Total Debt	346.3	352.7	394.9	392.6	385.3	427.6	425.2	417.9	416.4	457.8	450.6	424.0	373.8
Shareholders' Equity													
Share Capital/ Premium	1,535.7	1,535.7	1,535.7	1,535.7	1,535.7	1,535.7	1,535.7	1,535.7	1,535.7	1,535.7	1,535.7	1,535.7	1,535.7
Reserves & Surplus	(167.9)	(173.9)	(181.6)	(197.7)	(221.2)	(233.7)	(236.0)	(229.1)	(234.9)	(241.1)	(249.8)	(273.1)	(111.8)
f-Total Shareholders' Equity	1,367.8	1,361.9	1,354.1	1,338.0	1,314.5	1,302.0	1,299.7	1,306.6	1,300.8	1,294.7	1,286.0	1,262.6	1,424.0
g-Total Liabilities (d+e+f)	2,205.5	2,230.0	2,290.2	2,308.6	2,334.1	2,424.2	2,486.8	2,441.7	2,376.4	2,355.4	2,302.3	2,244.6	2,175.2
Difference (g-c)	(43.61)	(34.45)	(29.60)	(24.33)	(21.15)	(27.43)	(31.16)	(24.87)	(19.07)	(13.61)	(9.02)	(5.07)	0.00

*Difference between assets and liabilities should be zero. In an estimated balance sheet, if there is any difference, the reasons are identified and addressed by accounts.

BALANCE SHEET RATIOS

			Projected Current Year	Actual Previous year
1	Debt Equity Ratio	Total Debt / (Share Cap+Reserves)	0.34	0.19

2	Current Ratio	$\dfrac{\text{Current Assets}}{\text{Current Liabilities}}$	1.39	1.59
3	Debtors Turnover Ratio	$\dfrac{\text{Debtors}}{\text{Turnover}}$	0.23	0.27
4	Inventory Turnover Ratio	$\dfrac{\text{Cost of Sales}}{\text{Inventory}}$	9.27	3.13
Working Capital Ratios				
5	AR Days Outstanding	$\dfrac{\text{Sundry Debtors * Number of Days}}{\text{Turnover}}$	85 Days	71 Days
6	AP Days Outstanding	$\dfrac{\text{Sundry Creditors * Number of Days}}{\text{Cost of Sales}}$	94 Days	54 Days
7	Finished Goods Inventory Days	$\dfrac{\text{Inventory * Number of days}}{\text{Cost of Sales}}$	42 Days	47 Days
8	Raw material inventory days	$\dfrac{\text{Inventory * Number of days}}{\text{Raw material consumption}}$	35 Days	37 days
9	WIP days	$\dfrac{\text{Inventory * Number of days}}{\text{Cost of production}}$	28 Days	23 days

* AR-Account Receivable, AP-Account Payable

The Debt-Equity ratio is a measure of the relative contribution of the creditors and shareholders or owners in the capital employed in business. Simply stated, ratio of the total long term debt and equity capital in the business is called the debt-equity ratio.

The current ratio indicates a company's ability to meet short-term debt obligations. The current ratio measures whether or not a firm has enough resources to pay its debts over the next 12 months. Potential creditors use this ratio in determining whether or not to make short-term loans. The current ratio can also give a sense of the efficiency of a company's operating cycle or its ability to turn its product into cash. The current ratio is also known as the **working capital ratio.**

ESTIMATED FUND FLOW

Estimated Fund Flow Statement														
All figures in Rupees millions unless otherwise specified														Budget
Current year	Apr	May	Jun	Jul	Aug	Sep	Oct	Nov	Dec	Jan	Feb	Mar	Total	PY
Sources of Funds														
Profit before Depreciation	(9.43)	3.50	1.95	(6.44)	(13.76)	(2.59)	7.71	16.85	4.16	4.09	1.55	(13.13)	(5.53)	24.13
Increase in I/L (Term Loan)	-	13.80	43.80	-	-	43.80			-	43.80	-	(25.00)	120.00	130.00
Increase in C/C (Cash Credit)	10.31	-	-	-	-	-	-	-	-	-	-	-	10.31	127.99
Increase in Current Liabilities	54.14	24.08	25.63	36.96	56.23	60.36	67.26	(44.77)	(57.90)	(56.28)	(37.21)	(7.75)	120.74	(17.16)
	55.02	41.18	71.39	30.52	42.47	101.57	74.97	(27.92)	(53.74)	(8.39)	(35.66)	(45.89)	245.52	264.96
Applications of Funds														
Capital Expenditure	90.00	-	43.80	-	-	43.80	-	-	-	43.80	-	-	221.40	172.00
Repayment of Loans	2.37	7.24	1.53	2.38	7.25	1.53	2.39	7.25	1.53	2.39	7.25	1.53	44.64	44.74
Increase in Sundry Debtors	(11.98)	(12.18)	(11.92)	9.34	80.88	109.75	92.39	41.91	(39.14)	(86.21)	(74.20)	28.28	126.93	125.13
Increase in Stocks	64.04	46.95	23.13	33.53	(28.84)	(47.24)	(16.08)	(83.36)	(21.94)	6.16	26.71	(49.65)	(46.58)	(43.08)

Increase in Advance. & Deposits	36.96	10.00)	10.00	(20.00)	(20.00)	-	-	-	-	20.00	-	(30.00)	(13.04)	(34.46)
Increase in Bank & Cash balance	(7.90)	-	-	-	-	-	-	-	-	-	-	-	(7.90)	0.63
	173.50	32.01	66.54	25.25	39.29	107.84	78.70	(34.20)	(59.54)	(13.86)	(40.25)	(49.84)	325.45	264.96
	118.47	(9.17)	(4.85)	(5.27)	(3.18)	6.28	3.73	(6.28)	(5.80)	(5.46)	(4.59)	(3.95)	79.93	0.00

SCC limit or Cash Credit is a short-term source of financing for a company from the bank. Businesses need regular cash inflows to take care of their day-to-day operations. CC limit enables a company to withdraw money from its account without keeping credit balance upto a borrowing limit sanctioned by the bank to fulfil their working capital requirement. CC limit holders offer stock and debtors as primary security to the bank. Interest is levied only on the loan amount availed by the borrower and not on the credit amount sanctioned. The borrower has the option of repaying the loan on either a daily or a weekly basis, or as per the repayment structure designed by the lender.

Term Loan (T/L) is a loan issued by a bank for a fixed amount and fixed repayment schedule with either a fixed or a floating interest rate. Companies often use a term loan's proceeds to purchase fixed assets, such as equipment or a new building for its production process. Term loans can be long-term facilities with fixed payments, while short and intermediate-term loans might require balloon payments.

INTEREST AND DEPRECIATION CALCULATION

All figures in Rupees millions unless otherwise specified

Interest calculations for Current Year	Rate	Apr	May	Jun	Jul	Aug	Sep	Oct	Nov	Dec	Jan	Feb	Mar	Total
Bank C/C (Cash Credit)		200.00	200.00	200.00	200.00	200.00	200.00	200.00	200.00	200.00	200.00	200.00	200.00	
Interest	13.50%	2.25	2.25	2.25	2.25	2.25	2.25	2.25	2.25	2.25	2.25	2.25	2.25	27.00
Bank T/L (Term loan) Existing		69.52	62.30	60.80	58.45	51.23	49.73	47.37	40.15	38.65	36.29	29.07	27.57	
Interest	15.00%	0.87	0.82	0.77	0.75	0.69	0.63	0.61	0.55	0.49	0.47	0.41	0.35	7.40
Bank Capex T/L (For Looms)		76.40	90.00	90.00	90.00	90.00	90.00	90.00	90.00	90.00	90.00	90.00	65.00	
Interest	14.50%	0.92	1.01	1.09	1.09	1.09	1.09	2.18	1.09	1.09	1.09	1.09	0.94	13.74
Capex term loan 1		-	-	32.15	32.15	32.15	32.15	32.15	32.15	32.15	32.15	32.15	32.15	
Interest	14.50%	-	-	0.39	0.39	0.39	0.39	0.39	0.39	0.39	0.39	0.39	0.39	3.88
Capex term loan 2		-	-	-	-	-	32.15	32.15	32.15	32.15	32.15	32.15	32.15	
Interest	14.50%	-	-	-	-	-	0.39	0.39	0.39	0.39	0.39	0.39	0.39	2.72
Capex term loan 3		-	-	-	-	-	-	-	-	-	32.15	32.15	32.15	
Interest	14.50%	-	-	-	-	-	-	-	-	-	0.39	0.39	0.39	1.17
Capex term loan 1(retail store)		-	-	11.65	11.65	11.65	11.65	11.65	11.65	11.65	11.65	11.65	11.65	
Interest	14.50%	-	-	0.14	0.14	0.14	0.14	0.14	0.14	0.14	0.14	0.14	0.14	1.41
New capex term loan 2 (retail store)		-	-	-	-	-	11.65	11.65	11.65	11.65	11.65	11.65	11.65	
Interest	14.50%	-	-	-	-	-	0.14	0.14	0.14	0.14	0.14	0.14	0.14	0.99
New capex term loan 3 (retail store)		-	-	-	-	-	-	-	-	-	11.65	11.65	11.65	
Interest	14.50%	-	-	-	-	-	-	-	-	-	0.14	0.14	0.14	0.42
Vehicle Loan		0.38	0.36	0.33	0.30	0.27	0.24	0.21	0.18	0.15	0.12	0.09	0.06	
Interest	EMI	-	-	0.01	-	-	0.01	-	-	0.01	-	-	0.01	0.04
TD		40.00	30.00	30.00	30.00	30.00	30.00	30.00	30.00	30.00	30.00	30.00	30.00	
Interest	11.5	0.38	0.29	0.29	0.29	0.29	0.29	0.29	0.29	0.29	0.29	0.29	0.29	3.55
Payment to Debtors on Advance. Received		1.00	1.00	0.95	0.95	0.95	0.95	1.00	1.00	1.00	1.00	1.00	1.00	11.80
Bank Charges		0.40	0.40	0.30	0.40	0.30	0.40	0.45	0.45	0.45	0.40	0.40	0.40	4.75
Gross Interest		5.83	5.77	6.18	6.25	6.09	6.67	7.83	6.68	6.64	7.08	7.02	6.83	78.86
Interest Recovery														
Recovery on FD		19.75	19.75	19.75	19.75	19.75	19.75	19.75	19.75	19.75	19.75	19.75	7.75	
	9.5	0.16	0.16	0.16	0.16	0.16	0.16	0.16	0.16	0.16	0.16	0.16	0.06	1.78
Recovery from Drs. ***		0.67	0.67	0.67	0.93	1.13	1.13	1.13	0.93	1.00	1.20	1.34	1.20	12.00
Total Recovery		0.83	0.83	0.83	1.09	1.29	1.29	1.29	1.09	1.16	1.36	1.50	1.26	13.78

	Value	Apr	May	Jun	Jul	Aug	Sep	Oct	Nov	Dec	Jan	Feb	Mar	Total
Interest recoverable on Debtors														
Avg.Interest bearing domestic debtors		200.00												
Interest @ 12 %		24.00												
50 % recovery expected		12.00												
Repayment. of Bank Loan existing	71.87	2.35	7.22	1.50	2.35	7.22	1.50	2.36	7.22	1.50	2.36	7.22	1.50	44.30
Repayment. of Auto Loan	0.40	0.02	0.02	0.03	0.03	0.03	0.03	0.03	0.03	0.03	0.03	0.03	0.03	0.34
New capex loan (Production)		-	-	32.15	-	-	32.15	-	-	-	32.15	-	-	96.45
New T/L Looms	76.40		13.60	-	-	-	-	-	-	-	-	-	(25.00)	(11.40)
New capex loan (retail stores)				11.65			11.65			11.65				34.95
TL Outstanding	148.67	146.30	152.66	194.93	192.55	185.30	227.57	225.18	217.93	216.40	257.81	250.56	224.03	224.03
Depreciation calculations	Value													
Existing Block 1	1295.46	8.01	8.01	8.01	8.01	8.01	8.01	8.01	8.01	8.01	8.01	8.01	8.01	96.12
Existing Block Building 2	280.50	0.78	0.78	0.78	0.78	0.78	0.78	0.78	0.78	0.78	0.78	0.78	0.78	9.37
Existing Block Furniture3	14.43	0.08	0.08	0.08	0.08	0.08	0.08	0.08	0.08	0.08	0.08	0.08	0.08	0.91
Addn- Weaving looms	90.00	0.56	0.56	0.56	0.56	0.56	0.56	0.56	0.56	0.56	0.56	0.56	0.56	6.68
Addition in Production 2nd quarter	32.15	-	-	0.20	0.20	0.20	0.20	0.20	0.20	0.20	0.20	0.20	0.20	1.99
Addition in Production 3rd quarter	32.15	-	-	-	-	-	0.20	0.20	0.20	0.20	0.20	0.20	0.20	1.39
Addition in Production 4th quarter	32.15	-	-	-	-	-	-	-	-	-	0.20	0.20	0.20	0.60
Addition in 2nd quarter (Retail stores)	11.65	-	-	0.07	0.07	0.07	0.07	0.07	0.07	0.07	0.07	0.07	0.07	0.72
Addition in 3rd quarter (Retail stores)	11.65	-	-	-	-	-	0.07	0.07	0.07	0.07	0.07	0.07	0.07	0.50
Addition in 4th quarter (Retail stores)	11.65	-	-	-	-	-	-	-	-	-	0.07	0.07	0.07	0.22
Total Addition	221.40	9.42	9.42	9.69	9.69	9.69	9.97	9.97	9.97	9.97	10.24	10.24	10.24	118.50
Capex-P/M (Others)				32.15			32.15				32.15			
		-	-	32.15	32.15	32.15	64.30	64.30	64.30	64.30	96.45	96.45	96.45	96.45
Capex-Looms	90.00	-	-	-	-	-	-	-	-	-	-	-	-	90.00
			90.00	90.00	90.00	90.00	90.00	90.00	90.00	90.00	90.00	90.00	90.00	
Capex- Retail stores				11.65			11.65			11.65				34.95
			-	-	11.65	11.65	11.65	23.30	23.30	23.30	23.30	34.95	34.95	
Total		90.00	-	43.80	-	-	43.80	-	-	-	43.80	-	-	221.40
Cumulative		90.00	90.00	133.80	133.80	133.80	177.60	177.60	177.60	177.60	221.40	221.40	221.40	221.40

*T/L –Term loan, Adv- Advance, Recd-Received

DEBTORS, CREDITORS AND STOCKS

Debtors, Stocks and Creditors														
All figures in Rupees millions unless otherwise specified														
Sundry Debtors		Apr	May	Jun	Jul	Aug	Sep	Oct	Nov	Dec	Jan	Feb	Mar	Total
Opening	Domestic	251.65	251.08	194.81	151.83	156.78	218.98	325.75	431.90	469.26	426.33	366.46	317.83	251.65
	Govt/Inst	33.88	52.83	84.12	100.81	101.90	99.01	102.77	103.72	102.26	94.11	89.01	87.01	33.88
	Export	72.45	78.55	91.34	105.71	109.01	130.59	129.81	115.11	121.10	133.05	111.81	88.24	72.45
		357.98	382.45	370.27	358.35	367.70	448.58	558.33	650.72	692.63	653.49	567.28	493.08	357.98
Dispatches	Domestic	57.11	50.47	41.66	62.06	112.67	148.43	168.21	150.03	105.50	108.34	101.40	131.12	1236.98
	Govt/Inst	22.95	39.80	39.64	40.89	36.74	44.65	37.69	43.20	29.53	38.10	27.54	22.84	423.56
	Export	32.20	42.62	46.58	45.92	68.16	45.14	53.46	51.14	65.40	29.90	41.83	47.83	570.17
		112.26	132.89	127.87	148.87	217.57	238.22	259.35	244.37	200.43	176.34	170.76	201.78	2230.72

Realisation	Domestic		57.69	106.74	84.64	57.11	50.47	41.66	62.06	112.67	148.43	168.21	150.03	105.50	1145.19
	Govt/Inst		4.00	8.50	22.95	39.80	39.64	40.89	36.74	44.65	37.69	43.20	29.53	38.10	385.69
	Export		26.10	29.84	32.20	42.62	46.58	45.92	68.16	45.14	53.46	51.14	65.40	29.90	536.45
			87.78	145.07	139.79	139.53	136.69	128.47	166.96	202.46	239.57	262.55	244.97	173.50	2067.33
Closing	Domestic		251.08	194.81	151.83	156.78	218.98	325.75	431.90	469.26	426.33	366.46	317.83	343.45	343.45
	Govt/Inst		52.83	84.12	100.81	101.90	99.01	102.77	103.72	102.26	94.11	89.01	87.01	71.75	71.75
	Export		78.55	91.34	105.71	109.01	130.59	129.81	115.11	121.10	133.05	111.81	88.24	106.16	106.16
Total			382.45	370.27	358.35	367.70	448.58	558.33	650.72	692.63	653.49	567.28	493.08	521.36	521.36
Finished Goods															
Opening			298.10	311.87	321.82	359.95	385.47	355.64	325.40	281.32	240.04	228.11	229.27	224.89	298.10
Incr/(Decr)			13.77	9.95	38.13	25.53	(29.84)	(30.24)	(44.08)	(41.28)	(11.94)	1.16	(4.38)	(45.65)	(118.86)
Closing*		298.10	311.87	321.82	359.95	385.47	355.64	325.40	281.32	240.04	228.11	229.27	224.89	179.24	179.24
Process Stock		112.86	125.00	157.00	140.00	140.00	150.00	140.00	155.00	145.00	135.00	140.00	155.00	155.00	155.00
RM Stock		113.20	113.20	113.20	113.20	113.20	113.20	113.20	113.20	113.20	113.20	113.20	113.20	113.20	113.20
Stores		32.78	40.00	45.00	47.00	55.00	46.00	39.00	52.00	19.92	19.92	19.92	36.00	32.00	32.00
Total Inventories		556.94	590.06	637.02	660.14	693.67	664.83	617.60	601.52	518.16	496.22	502.38	529.09	479.44	479.44
Cash & Bank Balance		13.90	6.00	6.00	6.00	6.00	6.00	6.00	6.00	6.00	6.00	6.00	6.00	6.00	6.00
Advance. & Deposits		63.04	100.00	90.00	100.00	80.00	60.00	60.00	60.00	60.00	60.00	80.00	80.00	50.00	50.00
Current Liabilities															
Suppliers Stores		39.59	37.24	38.08	39.59	41.05	42.31	43.46	45.25	46.07	45.11	44.02	42.76	41.56	41.56
Suppliers RM		109.48	279.72	291.55	337.49	384.11	437.02	477.41	523.91	414.44	420.84	407.90	381.46	349.31	349.31
Trade Deposits		29.76	40.00	30.00	30.00	30.00	30.00	30.00	30.00	30.00	30.00	30.00	30.00	30.00	30.00
Employees dues		16.90	11.80	11.80	13.80	9.57	17.30	19.00	19.05	19.80	10.80	9.80	13.00	14.00	14.00
Gratuity/Leaves		1.91	4.00	1.00	3.00	3.00	4.00	4.00	4.00	6.00	6.00	5.00	5.00	5.00	5.00
Power		3.22	3.62	5.00	5.00	4.00	4.00	5.00	5.00	10.00	5.81	4.00	4.00	5.00	5.00
Legal		53.90	38.33	34.45	42.59	44.00	40.00	28.00	20.00	35.00	29.00	23.00	27.00	32.43	32.43
Interest		23.22	8.00	10.04	5.00	5.00	4.00	4.00	4.00	8.00	4.00	4.50	3.50	10.00	10.00
Publicity		39.56	5.00	9.50	8.50	9.00	7.00	12.00	8.00	26.82	10.00	8.00	9.00	6.00	6.00
Commn, TOD		78.89	43.70	47.84	42.00	38.00	42.00	47.00	75.01	81.16	73.31	46.46	29.24	42.00	42.00
Others		40.82	19.98	36.20	14.12	10.32	6.65	24.77	27.68	39.84	24.37	20.28	20.78	22.69	22.69
		437.25	491.38	515.46	541.10	578.05	634.28	694.64	761.90	717.13	659.24	602.96	565.74	557.99	557.99
			43.61	34.45	29.60	24.33	21.15	27.43	31.16	24.87	19.07	13.61	9.02	5.07	5.07

DEBTOR AND CREDITOR WORKING

(Rs Millions)	Days	Apr	May	June	July	Aug	Sept	Oct	Nov	Dec	Jan	Fab	March
Domestic (Net)	82	57	50	42	62	113	148	168	150	105	108	101	131
Govt./Institutional	44	23	40	40	41	37	45	38	43	30	38	28	23
Export	54	32	43	47	46	68	45	53	51	65	30	42	48
Total		112	133	128	149	218	238	259	244	200	176	171	202
Add: Opening debtors													
Domestic		252	251	195	152	157	219	326	432	469	426	366	318
Govt./Institutional		34	53	84	101	102	99	103	104	102	94	89	87
Export		72	79	91	106	109	131	130	115	121	133	112	88
Total		358	382	370	358	368	449	558	651	693	653	567	493
Realisation													
Domestic		58	107	85	57	50	42	62	113	148	168	150	105
Govt./Institutional		4	9	23	40	40	41	37	45	38	43	30	38

Export		26	30	32	43	47	46	68	45	53	51	65	30
Total		88	145	140	140	137	128	167	202	240	263	245	173
Closing debtors													
Domestic		251	195	152	157	219	326	432	469	426	366	318	343
Govt./Institutional		53	84	101	102	99	103	104	102	94	89	87	72
Export		79	91	106	109	131	130	115	121	133	112	88	106
Total		382	370	358	368	449	558	651	693	653	567	493	521
RM Opening stock		113	113	113	113	113	113	113	113	113	113	113	113
Add: Purchases (RM and dyes)		82	84	103	110	122	127	131	114	115	108	102	100
Less: Consumption		82	84	103	110	122	127	131	114	115	108	102	100
Closing stock		113	113	113	113	113	113	113	113	113	113	113	113
Payment to creditors													
RM (Domestic)	90	11	9	14	16	17	16	15	136	14	14	16	16
RM (Imported)	180	48	59	41	45	48	66	65	82	90	102	108	113
Dyes	90	4	4	3	3	4	5	5	5	5	5	5	4
Total		62	72	57	64	69	86	85	223	108	121	128	133
Opening creditors		260	280	292	337	384	437	477	524	414	421	408	381
Add: purchases		82	84	103	110	122	127	131	114	115	108	102	100
Less: Payment		62	72	57	64	69	86	85	223	108	121	128	133
Closing (RM and dyes)		280	292	337	384	437	477	524	414	421	408	381	349
No. of days													
Opening creditors (Stores)		37	37	38	40	41	42	43	45	46	45	44	43
Add: Purchases		6	7	8	8	9	9	11	10	10	9	8	8
Less: payment		6	6	6	7	8	8	9	9	11	10	10	9
Closing	90	37	38	40	41	42	43	45	46	45	44	43	42

*RM-Raw material, Govt- Government, Inst-Institutional, TOD-Turn over discount

QUALITY WISE MONTHLY FABRIC PRODUCTION PLAN

Quality	Lac Mtrs	Apr	May	Jun	Jul	Aug	Sep	Oct	Nov	Dec	Jan	Feb	Mar	Total
Economy Class 1	0.20	0.04	0.06	0.06	-	-	-	-	-	-	-	0.04	-	0.20
2	0.13	-	0.05	-	0.05	-	-	-	-	-	0.03	-	-	0.13
3 Etc...	0.20	0.02	0.03	0.04	-	-	-	0.02	0.02	0.02	0.02	-	0.03	0.20
Sub-total	12.15	0.91	1.06	1.20	1.30	0.76	0.69	0.65	1.08	1.14	1.00	1.15	1.21	12.15
Standard Class														
1	0.40	0.05	0.06	0.05	0.03	-	-	-	0.03	0.03	0.05	0.05	0.05	0.40
2	0.11	-	-	-	-	0.01	0.03	0.04	-	0.02	0.01	-	-	0.11
3 Etc...	0.35	0.04	-	-	0.05	0.05	0.05	-	0.06	0.05	0.03	0.02	-	0.35
Sub-total	14.72	0.89	0.85	0.99	1.07	1.60	1.97	1.83	1.46	1.32	1.18	0.87	0.69	14.72
Premium Class														
1	0.09	-	-	-	0.03	-	0.02	-	-	0.02	0.02	-	-	0.09
2	0.03							0.02	0.01	-				0.03
3 Etc...	0.02	-	-	-	-	-	-	0.01	0.01				-	0.02

	Total	Apr	May	June	July	Aug	Sept	Oct	Nov	Dec	Jan	Feb	March	Total
Sub-total	0.75	0.02	-	0.02	0.04	0.07	0.16	0.20	0.11	0.06	0.05	0.02	-	0.75
Domestic Total	29.12	1.90	1.99	2.30	2.50	2.56	3.09	2.93	2.78	2.60	2.40	2.16	1.91	29.12
Govt./Institutional														
1	1.95	-	-	-	-	-	-	0.20	0.27	0.23	0.50	0.35	0.40	1.95
2	0.85	0.25	-	0.10	0.25	-	-	-	-	-	0.25	-	-	0.85
Etc...	0.25	-	-	-	-	-	-	-	-	-	-	0.10	0.15	0.25
RMG														
1	0.30	-	0.03	0.05	0.06	-	-	-	-	0.06	0.09	-	0.01	0.30
2	0.05	-	0.03	-	-	0.02	-	-	-	-	-	-	-	0.05
3 Etc.	0.03	-	0.02	-	0.01	-	-	-	-	-	-	-	-	0.03
Sub total	7.90	0.45	0.58	0.82	0.88	0.31	0.61	0.39	0.56	0.42	1.13	0.88	0.88	7.90
Export -Others														
1	0.20	-	-	0.05	-	-	0.05	-	-	0.05	-	-	0.05	0.20
2	0.60	-	-	0.10	-	0.05	0.10	0.05	0.05	0.10	-	0.05	0.10	0.60
3 Etc...	0.65	-	-	0.10	-	0.05	0.10	0.05	0.05	0.10	0.05	0.05	0.10	0.65
	8.25	0.43	0.64	0.88	0.51	0.80	0.70	0.89	0.86	0.95	0.32	0.56	0.71	8.25
Export - EU														
1	0.68	-	-	-	0.06	0.06	0.10	0.10	0.10	-	0.10	0.06	0.10	0.68
2 Etc.	0.22	-	-	-	0.02	0.03	-	-	-	0.03	0.05	0.04	0.05	0.22
Sub Total	2.15	-	0.10	-	0.23	0.34	0.10	0.20	0.10	0.38	0.15	0.25	0.30	2.15
Export - US														
1	1.50	0.19	0.21	0.20	0.08	0.04	0.04	0.12	0.14	0.21	0.08	0.10	0.09	1.50
2 Etc.	0.25	-	-	-	0.10	0.10	0.05	-	-	-	-	-	-	0.25
Sub Total	6.60	0.51	0.55	0.54	0.60	0.84	0.53	0.54	0.58	0.65	0.39	0.44	0.43	6.60
Total Export	17.00	0.94	1.29	1.42	1.34	1.98	1.33	1.63	1.54	1.98	0.86	1.25	1.44	17.00
	54.00	3.29	3.86	4.54	4.72	4.85	5.03	4.95	4.88	5.00	4.39	4.29	4.22	54.00

QUALITY WISE MONTHLY SALES PLAN (Lacs Mtrs)

Economy Class	Apr-12	May	June	July	Aug	Sept	Oct	Nov	Dec	Jan	Fab	March	Total	Rate/Mtr
1.00	0.04	0.05	0.05	0.02	0.02	-	-	-	-	-	0.03	0.04	0.25	270.84
2.00	-	0.03	0.03	0.03	0.01	0.01	0.01	0.01	-	-	-	-	0.13	282.79
3.00	0.10	0.10	0.05	0.10	0.10	-	-	-	-	-	0.15	0.15	0.75	157.33
Sub-total	0.97	0.93	1.11	1.45	1.91	1.64	1.45	1.21	1.08	1.30	1.88	2.97	17.90	264.00
Standard Class														
1	0.10	0.06	0.06	0.01	0.01	0.01	0.01	0.01	0.01	0.05	0.07	0.10	0.50	401.28
2	-	-	-	-	0.01	0.02	0.03	0.02	0.02	0.01	-	-	0.11	560.60
3 Etc.	0.02	0.02	0.02	0.01	0.01	0.06	0.06	0.06	0.06	0.03	0.03	0.02	0.40	353.49
Sub-total	0.74	0.62	0.29	0.54	1.55	2.21	2.31	2.26	1.42	1.40	1.14	1.29	15.77	416.09
Premium Class														
1	-	-	-	-	-	0.01	0.02	0.02	0.02	0.02	-	-	0.09	583.50
2	-	-	-	-	-	-	0.01	0.01	0.01	-	-	-	0.03	649.22
3 Etc.	-	-	-	-	-	0.02	0.02	0.03	0.02	0.02	-	-	0.11	597.45

	Apr	May	Jun	Jul	Aug	Sept	Oct	Nov	Dec	Jan	Fab	March	Total	
Sub-total	0.08	0.07	0.02	0.01	0.09	0.20	0.24	0.22	0.19	0.21	0.14	0.12	1.59	598.26
Super Premium														
1	-	-	-	-	-	-	-	0.01	0.01	-	-	-	0.02	1,792.34
2	-	-	-	-	-	-	-	0.01	-	-	-	-	0.01	1,121.21
3 Etc.	-	-	-	-	-	0.01	0.01	0.02	0.02	0.02	0.01	-	0.09	838.41
Sub-total	0.02	0.02	0.02	0.04	0.04	0.06	0.14	0.13	0.09	0.10	0.06	0.03	0.75	1,161.63
Domestic Total	1.81	1.64	1.44	2.04	3.59	4.11	4.14	3.82	2.78	3.01	3.22	4.41	36.01	367.85
Govt/Institutional														
1	0.18	0.18	0.18	0.18	0.18	0.18	0.37	0.22	0.22	0.37	0.26	0.26	2.80	502.43
2	0.07	0.07	0.07	0.07	0.07	0.07	0.13	0.08	0.08	0.13	0.09	0.09	1.00	502.43
3 Etc...	-	-	-	-	0.10	0.15	-	-	-	-	-	-	0.25	376.82
RMG														
1	-	0.03	0.05	0.03	0.03	-	-	0.15	-	-	-	0.01	0.30	678.52
2	-	0.03	-	-	0.02	-	-	-	-	-	-	-	0.05	643.10
3 Etc.	-	0.02	-	0.01	-	-	-	-	-	-	-	-	0.03	663.20
	0.44	0.83	0.85	0.89	0.80	0.97	0.78	0.84	0.64	0.79	0.58	0.48	8.89	476.54
Export - Others														
1	-	-	0.05	-	-	0.05	-	-	0.05	-	-	0.05	0.20	307.94
2	-	-	0.10	-	0.05	0.10	0.05	0.05	0.10	-	0.05	0.10	0.60	307.94
3 Etc.	-	-	0.10	-	0.05	0.10	0.05	0.05	0.10	0.05	0.05	0.10	0.65	307.94
	0.43	0.64	0.88	0.51	0.80	0.70	0.89	0.86	0.95	0.32	0.56	0.71	8.25	287.42
Export - EU														
1	-	-	-	0.06	0.06	0.10	0.10	0.10	-	0.10	0.06	0.10	0.68	391.04
2 Etc.	-	-	-	0.02	0.03	-	-	-	0.03	0.05	0.04	0.05	0.22	391.04
	-	0.10	-	0.23	0.34	0.10	0.20	0.10	0.38	0.15	0.25	0.30	2.15	348.30
Export - US														
1	0.19	0.21	0.20	0.08	0.04	0.04	0.12	0.14	0.21	0.08	0.10	0.09	1.50	391.04
2 Etc.	-	-	-	0.10	0.10	0.05	-	-	-	-	-	-	0.25	391.04
	0.51	0.55	0.54	0.60	0.84	0.53	0.54	0.58	0.65	0.39	0.44	0.43	6.60	361.31
Total Export	0.94	1.29	1.42	1.34	1.98	1.33	1.63	1.54	1.98	0.86	1.25	1.44	17.00	323.81
G. Total	3.19	3.76	3.71	4.27	6.37	6.41	6.55	6.20	5.40	4.66	5.05	6.33	61.90	371.30

SALES AND PRODUCTION QUANTITY RECONCILIATION (Lacs Mtrs)

Sales	Apr	May	Jun	Jul	Aug	Sept	Oct	Nov	Dec	Jan	Fab	March	Total
Domestic	1.81	1.71	1.80	2.39	3.59	4.11	4.14	3.82	2.78	3.07	3.22	3.56	36.00
Less: O/S Purchase special	0.22	0.10	0.17	0.17	0.37	0.27	0.07	0.02	0.02	0.11	0.38	0.45	2.35
Less: O/S Purchase (PV)	0.10	0.10	0.15	0.15	0.25	0.20	0.20	0.05	0.02	0.25	0.25	0.28	2.00
Net Domestic Sale	1.49	1.51	1.48	2.07	2.97	3.64	3.87	3.75	2.74	2.71	2.59	2.83	31.65
Govt/Institutional	0.44	0.83	0.85	0.89	0.80	0.97	0.78	0.84	0.64	0.79	0.58	0.48	8.89
Export -Others	0.43	0.64	0.88	0.51	0.80	0.70	0.89	0.86	0.95	0.32	0.56	0.71	8.25
Export -EU	-	0.10	-	0.23	0.34	0.10	0.20	0.10	0.38	0.15	0.25	0.30	2.15
Export -US	0.51	0.55	0.54	0.60	0.84	0.53	0.54	0.58	0.65	0.39	0.44	0.43	6.60
Total	3.19	3.83	4.07	4.62	6.37	6.41	6.55	6.20	5.40	4.72	5.05	5.48	61.90

Production													
Domestic	1.90	1.99	2.30	2.50	2.56	3.09	2.93	2.78	2.60	2.40	2.16	1.91	29.12
Govt/Institutional	0.45	0.58	0.82	0.88	0.31	0.61	0.39	0.56	0.42	1.13	0.88	0.88	7.90
Export -Others	0.43	0.64	0.88	0.51	0.80	0.70	0.89	0.86	0.95	0.32	0.56	0.71	8.25
Export -EU	-	0.10	-	0.23	0.34	0.10	0.20	0.10	0.38	0.15	0.25	0.30	2.15
Export -US	0.51	0.55	0.54	0.60	0.84	0.53	0.54	0.58	0.65	0.39	0.44	0.43	6.60
Total	3.29	3.86	4.54	4.72	4.85	5.03	4.95	4.88	5.00	4.39	4.29	4.22	54.00
Stock Inc/(Dec) in Mtrs													
Domestic	0.41	0.48	0.82	0.43	(0.41)	(0.55)	(0.94)	(0.97)	(0.14)	(0.31)	(0.43)	(0.92)	(2.53)
Govt/Inst	0.01	(0.25)	(0.03)	(0.01)	(0.49)	(0.36)	(0.39)	(0.28)	(0.22)	0.34	0.30	0.40	(0.99)
Export -Others	-	-	-	-	-	-	-	-	-	-	-	-	-
Export -EU	-	-	-	-	-	-	-	-	-	-	-	-	-
Export -US	-	-	-	-	-	(0.00)	-	-	-	-	(0.00)	(0.00)	(0.01)
Total	0.42	0.23	0.79	0.42	(0.90)	(0.91)	(1.33)	(1.25)	(0.36)	0.02	(0.13)	(0.53)	(3.53)
Rate	331.82	331.82	331.82	331.82	331.82	331.82	331.82	331.82	331.82	331.82	331.82	331.82	331.82
Stock Inc/(Dec) in Lac Rs.													
Domestic	136.05	159.27	272.09	142.68	(136.05)	(182.50)	(311.91)	(321.86)	(46.45)	(102.86)	(142.68)	(305.27)	(839.50)
Govt/Institutional	1.73	(82.88)	(9.95)	(3.32)	(162.59)	(119.45)	(129.27)	(91.25)	(73.00)	111.16	99.55	131.07	(328.22)
Export -Others	-	-	-	-	-	-	-	-	-	-	-	-	-
Export -EU	-	-	-	-	-	-	-	-	-	-	-	-	-
Export -US	-	-	-	-	-	(0.66)	-	-	-	-	(0.66)	(0.66)	(1.99)
Total	137.77	76.39	262.14	39.36	(298.64)	(302.62)	(441.18)	(413.11)	(119.45)	8.30	(43.80)	(174.87)	(1,169.71)
Million Rs.	13.78	7.64	26.21	13.94	(29.86)	(30.26)	(44.12)	(41.31)	(11.95)	0.83	(4.38)	(17.49)	(116.97)

*Inc- Increase, Dec-Decrease, O/S-outsourced

SALES COST ELEMENTS

Category	Lac Mtrs	Price	Lac Rs.	Def %	Lac Rs.	Cash Disc.	Sales Value	TOD	Net Value	Commn.	Samples	Freight	Total	Royalty @8 %	Pkg Exp
Domestic															
Economy	17.11	281.16	4,809.24	.16	247.97	45.61	4,515.66	270.94	4,244.72	163.92	47.41	17.11	228.44	-	44.47
Standard	16.46	447.12	7,359.63	6.00	441.58	69.18	6,848.87	410.93	6,437.94	248.61	71.91	16.46	336.99	-	42.80
Premium	1.68	642.87	1,080.03	6.00	64.80	10.15	1,005.07	60.30	944.77	36.48	10.55	1.68	48.72	-	25.20
Luxury	0.64	992.89	635.45	6.00	38.13	5.97	591.35	35.48	555.87	21.47	6.21	0.64	28.32	-	99.20
Imported	0.11	2,734.04	300.74	6.00	18.04	2.83	279.87	16.79	263.08	10.16	2.94	0.11	13.21	25.16	17.05
Sub Total	36.00	394.09	14,185.09	5.71	810.52	133.75	13,240.83	794.45	12,446.38	480.64	139.03	36.00	655.67	25.16	228.72
Govt/ Institutional.	8.89	481.31	4,278.83	1.00	42.79	-	4,236.04	-	4,236.04	30.77	44.48	8.89	84.14	-	17.78
Export	8.25	305.77	2,522.57	6.00	151.35		2,371.22	-	2,371.22	82.99	24.90	-	107.89	-	16.50
Export -EU	2.15	370.53	796.64	6.00	47.80		748.84	-	748.84	37.44	7.86	-	45.30	-	4.30
Export -US	6.10	416.00	2,537.60	6.00	152.26		2,385.34	-	2,385.34	119.27	25.05	-	144.31	-	9.70
Sub Total	16.50	354.96	5,856.81	6.00	351.41		5,505.40	-	5,505.40	239.70	57.81	-	297.51	-	30.50
	61.39	396.20	24,320.73	4.95	1,204.72		22,982.27	794.45	22,187.82	751.12	241.31	44.89	1,037.32	25.16	277.00
	Kgs	Rate/ count													
Job Work (Yarn)	6.60	103.50	683.10	-	683.10		683.10			6.83					
										757.95					

*TOD- Turn over discount, Disc- Discount, Def-Defective, Commn-Commission, Pkg-Packaging, Exp-Expenses

MONTHLY MANUFACTURING EXPENSES

			April	May	June	July	August	Sept	Oct	Nov	Dec	Jan	Fab	March	Total		
	Production																
	Domestic	Lac Mtrs	1.90	1.99	2.30	2.50	2.56	3.09	2.93	2.78	2.60	2.40	2.16	1.91	29.12		
	Government	Lac Mtrs	0.45	0.58	0.82	0.88	0.31	0.61	0.39	0.56	0.42	1.13	0.88	0.88	7.90		
	Export	Lac Mtrs	0.43	0.64	0.88	0.51	0.80	0.70	0.89	0.86	0.95	0.32	0.56	0.71	8.25		
	Export -(EU)	Lac Mtrs	-	0.10	-	0.23	0.34	0.10	0.20	0.10	0.38	0.15	0.25	0.30	2.15		
	Export - (US)	Lac Mtrs	0.51	0.55	0.54	0.60	0.84	0.53	0.54	0.58	0.65	0.39	0.44	0.43	6.60		
1	Total	Lac Mtrs	3.29	3.86	4.54	4.72	4.85	5.03	4.95	4.88	5.00	4.39	4.29	4.22	54.0		
2	Power Cost																
	Power for Job Spg.																
	Production	Lac Kgs	0.55	0.55	0.55	0.55	0.55	0.55	0.55	0.55	0.55	0.55	0.55	0.55	6.60		
	Unit Consumed	kwh/Kg.	7.00	7.00	7.00	7.00	7.00	7.00	7.00	7.00	7.00	7.00	7.00	7.00	7.00		
	Total Units	Lac Kwh	3.85	3.85	3.85	3.85	3.85	3.85	3.85	3.85	3.85	3.85	3.85	3.85	46.20		
	Power for Fabric																
	Production	Lac Mtrs.	3.29	3.86	4.54	4.72	4.85	5.03	4.95	4.88	5.00	4.39	4.29	4.22	54.02		
	Unit Consumed	kwh/Mtr.	4.39	4.39	4.39	4.39	4.39	4.39	4.39	4.39	4.39	4.39	4.39	4.39	4.39		
	Total Units	Lac Kwh	14.45	16.96	19.94	20.74	21.30	22.09	21.75	21.44	21.97	19.28	18.84	18.55	237.30		
	Total Units required		18.30	20.81	23.79	24.59	25.15	25.94	25.60	25.29	25.82	23.13	22.69	22.40	283.50		
	Cost of Power																
	Source of Power																
	Own Generation (DG)	Lac kwh	0.29	0.34	0.40	0.42	0.43	0.45	0.44	0.43	0.44	0.39	0.38	0.38	4.80		
	Trading	Lac kwh	9.26	10.86	12.78	13.28	13.64	14.15	13.93	13.73	14.07	12.35	12.07	11.88	152.00		
	Govt Grid	Lac kwh	7.72	9.05	10.65	11.07	11.37	11.79	11.61	11.45	11.73	10.29	10.06	9.91	126.70		
	Total Units	Lac kwh	18.30	20.81	23.79	24.59	25.15	25.94	25.60	25.29	25.82	23.13	22.69	22.40	283.50		
	Rate per Unit																
	Own Generation (DG)	Rs. Unit	2.69	2.69	2.69	2.69	2.69	2.69	2.69	2.69	2.69	2.69	2.69	2.69	2.69		
	Trading	Rs. Unit	4.80	4.80	4.80	4.80	4.80	4.80	4.80	4.80	4.80	4.80	4.80	4.80	4.80		
	Govt Grid	Rs. Unit	6.46	6.46	6.46	6.46	6.46	6.46	6.46	6.46	6.46	6.46	6.46	6.46	6.46		
	Cost of Power																
	Own Generation (DG)	Lac Rs.		0.79	0.92	1.08	1.13	1.16	1.20	1.18	1.17	1.19	1.05	1.02	1.01	12.90
	Trading	Lac Rs.	44.44	52.14	61.32	63.75	65.48	67.91	66.86	65.91	67.54	59.28	57.92	57.04	729.60		
	Govt Grid	Lac Rs.	49.82	58.45	68.75	71.47	73.41	76.14	74.96	73.90	75.71	66.46	64.93	63.95	817.94		
	Cost of Power	Lac Rs.	95.04	111.51	131.15	136.35	140.05	145.25	143.00	140.98	144.44	126.79	123.87	122.00	1,560.44		
3	Cost of Coal																
	Quantity required																

															Total		
	Steam Coal	MT	325.86	382.31	449.66	467.49	480.17	498.00	490.27	483.34	495.22	434.71	424.70	418.27	5,350.00		
	Pet Coke	MT	-	-	-	-	-	-	-	-	-	-	-	-	-		
	Charcoal	MT	0.73	0.86	1.01	1.05	1.08	1.12	1.10	1.08	1.11	0.98	0.95	0.94	12.00		
	Rate per Unit																
	Steam Coal	Per MT	11,500	11,500	11,500	11,500	11,500	11,500	11,500	11,500	11,500	11,500	11,500	11,500			
	Pet Coke	Per MT	-	-	-	-	-	-	-	-	-	-	-	-			
	Charcoal	Per MT	15,000	15,000	15,000	15,000	15,000	15,000	15,000	15,000	15,000	15,000	15,000	15,000			
	Cost of Coal																
	Steam Coal	Lac Rs.	37.47	43.97	51.71	53.76	55.22	57.27	56.38	55.58	56.95	49.99	48.84	48.10	615.25		
	Pet Coke	Lac Rs.	-	-	-	-	-	-	-	-	-	-	-	-	-		
	Charcoal	Lac Rs.	0.11	0.13	0.15	0.16	0.16	0.17	0.16	0.16	0.17	0.15	0.14	0.14	1.80		
	Total	Lac Rs.	37.58	44.09	51.86	53.92	55.38	57.44	56.55	55.75	57.12	50.14	48.98	48.24	617.05		
4	Cost of HSD																
	Quantity required	Lac Ltrs.	0.45	0.52	0.61	0.64	0.66	0.68	0.67	0.66	0.68	0.59	0.58	0.57	7.31		
	Rate per Unit	Rs. Ltr	43.00	43.00	43.00	43.00	43.00	43.00	43.00	43.00	43.00	43.00	43.00	43.00	43.00		
	Total	Lac Rs.	19.15	22.46	26.42	27.47	28.21	29.26	28.81	28.40	29.10	25.54	24.95	24.57	314.33		
	Total Power & Fuel	Lac Rs.	151.77	178.07	209.44	217.74	223.64	231.95	228.35	225.12	230.66	202.47	197.81	194.81	2,491.82		
4	Repair & Maintenance	Lac Rs.	6.91	8.11	9.53	9.91	10.18	10.56	10.40	10.25	10.50	9.22	9.00	8.87	113.43		
	Fixed		5.87	6.89	8.10	8.43	8.65	8.97	8.84	8.71	8.93	7.83	7.65	7.54	96.42		
	Variable		1.04	1.22	1.43	1.49	1.53	1.58	1.56	1.54	1.58	1.38	1.35	1.33	17.02		
5	Stores & Spares	Lac Rs.	36.19	42.46	49.94	51.92	53.33	55.31	54.45	53.68	55.00	48.28	47.17	46.45	594.18		
	Fixed		27.14	31.85	37.46	38.94	40.00	41.48	40.84	40.26	41.25	36.21	35.38	34.84	445.63		
	Variable		9.05	10.62	12.49	12.98	13.33	13.83	13.61	13.42	13.75	12.07	11.79	11.61	148.54		
5a	Additional Spares & Major Repair	Lac Rs.													-		
6	Overhead	Lac Rs.															
	Stationery & Printing		1.67	1.67	1.67	1.67	1.67	1.67	1.67	1.67	1.67	1.67	1.67	1.67	20.00		
	Communication Exp. (Postage/ Courier/ Telephone)		3.77	3.77	3.77	3.77	3.77	3.77	3.77	3.77	3.77	3.77	3.77	3.77	45.28		
	Travelling & Conveyance (Incld.Vehicles running)		19.19	19.19	19.19	19.19	19.19	19.19	19.19	19.19	19.19	19.19	19.19	19.19	230.33		
	Rent, Rates & Taxes		5.80	5.80	5.80	5.80	5.80	5.80	5.80	5.80	5.80	5.80	5.80	5.80	69.61		
	Legal & Professional		31.75	31.75	31.75	31.75	31.75	31.75	31.75	31.75	31.75	31.75	31.75	31.75	381.06		
	Insurance		3.65	3.65	3.65	3.65	3.65	3.65	3.65	3.65	3.65	3.65	3.65	3.65	43.75		
	Director's fees		0.02	0.02	0.02	0.02	0.02	0.02	0.02	0.02	0.02	0.02	0.02	0.02	0.20		
	Audit Fees		1.00	1.00	1.00	1.00	1.00	1.00	1.00	1.00	1.00	1.00	1.00	1.00	12.00		
	General Chgs.		0.67	0.67	0.67	0.67	0.67	0.67	0.67	0.67	0.67	0.67	0.67	0.67	8.00		
	Fees & Subscription		1.55	1.55	1.55	1.55	1.55	1.55	1.55	1.55	1.55	1.55	1.55	1.55	18.59		
	Other Misc.		0.69	0.69	0.69	0.69	0.69	0.69	0.69	0.69	0.69	0.69	0.69	0.69	8.25		

			69.76	69.76	69.76	69.76	69.76	69.76	69.76	69.76	69.76	69.76	69.76	69.76	837.07
7	Sales Expenses														
	Publicity		68.18	68.18	68.18	136.36	204.55	204.55	136.36	136.36	136.36	109.09	95.45	136.36	1,500.00
8	Salary														
	Salary	Lac Rs.	97.02	97.02	97.02	97.02	97.02	97.02	97.02	97.02	97.02	97.02	97.02	97.02	1,164.24
	Add: Yearly Incre.@ 10%	Lac Rs.	9.70	9.70	9.70	9.70	9.70	9.70	9.70	9.70	9.70	9.70	9.70	9.70	116.42
	Add: Welfare Expenses		1.73	1.73	1.73	1.73	1.73	1.73	1.73	1.73	1.73	1.73	1.73	1.73	20.72
			108.45	108.45	108.45	108.45	108.45	108.45	108.45	108.45	108.45	108.45	108.45	108.45	1,301.38
	Add: New Appointment		-	-	-	-	-	-	-	-	-	-	-	-	-
			108.45	108.45	108.45	108.45	108.45	108.45	108.45	108.45	108.45	108.45	108.45	108.45	1,301.38
	Worker														
	No. of Hands	Nos.	1,214	1,214	1,214	1,214	1,214	1,214	1,214	1,214	1,214	1,214	1,214	1,214	1,214
	Wages / Day	Rs.	287.72	287.72	287.72	287.72	287.72	287.72	287.72	287.72	287.72	287.72	287.72	287.72	287.72
	Working Days	Nos.	27	29	30	31	29	30	29	28	30	28	27	30	348
	Cost	Lac Rs.	94.31	101.29	104.79	108.28	101.29	104.79	101.29	97.80	104.79	97.80	94.31	104.79	1,215.54
	Add: Yearly Increment @ 6%	Lac Rs.	-	-	-	-	-	-	-	-	-	-	-	-	-
	Mending Charge @1.25 per mtr	Lac Rs.	4.11	4.83	5.68	5.90	6.06	6.29	6.19	6.10	6.25	5.49	5.36	5.28	67.52
	Total Cost	Lac Rs.	98.42	106.12	110.46	114.18	107.35	111.07	107.48	103.90	111.04	103.29	99.67	110.07	1,283.06
	Salary & Wages		206.87	214.57	218.91	222.63	215.80	219.52	215.93	212.35	219.49	211.74	208.12	218.52	2,584.44

RAW MATERIAL REQUIREMENT

Yield % Upto Top		94	92	Wool Rate	Com Charges	Dyeing cost	RM Cost	Scou/Comb Charges	Dyeing cost
WOOL	Cloth Wt.	Yarn Wt.	Top Wt.	Rs Per Kg.	Rs Per Kg.	Rs Per Kg.	Lac Rs.	Lac Rs.	Lac Rs.
17.5 & 16.5 M	1,455	1,547	1,682	1,174	49.00	26.00	19.74	0.82	0.44
10.5M	44,336	47,166	51,267	975	45.00	26.00	499.77	23.07	13.33
19.5M	2,62,471	2,79,224	3,03,505	932	42.00	26.00	2,829.67	127.47	78.91
20.5M	45,494	48,398	52,607	899	38.00	26.00	472.90	19.99	13.68
21.5M	-	-	-	877	35.00	26.00	-	-	-
22.5M	3,16,581	3,36,788	3,66,074	826	33.00	26.00	3,024.24	120.80	95.18
22.5M Med	1,55,814	1,66,760	1,80,173	596	33.00	26.00	1,073.64	59.46	46.85
24.5M	33,645	35,792	38,905	746	31.00	26.00	290.20	12.06	10.12
28.0M	92,467	98,369	1,06,923	363	28.00	26.00	387.84	29.94	27.80
	9,52,262	10,13,044	11,01,135				8,597.99	393.62	286.30
Polyester	5,43,333	5,78,014	6,28,276	132	9.25	24.00	829.55	58.12	150.79
Low Pill	-	-	-	163	9.25	24.00	-	-	-
TBL Polyester	1,27,684	1,35,834	1,47,646	144	9.25	24.00	212.64	13.66	35.43
Viscose	6,132	6,523	7,091	335	9.25	24.00	23.74	0.66	1.70
	6,77,149	7,20,371	7,83,012				1,065.93	72.43	187.92
SILK	1,768	1,881	2,045	2,800	-	24.00	57.26	-	0.49
Cashmere	20	21	23	7,000	-	24.00	1.58	-	0.01
Lycra	39	41	45	1,200	-	-	0.54	-	-
Outsourced Fabric	1,47,718	1,57,147	1,57,147	275	-	-	432.15	-	-
	1,49,545	1,59,091	1,59,260				491.53	-	0.50
	17,78,956	18,92,506	20,43,407				10,155.45	466.05	474.71

*M-Micron, Wt-Weight, Com-Commissioning, Scouring and combing charges – Outsourced, Scou- Scouring, Comb-Combing

RAW MATERIAL COSTING

Items	Basic Rate	Exchange	CIF	Custom Duty	Cost Including duty and	Yield Upto Combed	Total Cost of	Scouring / Combing Charges	Cost of Combed
Wool	A$	Rs/A$	Rs/Kg	5.25	Freight	Top	Top		Top
17M	18.00	52.50	945.00	5.25	997.61	0.85	1,173.66	49.00	1,222.66
18.5M	16.00	52.50	840.00	5.25	887.10	0.91	974.84	45.00	1,019.84
19.5M	15.30	52.50	803.25	5.25	848.42	0.91	932.33	42.00	974.33
20M	14.75	52.50	774.38	5.25	818.03	0.91	898.93	38.00	936.93
21.5M	14.70	52.50	771.75	5.25	815.27	0.93	876.63	35.00	911.63
22.5M	13.85	52.50	727.13	5.25	768.30	0.93	826.13	33.00	859.13
22.5M	9.98	52.50	523.69	5.25	554.18	0.93	595.89	33.00	628.89
24.5M	12.50	52.50	656.25	5.25	693.70	0.93	745.92	31.00	776.92
28M	5.99	52.50	314.21	5.25	333.71	0.92	362.73	28.00	390.73
M-Micron					Tow				
Polyester 2.5D	-	-	-	-	108.75	0.99	109.85	9.25	119.10
Poly 3DN	-	-	-	-	108.75	0.99	109.85	9.25	119.10
Poly (2.5D) L.P	-	-	-	-	135.00	0.99	136.36	9.25	145.61
TLB Poly	-	-	-	-	118.75	0.99	119.95	9.25	129.20
Viscose	-	-	-	-	-	-	-	0.25	290.00
Flex Tops	-	-	-	-	-	-	-	0.25	650.00
Cashmere	-								7,000.00
Lycra									1,200.00

* Duty + Local Freight = Rs 18.63/Kg, A$- Australian Dollar, CIF- Cost, Insurance and Freight, Scouring and Combing Charges- Greasy wool is scoured and converted into sliver form after carding, combing and gilling process, Yield and Scouring /Combing charges depends on fineness of wool.

POWER & FUEL CONSUMPTION

		Qnty.	Avg. Rate	Lac Rs.
Own	Units in Lacs	4.80	2.69	12.90
Trading		152.00	4.80	729.60
Govt supply /Grid	Units	126.70	6.46	817.94
	Total	283.50	5.50	1560.44
Job Spinning	Units	46.20		
Own Spinning	Units	121.34		
		167.54		
Fabric	Units	115.96		
	Total	283.50		
Production				
Job Spinning	Lac Kgs.	6.60		
Own Spinning	Lac Kgs.	17.33		
		23.93		
Fabric- Own	Lac Mtr's	54.02		
Fabric- Out Source	Lac Mtr's	4.35		
		58.36		
Power Consumed for Spg.				
-Total Units Consumed		167.54		
-Per Kg.		7.00		

Units for Job Spg.		46.20		
Power Consumed for Fabric				
Total Units Consumed		283.50		
Less: Power for Job Spg.		46.20		
Net Power Units for Fabric		237.30		
Fabric Production		54.02		
Power Consumption/Mtr.		4.39		
Cost of per unit (DG set)				
Quantity Consumed		0.30		
Diesel Cost		12.90		
Units Generated		4.80		
Units Generated in one Ltr.		16.00		
Per Unit Cost		2.69		
Per Mtr. Consumption				
Power units		4.39		
Coal kg		0.99		
Furnace Oil Kg		0.14		
Furnace Oil/Thermopack	Lac Ltrs	7.31	43.00	314.33
Steam Coal	MT	5350.00	11500.00	615.25
Pet Coke	MT	0.00	0.00	0.00
Charcoal	MT	12.00	15.00	1.80
Diesel for DG	Lac Ltrs	0.30	43.00	12.90
				944.28

Govt Power Rate	Rs.
Basic Rate/unit	5.45
Less P.F. rebate@ 2%	0.11
Net Rate	5.34
Excise Duty Add 13%	0.69
Octroi @ 0.10/unit	0.10
Other Tax	0.24
Fuel surcharge	0.08
Rate/Unit	6.46

DETAILS OF OVERHEADS EXPENSES

(Current and last three consecutive Years)

	Previous	Previous	Previous	Proposed
(Figs Lacs Rs)	FY 3	FY2	FY1	Current FY
Stationery & Printing	16.90	20.48	17.95	19.95
Postage	0.53	0.45	0.50	0.49
Courier Charges	23.37	27.00	27.00	25.00
Telephone & FAX Charges	16.70	20.00	15.00	19.79
Travelling Exps				
Domestic	78.92	85.00	90.00	120.76
Overseas	20.00	30.00	35.00	34.61
For Export	17.25	20.00	25.00	26.12
Conveyance Exps.	25.15	28.00	30.00	30.00

Vehicle Running	7.08	9.50	11.00	18.84
Rates & Taxes	17.03	22.93	24.00	26.95
Rent	44.14	42.83	40.43	42.66
Books & Periodical	2.05	5.50	5.90	6.00
Fees & Subscriptions	14.21	17.68	17.79	18.59
Legal & Professional				
Normal	56.86	90.76	342.62	381.06
One time	3.87	65.57	2.50	0.00
Gardening Expenses	1.07	0.75	1.00	0.75
Insurance Charges	0.00	27.72	33.73	43.75
Charity & Donation	1.36	1.50	1.50	1.50
Audit Fee	8.97	12.00	12.00	12.00
Directors' Fee	0.11	0.05	0.20	0.20
General Charges	6.85	7.00	7.80	8.00
Total	362.42	534.72	740.95	837.06
Say	362.00	535.00	741.00	837.00
Rates & Taxes				
Water Supply & Sewerage Authority	10.97	10.96	11.00	6.62
House Tax Old buildings	1.38	1.38	1.38	2.68
House Tax	0.00	0.85	0.85	0.00
Renewal of Factory License	0.15	0.15	0.15	0.31
Weight & Measures	0.14	0.14	0.15	0.14
Pollution Control Board	1.90	1.90	1.90	0.00
Pollution Consent Fee	0.00	3.00	3.00	5.02
Road Tax	0.05	0.05	0.05	0.26
Wealth Tax/Environment Fees	1.65	2.00	2.00	3.60
Boards Expenses	0.20	2.00	3.00	3.04
Misc. (Land Revenue, etc.)	0.59	0.50	0.52	5.28
Total	17.03	22.93	24.00	26.95
Rent				
Corporate Office	33.39	30.70	30.70	33.86
Corporate Office Furniture Rent	8.64	8.64	8.64	8.64
Outside Office 1	1.62	1.80	0.00	0.00
Labour Resident colony	0.07	0.07	0.07	0.03
Outside Office 2	0.12	0.12	0.12	0.13
Show Room	0.30	1.50	0.90	0.00
Other Expenses	0.00	0.00	0.00	0.00
Total	44.15	42.85	40.45	42.65
Books & Periodicals				
Design Books	1.35	4.78	5.00	4.90
Law, Tax, Excise, Custom Books	0.24	0.35	0.40	0.55
Newspapers, Magazines, etc.	0.46	0.37	0.50	0.55
Total	2.05	5.50	5.90	6.00
Fees & Subscription				
P.F. Inspection Fee	2.15	2.25	2.25	2.20
Electrical Inspectorate	0.17	0.23	0.23	0.40
Boiler Inspection Fee	0.08	0.03	0.03	0.06

ISO Certification	0.22	0.26	0.26	0.32
Chief Controller of Explosive	0.00	0.17	0.17	0.58
Woolmark License Fee	4.09	5.00	4.75	5.39
Sample Testing Charges	2.53	4.03	4.00	3.78
Consulting agency	0.10	0.45	0.00	0.18
Wool Export Promotion Council	0.80	1.00	1.20	0.00
Agency in Mumbai	0.16	0.16	0.00	0.11
Tender Forms / Fees	0.23	0.25	0.30	0.17
Hazardous Waste Disposal Charges	2.69	1.25	2.00	0.88
Environment Control	0.50	0.50	0.50	0.66
Indian Energy Exchange Annual Fees	0.00	1.10	1.10	0.00
Other Agency	0.00	0.00	0.00	1.37
Quality Certificate Fee	0.00	0.00	0.00	1.84
Misc.	0.49	1.00	1.00	0.65
	14.22	1770	17.80	18.60
Legal & Professional Expenses				
Normal Expenses				
Marketing Consultant	0.00	34.75	119.12	119.12
Legal Company 1	1.55	1.55	1.55	0.00
Legal Company 2	0.40	0.40	0.50	0.50
Tax consultant	0.00	0.00	0.00	1.44
	0.00	0.00	0.00	0.20
Internal Auditors and Consultant	1.71	2.76	8.82	8.00
DO	0.33	0.33	0.30	0.60
DO	0.50	0.55	0.60	0.60
DO	0.00	0.00	0.00	1.80
DO	1.67	1.87	2.00	0.00
Corporate Advisor	0.00	0.00	0.00	0.60
Legal Charges for Documents	0.40	0.60	0.60	0.70
Expenses during Court Cases/Stamp Papers, etc.	0.25	0.35	0.50	0.45
Rating Agency	1.32	1.50	1.50	1.50
Computer Services	0.00	0.00	0.00	14.20
Software Service	6.11	5.00	5.00	0.00
Outsourced designer fees	38.76	38.05	0.00	0.00
Outsourced services 1	0.00	0.00	196.62	225.84
Outsourced services 2	0.32	0.05	0.50	0.50
Other Advocates & Misc. Expenses	3.54	3.00	5.00	5.00
	56.5	90.75	342.65	381.05
One-time expenses for different heads				
A	1.21	0.00	0.00	0.00
B	0.00	1.10	0.00	0.00
C	0.00	4.15	0.00	0.00
D	0.00	1.88	0.00	0.00
E	2.66	0.36	2.50	0.00
F	0.00	58.08	0.00	0.00
G	3.87	65.57	2.50	0.00
Total	60.75	156.35	345.15	381.05

Insurance				
Premium for the year		54.72	60.73	75.00
Add: ECGC Premium on Export to EU		0.00	3.00	3.75
Less: Transit Insurance Recovery		27.00	30.00	35.00
Net Expense	0.00	27.75	33.75	43.74
General Charges				
Guest Entertainment Expenses	2.09	2.30	2.50	2.25
Statutory & Other Advertisements	0.25	0.23	0.50	0.50
Other, etc.	0.11	0.20	0.25	0.15
Traveling accessories	0.16	0.10	0.25	0.15
	0.13	0.02	0.10	0.10
Auditors Expenses	0.13	0.23	0.25	0.25
Audit Committee Charges	0.11	0.05	0.10	0.10
Corp. Office Maintenance Charges	2.52	2.60	2.60	2.25
Misc.	1.35	1.27	1.25	2.25
Total	6.85	7.00	7.80	8.00
Management Cost				
Corporate Expenses	-	-	-	-
Management Fee	-	-	255.92	-
Travel Expenses	-	-	27.72	-
Consultancy Fees	-	-	-	-
Legal Advisory Fees	-	-	-	-
Misc.	-	-	20.72	-
	-	-	-	-
Total	-	-	304.36	-
Details of Welfare Expenses				
Canteen Subsidy	11.37	13.99	15.00	15.00
Uniform Expenses	2.73	3.07	3.20	6.07
Food Bills	8.47	10.74	12.00	9.30
Festival Gifts	1.69	2.37	3.00	2.85
Festival Expenses	2.72	1.91	2.00	3.75
Cycle/car Stand Charges	1.23	1.31	1.40	1.35
Cable TV Charges	0.39	0.51	0.50	0.30
Gift on Employees Retirement/Marriages	0.71	1.07	1.25	0.35
Club & Sports activities	0.22	0.28	0.30	0.25
Gratuity Insurance Charges	2.66	2.82	3.00	3.50
Workers Tour	0.15	0.85	1.00	-
Water Purifier	0.03	0.72	0.50	-
Misc. Expenses	2.74	3.36	4.00	3.00
Total	35.11	43.00	47.15	45.72
Workers	20.00	25.00	25.00	25.00
Staff	15.12	18.00	22.15	20.75

MARKETING BUDGET

		Lac Rs.	Lac Rs.
Year-Round Activity	Brand Strategy	35.00	47.73
	Celebrity Payments	104.00	141.82
	Creative Agency Fee	14.00	19.09
	Media Agency Fee	14.00	19.09
	Packaging, etc.	25.00	34.09
	Polybags	35.00	47.73
		0.00	0.00
Summer season campaign	Print media	60.00	81.82
	Billboards, etc.	6.00	8.18
	Internet / Digital	22.00	30.00
		0.00	0.00
Winter season campaign	Print and Electronics media	360.00	490.91
	Billboards, etc.	90.00	122.73
	Internet / Digital	25.00	34.09
	Special expenses	140.00	190.91
	Website	10.00	13.64
	Photoshoot	30.00	40.91
	Exhibitions & Conferences	90.00	122.73
	Miscellaneous	20.00	27.27
		0.00	0.00
	Photoshoot	20.00	27.27
Total		**1,100.00**	**1,500.00**

BUDGET FOR COMPANY RETAIL SHOP Rs Lacs	
Retail Shops Expenses	213.00
One Time Expenses	
Civil Work	11.50
Gensets + Air Conditioners	49.00
Aesthetic Environment	2.50
Fire Fighting /Safety Equipment	2.50
Sub Total:	65.50
Architect and Interior Designer Fee, etc.	49.52
Miscellaneous	21.48
Total	**349.50**

D. ANNUAL BUDGET AND PROFIT/LOSS PROJECTION AT A GLANCE

(High value suitings manufacturing company)

Example of summarised annual projection, for the review of senior management

Particulars	Monthly	Yearly	Per Mtr	%
Fabric Production (Lac Mtrs)				
Domestic:				
In house	1.25	15		
Outsourced	2.08	25		

Export	0.17	2		
Total	3.5	42		
Fabric Sales (Lac Mtrs)				
Domestic				
In house Production	1.25	15		
Outsourced	2.08	25		
Export	0.17	2		
Total	3.5	42		
Average Sales Realisation (Rs./Mtr)				
Fabric				
In house Production	614	614		
Outsourced materials	273	273		
Average Domestic	382	382		
Export	564	564		
Overall Average	408	408		
Income Rs Lacs				
Sales				
Fabrics				
Domestic				
In house Production	767	9206		
Outsourced material	569	6822		
Total (Domestic)	1336	16028		
Export	94	1127		
Total (Domestic & Export)	1430	17155		
Export Benefit	5	56		
Expenditure Rs Lacs				
Material Cost				
Raw Materials	250	2999		
Fabric Purchase	307	3684		
Decrease/ (Increase) in WIP				
Decrease/ (Increase) in Finished Goods	–	–		
Sub-total	557	6683	159.12	41.7
Stores Consumptions	48	571	13.61	3.57
- Mending Charges	12	138	3.29	0.86
Utility				
- Power	86	1028	24.48	6.42
- Coal	21	257	6.12	1.6
- Water	1	12	0.29	0.07
Repairs and Maintenance	15	175	4.16	1.09
Salaries and Wages	196	2348	55.9	14.65
Administrative Expenses	45	540	12.86	3.37
Advertising & Sales Promotion	35	415	9.88	2.59
Selling Expenses	190	2284	54.39	14.25

Sub-total	647	7769	184.97	48.47
Total Expenditure	1204	14452	344.09	90.16
EBITDA	230	2760	65.71	17.22
Interest	103	1235	29.41	7.71
Financial Charges	10	115	2.74	0.72
Less: Interest Income	-4	-50	-1.19	-0.31
Sub-total	108	1300	30.95	8.11
PBDT	122	1460	34.75	9.11
Depreciation	17	198	4.71	1.24
PAT (Profit after tax)	105	1262	30.04	7.87

E. ANNUAL BUDGET FOR REPAIR AND MAINTENANCE OF A SPINNING UNIT

Every machine and its spare parts have a natural life depending on their use. Regular wear and tear and spare parts consumption is a regular phenomenon for any machine. Machine manufacturers recommend the spare parts consumption based on trial running hours at their plants and collect data from few manufacturing companies that run in ideal conditions. However, their technical service team collects the data from all the users and makes necessary changes and improvements in the materiality and design of machines and spare parts time to time to remain efficient and viable. To achieve the optimum life of machineries and spare parts, it is essential that the users review the recommended preventive maintenance schedule and checklist and customize it according to their working conditions, working speed, utilisation and loading of machineries and have a holistic approach in its implementation.

User departments are asked to submit the annual spare parts consumption to incorporate in annual Budget.

In a new set up, the maintenance department may follow the recommendation of machine manufacturers, get the data of other competitor companies using same machines, etc. to fix the cost of spare parts consumption. Alternatively, the break down and spare parts' consumption data may be collected for three to five years in terms of quantity and its amount, to determine the life of individual spare parts and cost of spare parts per unit of production.

Records must also be maintained for running hours of individual machines, climatic conditions maintained (humidity and temperature), productivity level achieved, quantum of rework, reason wise machine stoppages, waste generation, quality of product produced, rejection percentage, analysis of defects and customer complaints related to machines fault during the period. Analysis should be done for expenses for spare parts consumed and losses caused due to machine stoppages, cost of rework, claim amount for market rejections and market feedback about the quality of products, etc. on account of spare parts. Management must compare the cost of spare parts consumption and cost of above losses and should work out its economy. A reasonable value should be fixed for spare parts consumption, and it is advisable to not compromise in it for the best interest of the business.

Example (Worsted Spinning Unit)

SUMMARY FOR ANNUAL BUDGET:

Sr No.	ITEM	Per Month Rs Lacs	Per Year
A	REGULAR CONSUMABLE ITEMS	18.46	221.5
B	MACHINES SPARES	17.31	207.7
C	MACHINES LUBRICANTS	0.69	8.3
D	ANNUAL MAINT. CONTRACT	0.22	2.7
	SUB TOTAL	36.68	440.2
E	FIBRE LUBRICANTS	5.63	67.6
	TOTAL	42.32	507.8

MACHINE WISE CONSUMABLE ITEMS

M/C NAME	SR. NO.	CONSUMABLE ITEMS MATERIAL	BUDGET FOR YEAR		PER MONTH	
			QTY.	Rs. In Lacs	QTY.	Rs. In Lacs
GILL BOX [Tow to Tow convertor]	1	CUTTER 88 MM (REPAIRING & NEW)	3	9.60	0.25	0.80
GILL BOX	1	PIN STRIPE	10503	29.04	875	2.42
	2	COTS	214	2.76	18	0.23
	3	FALLER CHAIN	92	4.59	8	0.38
	4	SCRAPPER TUBE 14x265x1.1mm	324	0.29	27	0.02
COMBER	1	SYN. LEATHER APRON	336	18.94	28	1.58
	2	CIRCULAR BRUSH DIA-160	186	8.37	16	0.70
	3	VARIO BARS	535	62.28	45	5.19
	4	TOP COMB	414	14.51	35	1.21
	5	FEED COMB	755	6.72	63	0.56
	6	NIPPER BRUSH (WHITE)	372	0.52	31	0.04
	7	TOP COMB BRUSH (BLACK)	248	0.37	21	0.03
RUBBING FRAME	1	APRONS	832	3.05	69	0.25
	2	RUBBING APRON	147	2.20	12	0.18
	3	COTS *	275	3.00	23	0.25
	4	SCRAPPER TUBE	118	0.11	10	0.01
RING FRAME	1	TOP APRONS R/F	23332	4.51	1944	0.38
	2	BOTTOM APRONS R/F	23332	4.57	1944	0.38
	3	BOTTOM APRONS SKIVED R/F	7500	1.50	625	0.13
	4	COTS R/F	14388	7.77	1199	0.65
	5	RING TRAVELLER	2111	9.63	176	0.80
	1	BEARINGS	LUMPSUM	9.0		0.75
	2	TANGENTIAL/CONVENOR BELTS *	LUMPSUM	7.5		0.63
	3	TIMING & V-BELTS	LUMPSUM	5.4		0.45
	4	MISC.	LUMPSUM	5.3		0.44
B-				221.53		18.46

Items of Longer Life

Sr. No.	ONE TIME ITEMS	Qty.	Cost in Rs./pc. Approxi.	Estimated Cost in Lacs A
TOP MAKING				55.23
	WILOW/OPENER			2.50
1	FEED LATTICE	1		1.00
2	SPIKED LATTICE	1		1.50
	CARDING			18.45
1	HOPPER RESERVER LATTICE	1		0.80
2	SPIKED LATTICE	1		1.75
3	FEED LATTICE	1		0.70
4	PNEUMATIC JACK HP 100K2	2		0.70
5	BANDE FREIN 50 BRAKING BAND - UTC44886	2		1.20
6	ROULEMENT A AIGUILLES XF 36423 NEEDLE BEARING XF 36423 - A422T026	4		1.00

7	REVOLVING PLATE - UGN20675AD	2		0.30
8	RING WIRE DIA 457.05 425 A 030 H	10		
9	RING WIRE DIA 457.05 425 A 031 K	10		
10	TAPE WI008 PE HD 500 425 A 231 H	30 MTS		
11	BALL DIA 9.525 MM 425 A 033 P	2000		3.50
12	BURR BEATER ROLLER RESHARPING	5		8.50
	GILL BOX			**19.33**
1	FLANGES 4 M/CS	16		3.20
2	FALLER BODY HIGH WALL - 270MM 3 M/CS.	432		2.16
3	VARIABLE PULLEYS FOR GILL BOX. 4 M/CS.	4		7.00
4	DRAFTING HOUSING, CARTERS 4 M/CS	4 Sets		1.44
5	PROTECTION GN 33963 AT	3		0.93
6	PROTECTION BOTTOM GN 23272 AF	3		0.18
7	TESTER TOP GN 35794 AP	3		0.04
8	TESTER BOTTOM GN 35793 AC	3		0.14
9	HEAD PARTS, FUNNELS	2 Sets		3.00
10	ROLLER DIA- 80 6 M/CS.	6 Sets		0.48
11	CREEL FEED ROLLERS 3 M/CS	30		0.75
	COMBER			**14.95**
1	DELIVERY ROLLERS 2.5 4 M/CS	4 Sets		3.20
2	TOP & BOTTOM JAWS 7 M/CS.	7 Sets		2.59
3	CARRIAGE CAM & RUNNERS 3 M/CS.	3 Sets		1.80
4	FEED TOP & BOTTOM GRIDS 5 M/CS.	5 Sets		1.20
5	FEED COMB BODY 3 M/CS	3 Sets		2.70
6	DRAWING OFF ROLLERS DIA 25	11		0.66
7	COARSE & FINE SEGMENTS 2 M/CS	2 Sets		1.80
8	TOP COMB BODY 2 M/CS.	2 Sets		1.00
PREPARATORY				**62.27**
	Gill Boxes GC			**14.06**
1	FLANGES	4 M/CS.		3.20
2	VARIABLE PULLEYS	2 NOS.		3.50
3	FALLER BODY 270MM	4 M/CS.		2.88
4	DRAFTING HOUSING CARTER, CAP	4 M/CS.		2.00
5	PROTECTION GN 33963 AT	1		0.31
6	PROTECTION BOTTOM GN 23272 AF	1		0.06
7	SCRAPPER BODY GN 36956 AV	12		1.35
8	DRAFT ROLL DIA - 30.MM	10		0.70
9	TESTER TOP GN 35794 AP	1		0.01
10	TESTER BOTTOM GN 35793 AC	1		0.05
	COMBER			**11.49**
1	DELIVERY ROLLERS	2 Sets		1.60
2	TOP & BOTTOM JAWS	1 Set		0.84
3	CARRAIGE CAM & RUNNERS 3 M/CS.	3 Sets		1.80
4	FEED TOP & BOTTOM GRIDS 5 M/CS.	5 Sets		1.20
5	FEED COMB BODY 5 M/CS	5 Sets		4.50

6	DRAWING OFF ROLLERS DIA 25	5		0.55
7	TOP COMB BODY	2		1.00
	Gill Boxes GV – 11			**5.22**
1	SLIDE BLOCK GN 35978 AC	8		0.86
2	TIGHTNER GN 35979 AP	8		0.82
3	SPROCKET GN 32066 AV	9		1.14
4	HOUSING GN 24984 AA	2		1.20
5	HOUSING 25686 AE	2		1.20
	Gill Boxes GV -20			**2.40**
1	BEARING GN 43978 AS	1		1.2
2	BEARING GN 25686 AE	1		1.2
	RUBBING FRAME FMV-36			24.59
1	SHAFT RH THREAD FN 4645 AY	2		1.08
2	SHAFT LH THREAD FN 4646 AK	2		1.08
3	ROD RH THREAD FN 9584 AC	4		2.67
4	ROD LH THREAD FN 9585 AP	4		2.67
5	SHAFT FN 4647 AX	4		0.26
6	SPACER FN4648 AJ	8		0.28
7	PLATE FN 14785 AC	24		2.41
8	EYELET DIA 5 FN 15133 AJ	120		0.52
9	CONDENSER WIDE 16 FN 1935 AS	32		0.33
10	CONDENSER 9 - 25 FN 14901 AC	7		0.05
11	CONDENSER 10 FN 14714 AZ	22		0.07
12	SLIVER GUIDE 12 502 A 379 B	96		0.31
13	SLIVER GUIDE 21 502 A 381 J	48		0.13
14	FEED ROLLER FN 9388 AM	24		3.36
15	JOINT BUSHING FN 8514 AY	24		1.32
16	ASSEMBLY BEARING RH FN 9715 AL	26		3.59
17	BALL BEARING RLF 20X40X50	54		3.50
18	CONNECTING SHAFT RH FN 8816 AD	24		0.48
19	CONNECTING SHAFT LH FN 8817 AR	24		0.48
	RUBBING FRAME SANT ANDREA RF 2A			4.51
1	SMALL PLAT 3 4539550001	24		0.63
2	WIRE GUIDE 4 4519730004	24		0.58
3	WIRE GUIDE 5 4519731004	24		0.58
4	CAM 45395180	2		0.03
5	STUD FOR GEAR 45325390	6		0.09
6	HELICAL GEAR Z = 25 (EX 45396430) – 4539643001	12		1.07
7	GEAR Z=30 (EX 45395690) – 4539569001	12		0.90
8	25 JOINT RSP0010 S308 L398, PART NO. C2110003	2		0.64
SPINNING				**42.31**
	RING FRAME ITEMS			**42.31**
1	SUCTION TUBE 4 M/CS.	1752		0.18
2	ABC RINGS 7 M/CS. (1,3,4,7,10,11,12)	4380		2.41

3	LAPPET HOOK DIA - 5.0MM 5 M/CS. (5,7,9,11,19)	3504		0.45
4	KNEE BRAKES DIA 30.0MM 2 M/CS.	7008		4.20
5	SUCTION TUBE SLEEVE 2 M/CS.	1752		0.19
6	BEVEL SHAFT	48		0.58
7	BEVEL WHEELS SETS	16		1.09
8	FLAT SEPARATORS 3 M/CS.	1752		0.57
9	FLOTTING CONDENSERS 75.0MM 4 M/CS.	1314		0.86
10	FLOTTING CONDENSERS 82.5.0MM 1.5 M/CS.	400		0.20
11	TANGENTIAL BELT 66720 X 35 MM 4M/CS. (6, 8, 19, 22)	7		3.82
12	SPINDLE PADS 0.3 MM 12 M/CS.	12000		0.67
13	SPINDLE PADS 0.4 MM 12 M/CS.	12000		0.67
14	BOBBIN CONVEYOR STEEL BELTS 82.5MM	2		0.31
15	BOTTOM ROLLER BEARINGS	500		1.63
16	CONICAL RINGS ID - 45.0MM 5 M/CS. (4, 6, 8, 10, 11)	6132		23.61
17	LATTICE APRON FOR COMPACT SPINNING	876		0.70
18	FILAMENT FLOTTING CONDENSERS 75.0 MM 1 M/C.	200		0.17
POST SPINNING				**46.95**
	AUTOCONER			**6.74**
1	RETAINER TUBES S SAW TOOTHED	30		0.63
2	INTERMEDIATE PIECE 36010 – 148021716	120		0.02
3	DISTRIBUTOR 146.013.323	60		0.54
4	O RING 6X2 903.770.002	720		0.04
5	O RING 5 X 1.5 41501 – 836460131	120		0.14
6	CUTTING KNIFE 146-015.415	120		2.04
7	CUTTING PLATE 41501 – 146015355	60		0.12
8	COLLER SCREW 41501 – 146004761	120		0.15
9	ROUND CORD RING 34201-836465058	120		0.16
10	HUB 14 - 34201, 146001834	120		0.78
11	BEARING 34201 – 138012554	120		1.92
12	ELECTRONIC CRANCK FOR SPINDLE 147655051	1		0.20
	VOLKMANN TFO			**8.49**
1	TANGENTIAL BELT ENDLESS 57200 X 50MM	2		1.36
2	BELT GUIDE ROLLER 400-43 – 0421966	120		1.56
3	SLEEVE 0444489 – 641662	500		1.15
4	BALLON LIMITOR 405 - 40 – 0406827	120		2.40
5	OIL TANK 510 - 01 – 0947721	1		2.02
	MURATA TFO			**23.24**
1	TANGENTIAL BELT ENDLESS 66150 X 50MM	4		3.20

2	MTN COMPLETE SET 364 - 600B	120		8.40
3	CAPSULE 1 R 364-6008-017	720		5.04
4	SPINDLE 364 - 310C-001.	60		6.60
	SSM & FADIS ASSEMBLY WINDER.			8.48
1	VARIABLE BELT 775 X 10 X 5 37220230	96		1.34
2	VARIABLE PULLEY 372 – 15441	48		6.72
3	SMOOTH CYLINDER (DRUM) 4204010	6		0.42
	C-	Total		206.76
				17.23

REQUIREMENT OF PIN STRIPES

M/C	Type of Pin Stripes	No of M/C.	Per M/C Reqd.	Total	Life (Months)	Consumption Nos /Month	Annual Consumption	Landed Cost / Pc Rs	Total Cost Rs Lacs
GC-14	3x15x7/8"-270mm	1	144	144	4	41	497	244.00	1.21
GC-14	4x16x7/8"-270mm	2	144	288	4	83	994	264.00	2.62
GC-15	5-15/21x1"-270mm	3	144	432	4	124	1490	280.00	4.17
GC-15	6-16/22x1"-270mm	3	144	432	5	99	1192	288.00	3.43
GC-14	4x16x1"-270mm	2	144	288	6	55	662	272.00	1.80
GC-14	5-15/21x1"-270mm	3	144	432	6	83	994	280.00	2.78
GC-14	6-16/22x1"-270mm	4	144	576	6	110	1325	288.00	3.81
GC-14	7-17/23x1"-270mm	7	144	1008	6	193	2318	304.00	7.04
GV 11	7-18/24*7/16"-84.3	2	224	448	12	37	448	240	1.07
GV-20	7-18/24*7/16"-84.3	1	448	448	12	37	448	240	1.07
GC-14	ROLLERS & RINGS	25	144	3600	36	100	1200	7.00	0.08
					Life (Month)				
PB-32	FEED COMB (TYPE-B) 19 / 26 / 5.0 / 410 mm	11	3	33	4	9	114	839.90	0.95
PB-32	FEED COMB (TYPE-B) 20 / 26 / 6.0 / 410 mm	11	3	33	4	9	114	861.82	0.98
PB-32	FEED COMB (TYPE-B) 20 / 26 / 7.0 / 410 mm	11	3	33	4	9	114	913.43	1.039
									2.98
					Life (Month)				
PB-31	FEED COMB (TYPE-B) 19 / 26 / 5.0 / 410 mm	7	3	21	6	4	48	839.90	0.41
PB-31	FEED COMB (TYPE-B) 20 / 26 / 6.0 / 410 mm	20	3	60	6	12	138	861.82	1.189
PB-31	FEED COMB (TYPE-B) 20 / 26 / 7.0 / 410 mm	20	3	60	6	12	138	913.43	1.26

PB-31	FEED COMB (TYPE-B) 20 / 26 / 8.5 / 410 mm	13	3	39	6	7	90	992.61	0.89
							756		3.75
					Life (Month)				
PB-32	S+U TOP COMB NEEDLE STRIPS NO. 25.82.470	9	1	9	1	10	124	2,416.48	3.001
PB-32	S+U TOP COMB NEEDLE STRIPS NO. 28.82.470	2	1	2	1	2	28	2,603.83	0.718
PB-31	S+U TOP COMB NEEDLE STRIPS NO. 32.82.470	6	1	6	1	7	83	3,057.72	2.53
PB-31	S+U TOP COMB NEEDLE STRIPS NO. 35.82.470	13	1	13	1	15	179	4,606.02	8.26
							414		14.51
					Life (Month)				
PB-32	S + U VARIO BARS NO. 090.440	11	1	11	6	2	22	8,807.64	1.93
PB-32	S + U VARIO BARS NO. 075.440	11	1	11	6	2	22	7,877.25	1.73
PB-32	S + U VARIO BARS NO. 065.440	11	1	11	6	2	22	8,571.51	1.88
PB-32	S + U VARIO BARS NO. 055.440	11	1	22	6	4	44	9,448.88	4.15
PB-32	S+U VARIO - BARS NO. 045.440	11	3	22	6	4	44	9,787.53	4.31
PB-32	S+U VARIO - BARS NO. 035.440	11	2	22	6	4	44	11,302.60	4.97
PB-32	S+U VARIO - BARS NO. 030.440	11	2	22	6	4	44	11,978.48	5.27
PB-32	Starting Bar 15x4mm, no-3623.6	11	1	11	12	1	11	3,380.10	0.37
							242		24.64
PB-31	S + U VARIO BARS NO. 075.440	20	1	20	9	2	27	7,877.25	2.10
PB-31	S + U VARIO BARS NO. 065.440	20	1	20	9	2	27	8,571.51	2.28
PB-31	S + U VARIO BARS NO. 055.440	20	1	20	9	2	27	9,448.88	2.52
PB-31	S+U VARIO - BARS NO. 045.440	20	2	40	9	4	53	9,787.53	5.22
PB-31	S+U VARIO - BARS NO. 035.440	20	2	40	9	4	53	11,302.60	6.03
PB-31	S+U VARIO - BARS NO. 030.440	20	2	40	9	4	53	11,978.48	6.38
PB-31	S+U VARIO - BARS NO. 025.440	20	2	40	9	4	53	14,650.18	7.81
PB-31	Starting Bar 15x4mm, no-3623.6	20	1	20	12	2	20	3,380.10	0.67
							293		33.03
								Total	107.95

REQUIREMENT OF APRONS

NAME	Specs	No Per M/C	Total M/CS	Spares Reqd	Life (Months)	Cons per month	cons per year	Cost / Pc RS.	Total cost Rs Lacs
HEAD APRON	290 X 1690 MM	1	1	1	3	0.33	4.6	1290.6	0.06
PB 32	580 X 540 X 4	1	11	11	1	11	152	5,640.89	8.56
PB 31, 30	640 X 530 X 4	1	20	20	1.5	13	184	5,640.89	10.37
FMV 36, FM-7N	468 X 166 X 4	40	4	160	24	7	92	1,562.44	1.43
RF-2A	459 X 117 X 4	96	1	96	24	4	55	1,378.62	0.76
DRAFTING APRON	146.5 X 160 X 1.6	24	2	48	6	8	110	551.45	0.60
CRADLE APRON	58.2 X 53 X 1.1	48	2	96	4	24	331	57.90	0.19
BOTTOM APRON	75.8X54.0X1.2	48	1	48	4	12	166	70.77	0.12
TOP APRON	51.0X117.0X1.3	24	1	24	4	6	83	137.86	0.12
DRAFTING APRON	146.5 X 145 X 1.65	16	1	16	4	4	55	574.43	0.32
DRAFTING APRON	159 X 145 X 1.65	16	1	16	4	4	55	588.21	0.33
HEAD APRON	120 X 140 X 1.2	2	1	2	1	2	28	390.61	0.11
PRESSURE ROLLER (Grey)	Dia-36-125	4	1	4	0.5	8	110	1,148.85	1.26
WHITE - 75 HARDNESS	L-53.0, TH-1.25	48	1	48	3	16	0.0	316.0	0.00
									24.25
TOP APRON	83.2 X 31.8 X 1.1	876	23	20148	12	1679	20148	19.0	3.82
TOP APRON	83.2 X 40 X 1.1	796	4	3184	12	265	3184	21.3	0.67
BOTTOM APRON	79.0 X 34 X 1.1	876	23	20148	12	1679	20148	19.6	3.94
BOTTOM APRON	79.0 X 42 X 1.1	796	4	3184	12	265	3184	19.6	0.62
BOTTOM APRON (SKIVED)	79.0 X 34 X 1.1	-	-	-	-	500	6000	20.0	1.20
BOTTOM APRON (SKIVED)	79.0 X 42 X 1.1	-	-	-	-	125	1500	20.0	0.30
									10.58
CARDING	64.5 X 80 X 190	3	2	6	12	0.50	6.0	1488.0	0.09
GILL BOX	64.5 X 80 X 265	2	27	54	12	5	54.0	1813.4	0.97
DELIVERY ROLLER (GB)	48 X 60 X 265	1	9	9	6	1.50	18.0	1314.3	0.23
TT 11 & GV11	64.5 X 80 X 115	12	2.5	30	9	3.33	40.0	885.1	0.35
FMV 36	40 X 60 X 135	48	2	96	12	8.00	96.0	905.3	0.86
RF-2A FRONT	40 X 55 X 130	24	1	24	12	2.00	24.0	749.1	0.18
RF-2A BACK	35 X 45 X 130	24	1	24	18	1.33	16.0	560.6	0.089
FM7N	48 X 70 X 133	32	1	32	18	1.78	21.3	963.2	0.205
RING FRAME 68 (Front)	19 X 50 X 32	876, 796	25	10192	36	283	3397	60.0	2.04
RING FRAME 85 (Front & back)	19 X 50 X 32	876, 796	25	32968	36	916	10989	51.0	5.60
									10.65
CIRCULAR BRUSH -160 (Nylon)	DIA - 160MM	1	31	31	2	16	186	4,500.00	8.37
SCRAPER TUBE (DIA.14x265X1.0)		3	27	81	2.2	37	442	90.6	0.400
FALLER CHAIN GC	432A 152M	4	25	100	10	10.0	120	4,620.00	5.54
TT 11 CUTTER RESHARPINIG	88.0MM	1	1	1	4	0.3	2	2,76,922.64	5.53

TT 11 CUTTER NEW	88.0MM		1	1	1	4	0.3	1	4,05,186.48	4.05
FALLER CHAIN GV	432A 319X	8 16	2	24	24	1.0	12		7,423.88	0.89
NIPPER BRUSH (WHITE)	PB17721AF		1	31	31	1	31	372	139.7	0.52
TOP COMB BRUSH (BLACK)	PB17722AT		1	31	31	1.5	21	248	149.8	0.37
										25.69
									Total	71.16
										5.93

ANNUAL MAINTENANCE CONTRACT

Sr. No.	Suppliers Name & Details	Department	Instrument	Rs. In Lacs (Approx.)
1	--- T....... (India) Pvt. Ltd.	Quality Control	---Tester 3B Model	0.76
			--- Tensorapid 3	
2	INTERWOOL LABS	Quality Control	--METER-AL100 & OFDA 100	1.25
3	An India Services Pvt. Ltd.	Production	Jointair Splicer with Air Track	0.67
			System and Splice Scanner	
D-			TOTAL	2.68

FIBRE LUBRICANTS

Textile material passes through different machines, contacts with various metal or synthetic or rubber surfaces, generates heat and static charges due to abrasion and frictions depending on material, area of contacts and speed of different machines during the process and climatic condition of surroundings. As a result, loss of natural moisture content in textile materials produces static charges due to heat generation, which results in fiber fraying, entanglements and ends breakages and ultimately creates quality and productivity issues. The selection of appropriate lubricants and its application is based on physical and chemical properties of fibers and type of machines. The quantity and stage of application of lubricants plays an important role in smooth working of machines and to obtain proper quality of product. Different lubricating and anti static agents are mixed or applied at different stages according to the fibers, its blend ratios, type and speed of machines and climatic conditions maintained which is determined after several trials and experience. Different lubricants are applied at fiber, sliver and yarn stage in spinning, winding, warping, weaving, finishing, etc.

Example:

| SR. | NAME | USED | CONSUMPTION | | | | | | QTY | | LANDED COST / UNIT IN Rs. | COST / MONTH (Rs. IN LACS) | Approved supplier |
| | OF | ON | PER MONTH | | | | | | PER | | | | |
NO.	LUBRICANT	M/C	TOP MAKING		PREP.		P.SPG.		MONTH				
1	Selbana --- Oil or Cerasol	Opener, GILL BOX	435	Kg		Kg			435	Kg			
2	Selbana N Oil or Cerasol	GILL BOX	750	Kg	300	Kg			1050	Kg			
3	Katex Oil	GILL BOX			50	Kg			50	Kg			
4	Spreitan	TFO					900	Kg	900	Kg			
5	Ring Sol or Spin Oil	RING FRAME					210	Ltr	210	Ltr			
E	Total												

MACHINE LUBRICANTS

Appropriate lubrication of machines is essential to avoid or minimise the friction and wear and tear of spare parts and optimise a machine's working life. Lubrication is like feeding machines. While less and untimely lubrication is harmful and can at times cause costly and unrepairable damage, excess lubrication is also not helpful and causes quality issues such as oil and black stains, etc. The quality (specifications) of lubricants must be as per the recommendation of machine manufacturers which are prescribed based on machine materiality, speed, friction, sizes of moving parts and loading capacity of machines. Timing of lubrication is also important and influenced by working conditions and type of material produced. Thus, the lubrication schedule recommended by machine manufacturers must be reviewed and customized as per individual conditions to achieve the optimum life of machines and output. It is also essential to replace the oil use in closed chambers, gear boxes, etc. as per the prescribed schedule because of the oil burned after a period which deposits carbon at the bottom. It reduces the lubricating property and viscosity, which is harmful for spares, gears, bearings, bushes, etc. It is also important to clean the lubricating points before lubrication. Mixing of fluffs with oil causes a hard tuft-like effect, which may not only create hindrance in allowing the lubrication upto the desired point but also in free movement of related parts. Mixing of oil with dust may create emulsion-like effect, which can cause harm to the related spare parts. Untimely insufficient lubrication causes frequent breakdowns, high wear and tear, breakdowns, loss of production, bad quality of product, delayed delivery, dissatisfaction among the workforce, and loss to the company as a whole. Compromise in lubrication of machineries and equipment is like 'Penny wise and pound foolish'.

The assessment of consumption of lubricants: Use one oil gun and grease gun full of oil and grease respectively and count the no. of machines lubricated. Multiply with the frequency of lubrication within a month. Accordingly, decide the annual consumption of different lubricants. Compare with actual consumption, and analyse the reasons for variations.

Example:

Name of Lubricants	Used on	Purpose	Consumption Per Month TOP MAKING		PREP.		SPG.		P.SPG.		Qty Per Month		Landed Cost / Unit (Rs.)	Total Cost/ Month (Rs.)	Approved supplier
SERVO SPIN -12	RING FRAME	SPINDLE OIL					22	Ltr			22	Ltr	150.0	3300.00	
SERVO SYSTEM -32	RF / PSPG	FADIS GEAR BOX							18	Ltr	18	Ltr	150.0	2700.00	
SERVO SYSTEM -68	RF / NSC	GEAR BOX / AUTOMATIC OILING TANK & DAILY OILING	30	Ltr	30	Ltr	20	Ltr	13	Ltr	93	Ltr	150.0	13950.00	
SERVOMESH - 150	TFO	GEAR BOX							37	Ltr	37	Ltr	150.0	5550.00	
SERVO 320	SPG / NSC M/C	DIFF. GEAR BOX / GEAR BOX	5	Ltr	6	Ltr	5	Ltr	7	Ltr	23	Ltr	150.0	3450.00	
SILICON OIL	AUTOCONER	ADOPTER							0.9	Ltr	0.9	Ltr	100.0	90.00	
AP-3 GREASE	ALL M/CS		11	KG	10	KG	9	KG	6	KG	36	Kg	250.0	9000.00	
SKF GREASE LGFT2	TFO MURATA	SPINDLE							2	KG	2	Kg	4500.0	9000.00	
KUBLER GREASE	TFO VOLKMAN	SPINDLE							2	KG	2	Kg	8500.0	17000.00	
AC-400 (450 ML)	AUTOCONER	SPLICER CLG.							10	BTL	10	Btl	500	5000.00	
														69,040	
Per Year														828480	

6

Production Planning and Control

Production planning and control (PPC) is also termed as system chain management (SCM). In a business, the PPC department controls the predetermined plans and processes that monitor allocation of human resources, raw materials, and machinery to achieve targets. The responsibility of this department is to effectively execute the annual budget or annual plan. The PPC head plays a key role for the organisation by coordinating with the raw material section and other functional heads and management.

Production planning is a sequence of steps that empowers manufacturers to work smarter and optimise their production process in the best possible manner by efficiently managing internal resources to meet customer demands.

The function of the PPC team includes arranging timely availability of raw materials to ensure required production within a defined period and satisfy the needs of the sales and marketing team for a consistent revenue generation cycle and the smooth functioning of the organisation.

The PPC team must have comprehensive knowledge and information about manufacturing capacity, utilisation, efficiency of individual machines, effective output and lead-time of individual machines, waste generation, rework, ratio of sound & defection percentage and production and revenue realisation cycle to assess the inventory levels at different stages.

Usually, the quantity of raw materials required is assessed and ordered based on its availability and arrival lead-time and accordingly, the inventory is maintained in raw material godowns.

SPECIFIC FUNCTIONS OF PRODUCTION PLANNING AND CONTROL (PPC)

These days organisations have reliable PPC software, which makes it easy to plan, monitor, and control including following :

- Maintaining optimum levels of raw material, working stock (WIP) and finished goods inventory
- Online follow-up of process and product status to ensure timely delivery with full assortments
- Taking timely corrective measures in case of any accident to avoid delay in commitment
- Ensure cost-efficient production process
- Minimise lead time
- Improve customer satisfaction
- Coordinate with departments about production, to ensure things are moving as per plan
- Collect and study the data for individual product performance at various stages of operations, compare with targeted results, and provide relevant information to the costing and sales department for appropriate measures.

PPC plays a key role in any manufacturing organisation. It includes material forecasting, master production scheduling, long term planning, demand management and more. The PPC process starts with a targeted plan, and thereafter designing the production plan according to sales demands to move forward.

Production planning is therefore, a strategy to plan a chain of operations that supports manufacturers to be at the right place at the right time so that they can achieve maximum efficiency from their resources.

The process of production planning and control may be classified into the following steps:

A. Routing:

Routing determines the path from which the raw materials flow within the factory. Once the sequence is followed, raw materials are transformed into finished goods. Setting up time for every step is important to measure the overall duration of the production process. In simple words, routing in manufacturing states the sequence of work and operations. Routing throws light on the quantity and quality of materials to be used, resources involved i.e., men, machine, and material, the series of operations and place of production in case of multiple options. It systematises the process and nurtures optimum utilisation of resources to get the best results.

B. Scheduling:

Scheduling is the second step in PPC that emphasises on the deadline of operations. It aims to make most of the time given for completion of the operation. It can be defined as determining the period required to complete the operation cycle as routed, considering all the allowances during the process. Organisations use different types of schedules to manage the time element. These include master schedule, operation schedule, and daily schedule, etc.

C. Dispatching:

The third step ensures that operations are done successfully. Dispatching includes the release of orders, in accordance with the scheduled charts. Here are the points that encapsulate "Dispatching":

- Issue of materials that are important for the production

- Issue of orders for initiating the work

- Maintain the records from start to end

- Initiate the control procedure

- Cascade the work from one process to another

D. Follow-up

Also known as expediting, follow-up is the final step that finds faults or defects, bottlenecks and loopholes in the entire production process. In this step, the team measures the actual performance from start until the end and then compares it with the expected one.

Expediters or stock chasers are responsible for performing follow-up process. It is obvious that any of the processes may undergo breakdowns or machine failure. Follow-up promotes smooth production by eliminating these defects.

RAW MATERIAL PROCUREMENT

Quality wise individual components of raw material (RM) requirement are incorporated in the annual budget. PPC also has the formula to calculate the components of RM per unit of the finished product and for required volume, considering the expected yield, waste generation during the process and percentage of sound quality production.

Manufacturing companies produce products as per the market demand, which is not always consistent due to various factors. Accordingly, sales plans need to be reviewed and altered in between. PPC department faces a great challenge to procure RM and fulfil the sales requirement on time. There is a lead-time from placing the order to arrival in the plant depending upon the type of RM, availability, price, import or indigenous, lead-time of arrival, etc. For example, if the RM is locally available, it may arrive within a week depending on distance. If the RM is to be imported, it may take thirty to ninety days based on distance of countries and other factors.

PPC has to ensure optimum level of RM inventory in godown. While they have to ensure RM is available timely to execute every sales order on time, at the same time there should not be any slow moving or dead inventory accumulation in godown to avoid blockage of liquidity. Following points may be considered before placing the order for RM:

- Specification of raw materials required and competitive prices from reliable suppliers
- Confirmed sales order in hand as on date and delivery date
- Work in progress as on date
- Finished goods inventory as on date
- Sales order under consideration and expected
- RM stock in RM godown
- RM in transit
- RM at supplier end, finalised
- Total exposure of RM as on date
- Excess/shortage exposure of RM as on date
- Average consumption of last three years and present trend

Despite the annual budget, there may be certain alterations in between because of inconsistency in demand and various other market factors. Considering the above facts, PPC department has to coordinate regularly with the sales department and adopt such alterations from time to time. Accordingly, the production plan is released monthly, considering the machine capacity and utilisation. After releasing the summarised production plan to concerned departments, shade wise dyeing plan, shade wise and count wise spinning plan, shade, design and quality-wise weaving and finishing plan, customer-wise folding, packing and dispatch plan is released by PPC department along with quantity and cut-off date of delivery for individual departments. Companies install different ERP systems, which provide online movement status of material and facilitates smooth monitoring and control. The system provides the position of individual orders, shades number, quality number, etc. with specific location, machine, time and date of input and output between and within the departments to calculate the lead time.

The following sheets indicate raw material order system according to the lead-time. It is evident that the order is placed in advance considering two to three months lead-time in case of import and in the same month in case of local availability.

Requirements and availability of LC (Letter of credit): Suppose there is 116 M Tons stock of raw material at plant, 54.5 Tons at converter place (Greasy wool scouring is outsourced) and 58.1 Tons is already ordered and expected in the month of September and October. Thus, total exposure of raw material is 229.1 Tons. Monthly requirement is 105.5 Tons, So, 123.5 Tons is balance. Thus, PPC department has to monitor continuously to review the monthly requirement, stock level and accordingly place the order for specific components. They should be in close coordination with finance department regarding availability of funds. Order should be placed in a distributed manner, synchronised with availability of fund to open LC, without delay in case of export and to release normal payment as per agreed terms in case of indigenous purchase.

RAW MATERIAL BOOKING FORMAT

RM Booking Month	March			June	April			May			June		
RM Required in Plant	May				June			July			August		
Particulars	Import Quantity	Proposed Price	Total Amount	Local	Import Quantity	Proposed Price	Total Amount	Import Quantity	Proposed Price	Total Amount	Import Quantity	Proposed Price	Total Amount

FOLLOW-UP CHART FOR RM IMPORT CONSIGNMENTS

Consignment Status	Arrival Month	ETA Plant	ETA Plant	RM Specs	Order Allocation	Supplier's Order No	Purchase Date	Supplier	Payment Terms	ETD Aus	B L Dt Aus	ETA Port India	Order Qty (Tons)	Invoice Qty (Tons)
At Indian Port Total														
In Transit from Scourer Total														

* ETA- Expected time of arrival, ETD-Expected time of departure

COTTON SPINNING PRODUCTION CAPACITY AND BALANCING FORMAT

Ring Frame								Total
	16	20	30	40	20	40	40	
Count Ne	16s K	20s K	30s C.	40s C.	20s K	40s K	40s K	
RF Make	SS	SS	LR	SS	NMM	SS	NMM	
No. of Splindles	952	4760	1536	4284	2470	8628	8770	31400
No. of Ring Frames	2	10	2	9	6.5	21	23.5	74
Speed	12000	13000	15500	15000	11500	14000	13000	
T.M	4.5	4.5	4.2	4.2	4.5	4.5	4.5	
TPI	18.00	20.12	23.00	26.56	20.12	28.46	28.46	
Production Grams / Spindle /Shift	300.36	232.83	161.90	101.77	205.97	88.65	82.32	
Total Prod./Day @ Eff.90% & Uti.96%	741	2873	645	1130	1319	1983	1871	10561
Speed Frame								
Hank	0.8	0.8	1.4	1.4	0.8	1.2	1.2	
Make	MMC	MMC	LR	LR	MMC	MMC/LR	MMC / LR	
Speed	625	625	850	850	625	750	750	
T.M	1.3	1.25	1.25	1.25	1.3	1.25	1.25	
TPI	1.16	1.12	1.48	1.48	1.16	1.37	1.37	
Kg / spindle /shift	4.84	5.04	2.96	2.96	4.84	3.29	3.29	
Spindles Reqd.@ 80% Efficiency	66	247	94	165	118	261	246	1199
Total Spindle available	MMC- 792 & LR- 456							1248
Total Production	771	2988	670	1175	1371	2062	1946	10983
Draw Frame								
Finisher								
Make	RSB 851	RSB 851	RSB 851	RSB 851	DO/2S	DO/2S	DO/2S	
Hank	0.12	0.12	0.12	0.12	0.12	0.12	0.12	
Speed	500	500	400	400	150	150	150	
Production/Shift	1181	1181	945	945	709	709	709	
M/C required. @85% Eff.	0.26	1.00	0.28	0.49	0.76	1.15	1.08	5.02

Total M/c Available	RSB 851 - 2 & LR Do/2S - 3							5.00
Total Production	775	3002	674	1181	1378	2072	1955	11036
Breaker								
Make	DO/6	DO/6	DO/6	DO/6	DO/6	DO/6	DO/2S	
Hank	0.12	0.12	0.12	0.12	0.12	0.12	0.12	
Speed	400	400	400	400	400	400	400	
Production/Shift	1890	1890	1890	1890	1890	1890	1890	13227
M/C required @85% Eff.	0.16	0.62	0.14	0.25	0.29	0.43	0.41	2.29
Total M/c Available	DO/6 M/C 2 & DO/2S M/C 1							3
Total Production	775	3002	674	1181	1378	2072	1955	11036
Comber								
Make			E7/4	E7/5				
Hank			0.12	0.12				
Nips/Min			180	300				
Efficiency			80%	80%				
Total M/c Available.	E7/4 -6 & E7/5 -1							7
Production/Day @80%Eff.			1574	286				1859
Card								
Make	DK-780	DK-780	DK-780	DK-780	DK-740	DK-740	DK-740	
Hank	0.11	0.11	0.11	0.11	0.1	0.1	0.1	
Speed	165	165	165	165	145	145	145	
Production/Day/M/C @90% Eff.	1148	1148	1148	1148	1110	1110	1110	
M/c Required.	0.68	2.64	0.68	1.19	1.25	1.88	1.78	10.10
Total Production	782	3031	777	1362	1391	2092	1974	11408
Total M/c Available	DK-740 - 5 M/C & DK - 780 - 5 M/c							10.00
Blow Room								
Hank	0.00123	0.00123	0.00123	0.00123	0.00123	0.0012	0.00123	
Speed	9.5	0.6	9.5	9.5	9.5	9.5	9.5	
Scutcher Reqd/Day @85% Eff.	0.20	0.76	0.19	0.34	0.35	0.52	0.49	2.85
Total M/c Available	2 Scutcher with CVT & 1 Scutcher without CVT							3.00
Production/Day	815	3160	806	1413	1451	2181	2058	11883
Mixing	H-4	Mech	Mech	Banny	H-4	Mech	Mech	
Winding								
Make	Murata	Murata	Murata	Murata	Murata	Autoconer	Winding M/C	
Speed/mts/minute	1200	1200	1200	1200	1200	800	800	
Prod./drum/day at 75 %Eff.	42.9	34.3	22.9	17.2	34.3	11.4	11.4	
Drum required	17.3	83.6	28.1	65.8	38.4	173.2	163.4	569.8
Drum available	Murata - 276 drum & AC138 - 168 drum							444
Doubling								
Count				2/40s C		2/40s K	2/40s K	
				20		20	20	
Make				SS		SS	SS	
Speed				10000		10000	10000	

T.M				5.09		5.09	5.09	
TPI				22.76		22.76	22.76	
Grams /spindle/shift.				158.36		158.36	158.36	
Grams /Spl/shift. @Eff.90 % & Uti.96 %				136.83		136.83	136.83	
Spindle. Required.				2698		4733	4468	11899
Spindle available				Super Doubler-11160 & NMM -880				12040
Ply Winding								
Make				Textool				
Speed/mts/minute				450	450	450	450	
Production/drum/day at 70 %Eff.				13.4	13.4	13.4	13.4	
Drums required				84	98	148	140	470
Drums available				540				
Dobble yarn Winding								
Make				PSMettler				
Speed/mts/minute				450	450	450	450	
Production/drum/day at 75 %Eff.				14.4	14.4	14.4	14.4	
Drums required				78	92	138	130	438
Drums available				480				

K-Carded, C-Combed, Eff-Efficiency, Uti- Utilisation, Spl-Spindle, Mts-Meters, TM-Twist multiplier, TPI-Twist per inch

WORSTED SPINNING CAPACITY (Polyester Tow and Wool Top to Yarn)

Machine	Model	No. of Machines	No. of Delivery	Feed Gms/ mtr	Speed Mtr/ minute	Efficiency %	Production /Mc. Per day Kg
Converter/ Tow to Top convertor		1	1		250	75	5400
G.B.-1		1	1	140	280	65	5242
G.B.-2		1	1	140	250	60	5184
Bump		1	1	24	1/1.5	60	5000
RECOMBING							
Defelter		1	1	250	200	50	4320
G.B.-1		1	1	150	230	60	5564
G.B.-2		1	1	150	250	60	5400
G.B.-3/Bl Coiler		1	1	200	250	65	5148
Comber		9	9	288	200	85	4315
G.B.-1		1	1	180	250	60	5184
G.B.-2/Ball Top		1	1	210	250	60	4752
PREPARATORY							
G.B.-1(Wool op)		1	1	160	70	45	1270
G.B.-2		1	1	175	70	50	1109
G.B.-3		1	2	176	70	55	1220
G.B.-4/Finisher		1	4	132	140	60	2419
G.B.-1		1	1	175	200	50	3168
G.B.-2		1	1	200	200	60	4838
G.B.-3		1	2	150	240	60	4147
G.B.-4/Finisher		1	4	132	250	65	3744

Rubbing Frame		1	20	4.8	70	60	968
Rubbing Frame		1	20	4.8	100	60	1382
Rubbing		1	24	4	100	65	1348
SPINNING							
Ring Frames		16	432	0.4	9500	85	150
Ring Frames		14	432	0.3	10000	88	86
POST SPINNING				Nm			
Auto Winding		10	24	56	900	67	372
Assembly Winding		1	24	56	400	70	346
T.F.O.		18	180	60	16000	90	156
Doffer							
Yarn Checker							
Soft Winding		1	16	56	600	80	197
Rewinding		1	60	56	400	70	432

*GB-Gill Box

WEAVING PRODUCTION CAPACITY (A PV Suitings manufacturing unit)

UNIT	TYPE OF LOOMS/ WIDTH	NO. OF LOOMS	PPI	PPM	TOTAL	AVG PPM	EFF. %	MTR IN LACS
UNIT-1	DORNIER PTV 190 CM	16	72	360	5760	360	80	0.70
	DORNIER HTVS 190 CM	24	72	360	8640	360	80	1.05
	DORNIER PTV 200 CM	8	72	330	2640	330	80	0.32
	DORNIER HTVS 200 CM	1	72	330	330	330	80	0.04
	DORNIER HI TECH 220 CM	18	72	290	5220	290	75	0.60
	DORNIER CIMMCO 200 CM	4	72	200	800	200	75	0.09
	DORNIER E.D.U. 210 CM	1	72	260	260	260	75	0.03
	TOTAL	72			23650	328		2.83
UNIT-2	DORNIER E.D.U.210 CM	16	72	260	4160	260	75	0.48
	DORNIER E.D.U. 190 CM	16	72	240	3840	240	75	0.44
	DORNIER E.D.U. 210 CM	2	72	260	520	260	75	0.06
	DORNIER E.D.U. 190 CM	2	72	240	480	240	75	0.05
	TOTAL	36			9000	250		1.03
	TOTAL LOOMS UNIT 1 & 2	108			32650			3.85
UNIT-3	SULZER (D.W) 390 CM	28	56	210	5880	210	82	1.89
	TOTAL	28						1.89
	G. TOTAL	136						5.74

*PPI-Picks/Inch, PPM-Picks per minute (Machine Speed)

DYEING DEPARTMENT PRODUCTION CAPACITY

(Top/Fiber/Yarn dyeing −Three shift)

Dyeing Mcs No	Make	No. of Mcs	Capacity	No. of Lots/day	Production/Day (Kg) @ 80% Efficiency
1		4	250	5	4000
2		3	200	5	2400
3		4	100	5	1600
4		3	50	5	600

5		1	50	5	200
6		2	25	5	200
7		2	25	5	200
8		1	12	5	48
9		1	6	5	24
10		1	5	5	20
Total		22			9292

FINISHING DEPARTMENT PRODUCTION CAPACITY

Mcs Name	Make	No. of Mcs	Production/Day (Mtrs) Three Shift working
Singing		1	56700
Crabing		1	11000
Rope Scouring		5	8000
Rope scouring		2	4500
Rope Scouring		1	3250
Rope scouring		1	3250
Milling cum Scouring		1	3250
Milling cum Scouring		1	1620
Open width scouring		1	20400
Jigger		1	900
Jigger		1	1800
JT 10		1	6000
Hydro Extractor		1	25200
Stenter		1	18500
Stenter		1	18500
Shearing		1	14000
Shearing		1	10500
Shearing		1	10500
Decatising		1	13000
Rotary Press		1	17000
Super Finish		1	19000

WEAVING MONTHLY PROGRAM SHEET (Cotton and blended materials)

PRODUCTION PROGRAMME										APRIL	200X		Working Days =		25.5	
							Mtrs/ day	Total	Total				No. OF		Summary	
Looms Group	Q.NO	Description	No of Looms	Picks/ Inch	Eff%	Width	Loom speed PPM	/Loom	Mtrs/ Day	Mtrs/ Month	Looms Group	Mtrs/ Day	Mtrs/ Month	Working Looms	Quality	Mtrs/ Day	Mtrs/ Month
130"	40401	Cotton Base	1	72	70	2	224	159	159	4062							
TW11	2096 2W	Bed Sheets	20	38	80	2	224	345	6899	175935							
	1362/2	Filament Suiting	2	38	80	2	224	345	690	17593	Sulzer						
	5760	Base Cloth	3	50	80	2	224	262	787	20057				60	Ankara Normal 3W	26654	679677
	49225/2	Shirting 2W	20	48	80	2	224	273	5462	139282					Ankara 2W		
	44712	School Check	4	50	80	2	224	262	1049	26742						26654	679677
	35192	Ankara- 2 W		44	80	2	224	298									

	81016	TOP DYED		42	80	2	224	312						3	Base cloth	787	20057
		S. TOTAL→	50						15046	383671	130"TW-11	15046	383671	4	School Check (2)	1049	26742
153" PU														35	45" Shirting 3W	13749	350598
	3519 3W	Ankara-Normal	12	44	80	3	235	469	5626	143465				20	45" Shirting 2W	5462	139282
		S. TOTAL→	12						5626	143465	153"PU	5626	143465			21046	536678
153"SH																	
	49225 3W	Shirting 3W	23	48	80	3	210	384	8833	225245				20	Bed Sheet 2W	6899	175935
	3519 3W	Ankara-Normal	21	44	80	3	210	419	8798	224354							
		S. TOTAL→	44						17631	449599	153"SH	17631	449599			6899	175935
153" MW																	
	3519 3W																
		Ankara-Normal	5	44	80	3	240	479	2394	61049							
		S. TOTAL→	5						2394	61049	153"MW	2394	61049		Cotton Drill		
														1	Cotton Base	159	4062
														2	Filament Suiting (2W)	690	17593
153" PU Dobby	3519 3W	Ankara-Normal		44	80	3	224	447							Total Dyed		
	49225 3W	SHIRTING 3W	12	48	80	3	224	410	4916	125354						849	21656
														145		55449	1413946
		S. TOTAL→	12						4916	125354	153" PU	4916	125354		TOTAL→	55449	1413946
153" PU D1	3519 3W	Ankara-Normal	8	44	80	3	240	479	3831	97678							
	49225 3W	SHIRTING 3W		48	80	3	240	439						4	Shirting Production	1536	39168
		S. TOTAL→	8						3831	97678	153" PU D1	3831	97678		(sample Looms)		
153" PU D2	3519 3W	Ankara-Normal	14	44	80	3	215	429	6005	153131							
	3710	DRILL															
		S. TOTAL→	14						6005	153131	153" PU D2	6005	153131				
		153" LOOMS	57											149	G TOTAL--->	56985	1453114
		GR TOTAL→	145						55449	1413946	Sulzer Net	55449	1413946				

*EFF-Efficiency

MONTHLY PRODUCTION PROGRAM AND TENTATIVE CONTRIBUTION SHEET

In certain businesses, monthly sales programs are not fixed as per the annual plan because of the fluctuating market condition. In such cases, a tentative contribution may be calculated to assess the profit or loss before releasing the production plan.

Contribution margin = revenue − variable costs

Fixed costs are business costs that remain the same, irrespective of the quantity or volume of production, for example, rent and administrative salaries, etc. Variable costs are those expenses that vary with the quantity of product produced, such as direct materials or sales commissions, etc.

Example:

Quality Name	Warp Ne	Weft Ne	EPI/ PPI	Warp Kg/Mtr	Weft Kg/Mtr	Mtrs/ Kg	Rate of Yarn Warp N	Weft N	Yarn /Mtr N	Expense /Mtr N	Net Sale/Mtr PRICE N	Cont /Mtr	Prod./Day /Loom	Cont Day/ Loom	No. of Looms	Total Prod	Exptd Cont.N=
ANKARA 3W	20/1 OE	20/1 OE	56/42	8.83	6.99	6.32	285	285	48.49	6.5	63	8.02	459	3679.03	60	716040	5739287
ANKARA 2W	20/1 OE	20/1 OE	56/42	8.83	6.99	6.32	285	285	48.49	6.5	63	8.02	306	2452.69	11	87516	701468

BUTTERFLY	24/1 C	24/1 C	56/44	7.49	5.86	7.49	315	315	45.39	6.81	53	0.8	368	294.83		0	-
DYED SHIRT 3W	40/1 P	150/1 Tex	64/48	4.95	4.21	10.9	302	202	26.42	7.42	45	11.2	377	4204.98	32	313664	3498545
DYED SHIRT 2W	40/1 P	150/1 Tex	64/48	4.95	4.21	10.9	302	202	26.42	7.42	45	11.2	251	2799.6	11	71786	800687
DYED SHIRT 2W	40/1 P	40/1 P	64/56	4.95	4.18	11	302	302	30.62	8.66	45	5.71	220	1257.23		0	-
Filament Shirting	150/1 TW	150/1 Tex	64/46	5.58	4.04	10.4	244	202	24.71	7.11	43	11.2	273	3050.47	0	0	-
SCHOOL CHECK2W	150/2 TW	150/1 TexC	52/50	9.02	4.35	7.48	264	222	36.64	7.73	65	20.6	262	5404.41	0	0	-
SCHOOL CHECK2W	40/2 PE	150/1 Tex	52/50	7.98	5.35	7.5	353	222	43.65	7.73	65	13.9	338	4704.76	3	26364	366971
COTTON DRILL -2W60"	12/1 C	12/1 C	64/38	22.14	13	2.84	225	225	83.17	8.82	80	-12	338	-4051	0	0	-
COTTON DRILL -2W45"	12/1 C	12/1 C	64/38	17.13	10.1	3.67	225	225	65.04	5.88	65	-5.9	530	-3135.5		0	-
FILAMENT SUITING2W	150/2 TW	150/2 Tex	56/42	13.69	10.1	4.2	244	202	57.46	9.74	80	12.8	338	4323.68	0	0	-
MERC SUITING	20/1 PE	150/2 Tex	56/42	10.88	9.28	4.96	285	202	53.25	9.74	70	7.01	276	1934.05		0	-
TOP DYED SUITING2W	12/1 C	20/1 PE	56/42	11.49	8.51	5	255	255	54.52	9.74	70	5.74	312	1789.66	1	8112	46531
PRINTING BED SHEET2W	20/1 OE	150/1 Tex	44/38	15.24	4.6	5.04	255	202	46.95	8.82	55	-0.8	340	-261.65	10	88400	-68029
COTTON RIB FUR.3W	150/2 TW	12/1 C	52/34	8.35	9.14	5.72	285	225	47.25	5.26	55	2.49	568	1415.24	1	14768	36796
DOBBY FURN.3W	40/1 C	150/1 Tex	56/50	9.97	4.63	6.85	264	284	42.76	7.73	65	14.5	386	5599.83	0	0	-
BASE CLOTH NSF	40/1 PE	40/1 C	74/72	6.8	6.5	7.52	310	310	44.55	16.7	100	38.7	159	6159.91	0	0	-
BASE CLOTH NSF	20/1 CE	150/1 TEX	64/50	6.2	5.2	8.77	302	202	32.31	11.6	60	16.1	262	4214.94	3	20436	328765
EXPORT	20/1C	20/1C	60/60	8.82	9.13	5.57	265	265	51.02	9.28	87	27	320	8623.52	20	166400	4484229

YARN PRICES/ Kg N			MIBS	Expenses N OOO		Total Picks Inserted per Loom			152	1513486	15935251
20/1 Cotton	285	150/1 Tw.	244	SALE & WAG.	4000	Per Month in 000		6589			
24/1 Cotton	315	150/1 Tex	202	REPAIRS & MAINT.	1000	2W LOOMS	48	96			
12/1 Cotton	225	20/1 PE	255	POWER&FUEL	4000	3W LOOMS	102	204			
16/1 Cotton	245	30/2 PE	310	ENG.OH	1500	TOTAL W		402	ACTUAL 2672771		
150/2 Tex	202	34/1 PE	280	TOTAL	10500	TOTAL PICKS		2648955			
150/2 Tw.	244	40/2 PE	333	COST PER PICK		AVG. PRODUCTION		1500			
		40/1 PE	302	FOR SINGKE W	0.155	AVG. PICKS PER INCH		44.86			

*C-Cotton, OE-Open end, Tex - Texturised, PE-1--%Spun Polyester, W-Width, N-Nigerian currency, Cont-Contribution, Exptd-Expected

MONTHLY PRODUCTION TARGET SUMMARY FOR A COMPOSITE WORSTED MILL

PRODUCTION PLAN		MONTH ---			
		Unit	Per Month	W. Days	Per Day
Dyeing	Fabric	Mtrs	22,500	31	726
	Top	kgs	85,000	27	3,148
	Yarn Dyeing - Own	kgs	3,000	27	111
	Cotton	kgs			
Recombing		kgs	93,500	31	3,016
Spinning Own		kgs	85,500	31	2,758
Yarn Delivery Own		kgs	80,400	31	2,594

Weaving		PW Export	Mtrs	15,000	31	484
		PW Domestic / st yarn	Mtrs	225,000	31	7,258
		PV	Mtrs	49,500	31	1,597
		Total	Mtrs	289,500	31	9,339
Finishing		PW Export	Mtrs	20,000	31	645
		PW Domestic / st yarn	Mtrs	230,000	31	7,419
		PV	Mtrs	5,000	31	161
		Total	Mtrs	255,000	31	8,226

Spinning: Average Count: 63 Nm Installed Efficiency - 88 % RF Spindle Speed - 8315 RPM TPM - 745

Weaving: Average Picks per inch: 71

SPINNING MONTHLY PROGRAM FORMAT

Worsted Spinning No of Spindles - 10512

Market	Order month	Sales Order No	Customer Name	Order Status	Week	Quality Specs	Shade	Blend	Blend Ratio	Spin factor	Qty (Kg)	Converted on 60 Nm Basis	Spindles / Day	TPM	Micron	Delivery date
Domestic	Oct-19			C	II											25-Nov-19
Domestic	Oct-19			C	IV											10-Nov-19
Export	Oct-19			C	II											15-Nov-19
Export	Oct-19			C	III											15-Nov-19
Export	Oct-19			C	II											20-Nov-19
Export	Oct-19			C	I											30-Nov-19
Export	Oct-19			C	I											30-Nov-19
Export	Oct-19			C	I											30-Nov-19
Export	Oct-19			C	I											30-Nov-19
Export	Oct-19			C	I											30-Nov-19
Export	Oct-19			C	I											30-Nov-19
Export	Oct-19			C	I											30-Nov-19
Domestic	Oct-19			C	I											15-Nov-19
Domestic	Oct-19			C	I											25-Nov-19
Domestic	Oct-19			C	I											25-Nov-19
Domestic	Oct-19			C	I											25-Nov-19
Domestic	Jun-19			UC	III											15-Apr-19

*C-Confirm, UC-Under confirmation

FABRIC DYEING DAILY FOLLOW-UP SHEET

Quality Details					Location			
Yarn Lot No	Dyeing Lot no	Fabric Q No	Balance Shade nos to Dye	Total shades dyed	Dyeing	Finishing	Mending	Weaving
			09	8		5	3	
			10	4	4			
			02	4	4			

				Code					
				09	4		4		
				04,05,06	12	4	4		4
				81	4	4			
				01,02	8	4		1	3
				05,	4	4			
				01,04,05,06,09	20	8	9	1	2
				04,10	8	4	3	1	
				02	4	4			
				02	4		4		
				302,303,39	12	4	5		3
				01	4		2	1	1
				07,08,09,10,11,12	12		3		9
				07,08,09,10,11,12	12		1		11
				01	8		2	1	5
				01	4				4
				01,03,07	12				12
				01,02,03,04,05	20			1	19
				02,05,07,04,03	20				20
				08,21	8		4		4
				06,143,81,10	16		3	1	12
				05,06	8			2	6
				03,04	8				8
				04	4				4
				08,136,50	12			3	9
					244	**44**	**49**	**15**	**136**

DAILY PRODUCTION FOLLOW-UP SHEET

	Dyeing				Spinning				Wvg Grey Perch			Mending	Finishing	Power	FO
Target	50000 Mtr	75400 Kg	2500 Kg		100500 Kg	95800 Kg	90052 Kg			365000 Mtrs		Mtrs	300000 Mtrs		
Date	Fabric	Top	Yarn dyeing (Own)	Yarn dyeing (Cotton)	Recombing	Spinning	Yarn Delivery	Ring frames running	Warping issue mtrs	Woven mtrs	Looms Running	Total mtrs	Total mtrs	Units	Liters
01-Nov	0	0	0	0	2249	2846	2781	23	11677	7619	65	7358	0	49950	4320
02-Nov	2653	2801	0	0	1300	2762	2066	27	10410	9431	74	9194	9413	55800	5490
03-Nov	1361	3435	0	0	3311	2766	2750	27	13934	9852	72	8938	7333	54000	4895
04-Nov	3058	1261	186	0	2542	2735	2805	26	7984	8811	68	8653	6469	54450	5420
05-Nov	2534	2959	124	0	3304	2648	2585	28	11536	9279	72	10082	9162	56700	5600
06-Nov	3587	2445	82	0	3293	3152	2200	28	11985	10129	75	10528	6041	54900	5400
07-Nov	2722	3345	91	0	3319	3003	2423	27	8074	11086	77	10272	8456	56250	5480
08-Nov	0	0	0	0	2324	3071	2701	26	6345	10766	77	10467	8351	53550	3690
09-Nov	2171	1904	49	0	2506	2742	2800	24	16649	9166	77	9333	6640	54450	5430
10-Nov	1799	1658	172	0	2819	2835	2050	25	9729	10583	77	9555	8845	53550	5480
11-Nov	2774	1517	116	0	2606	3004	3010	29	10472	11595	77	9192	8393	54450	5490
12-Nov														8550	450
Cumulative	22657	21322	819	0	29573	31564	28171	290	118795	108317	811	103572	79103	606600	57145
Cumu. Avg.	2517	2369	91	0	2688	2869	2561	26	10800	9847	74	9416	7910	50550	4762
Average Beam length				695	mtrs										
Average picks				71.41											

Average count Nm				55.59													
Manufacturing Program received				9.72	Lac mtrs												
Raw Material cleared				9.72	Lac mtrs												
Raw material																	
	Received	Stock															
Wool	6.6	13.7	Tons														
Polyester	13.7	55	Tons														

NP-Normal polyester, TBL-Trilobal, LP-Low Pill

MONTHLY PRODUCTION SUMMARY

Date- Year -	Dyeing		Spinning			Weaving				Finishing
Months	Fabric	Top	Spinning	Yarn Delivery	Avg count	Woven mtrs	Avg / day	Avg PPI	Avg. Per day Picks instreated 000	Total mtrs
April	16802	31713	45420	43035	57.6	132597	5100	68.4	13694	137904
May	28567	24070	42536	33505	54.6	115639	5507	64.4	13899	115984
June	31602	46858	49009	51330	58.3	150062	5772	57.9	13046	120515
July	25946	80893	84893	70404	58.8	219966	8147	66.6	21219	204922
August	43671	53652	81469	69863	59.5	230664	9227	64.5	23319	197366
September	36161	44537	76008	66989	57.5	240992	10041	63.9	25174	255194
October	43973	82962	84352	73186	58.7	280432	9348	70.0	25985	290222
November Till 18th	33946	**23262	44569	40629	56.7	161323	10083	70.2	27287	123889

** Low production due to shortage of program

RAW MATERIAL MONTHLY RECORD

Purchased Yarn Received		Raw material Received		
Month	2010	Month	Wool Kg	Polyester
		Apr-15	17191	27647
Jan	6,858	May	9161	41259
Feb	26,559	June	41169	42198
March	36,165	July	62762	83050
April	27,382	Aug	36349	48807
May	55,440	Sept	24663	38377
June	23,664	Oct	38758	76002
July	25,882	Nov	6783	13723
Aug	57,818	Total	236826	371063
Sept	79,563	Avg /Month	29603	46383
Oct	32,297			
Nov	41,731			
Dec	36,452			
Total	4,49,811			
Ave/Month	37484			

WORK IN PROGRESS (WIP)

MONTHLY WIP (WORK IN PROGRESS)								
Month Closing	April	May	June	July	Aug	Sep	Oct	As on 15th Nov
Dyeing Kgs	15.9	5.0	6.0	17.4	1.7	10.8	11.5	8.9
Spinning Kgs	27.2	34.4	35.2	52.2	58.9	62.6	65.7	64.8
Weaving Mtrs	64530	43948	75732	60587	85797	113508	115143	92170
Mending Mtrs	22068	13612	23958	31233	44217	20450	14862	23703
Finishing Mtrs	51178	53068	65968	56436	63548	59867	41322	62844
Total Equivalent Mtrs	**266433**	**228240**	**288640**	**408587**	**444046**	**416526**	**423022**	**412420**
Yarn Room Stock Tons								28

MONTHLY SALES PROGRAM AND SPINNING CAPACITY BALANCING

Domestic Market	Top Dyed	Piece Dyed		Spinning Production Capacity						
Month	Mtrs Recd		Total	Coarse - 25%		Medium - 40%		Fine - 35%	
	TD	PD		Mtrs	Percentage	Mtrs	Percentage	Mtrs	Percentage
Apr	648	0	648	0	0	648	100	0	0
May	73624	0	73624	34128	46	3024	4	36472	50
Jun	105948	14520	120468	24288	20	58488	49	37692	31
Jul	408996	74800	483796	79388	16	163956	34	240452	50
Aug	159840	62260	222100	80548	36	38792	17	102760	46
Sep	30240	5720	35960	10800	30	13448	37	11712	33
Oct	25920	9680	35600	18712	53	8008	22	8880	25
Total	805216	166980	972196	247864		286364		437968	
Export			**172330**	**48080**		**8600**		**115650**	
G. Total			**1144526**	**Actual %**	**26%**		**26%**		**48%**

Disbalancing the spinning capacity affects the spinning production. There should be synchronisation between sales order issued and available spinning capacity since the budget is prepared accordingly. However, because of market trends, spinning capacity may be disbalanced and may be outsourced for timely execution of sales order.

DAILY YARN STOCK MOVEMENT FORMAT

In a spinning mill, yarn is the final product for sale. However, in a fabric manufacturing company, yarn is used to produce fabrics. Daily yarn stock report is essential to track the yarn stock position and control the important inventory. It is obvious that despite proper monitoring, there are chances of variation in yarn production as compared to the actual requirement due to various reasons. In case of fancy fabric designs production, even if a single component of yarn is short or excess in quantity, it affects the other components proportionately. If one component is short for fabric production, the planning department may reduce the beam length, as a result other yarn components becomes remnant. There are many other reasons for remnant's generation during the process. Yarn stock report helps in monitoring and controlling the reasons for excess or short production, slow moving, and non-moving yarn inventory. The designing and development department should use remnants yarn regularly with the help of sales team and convert into asset from liability. Age wise yarn stock may also be generated for better control.

Example:

Quantity in Kg			DATE	31-Mar-03	
Opening	Recvd	Cumm	Issued	Cumm	Closing
Balance	Today	Recvd	Today	Issued	Balance

DAILY GREY FABRIC STOCK MOVEMENT FORMAT

Grey or unprocessed fabric is transferred in a godown after weaving and mending. It takes time to complete the total quantity of a specific quality in weaving, depending on weaver's beam length. Lot sizes required in Dyeing and Finishing are much larger compared to weaving. The fabrics are stored in Grey Godown, quality and shade wise. Lots are prepared for dyeing, finishing based on coarse, medium, fine, and light, medium, dark shade, type of finishes required, etc., and issued as per requirement. Daily grey fabric stock report is essential to monitor and control the inventory level of fabric and timely issuing for next process in a systematic manner (first in, first out basis).

Example:

Quantity in Mtrs	A	B		DATE	31-Mar-03	
QUALITY	Opening	Recvd	Cumm	Issued	Cumm	Closing
	Balance	Today	Recvd	Today	Issued	Balance
2520	500	100	3000	200	2000	1500
5570	61	430	1512			1573
5571	826					826
5572	1090					1090
5573						
5560						
SUB TOTAL→	2477	530	4512	200	2000	4989
EXP. SAMPLE 009	630					630
EXPORT 10	1304		544			1848
BASE CLOTH						
5760	2794		17067		16687	3174
40401/02	71		1670		965	776
SUB TOTAL→	2865		18737		17652	3950

RAW MATERIAL RECONCILIATION SUMMARY REPORT

The raw material daily stock and movement report is an important document for the management to monitor its inventory level. Sometimes few raw materials become slow moving or non-moving due to either excess order or cancelation of sales order for various reasons. Few costly special fibers or yarn are procured and remain in stock for long periods and block the liquidity. In daily reports, such items are highlighted for the management to take appropriate decision. Age wise raw material report is also helpful for the management to avoid blockage of inventory. At the month end, raw material reconciliation is essential for proper tracking, control, and accounting purposes.

Example:

Particulars	Imported			Indigenous			Total		
	Qty. (Kgs)/ Mtrs	Value (Rs)	Rate / Kg/ Mtrs	Qty. (Kgs)/ Mtrs	Value (Rs)	Rate / Kg/ Mtrs	Qty. (Kgs)/ Mtrs	Value (Rs)	Rate / Kg/ Mtrs
Opening Balance									
Wool / Top									
Viscose/Top									
Silk Sliver									
Silk / Wool									
Opening Balance Total									
Purchases									
Wool/ Top									
Viscose/Top									
Silk Sliver									
Silk / Wool									
Total									
Add / (Less):- Raw Material Freight & Other Charges									
Wool / Top									
Viscose / Top									
Silk Sliver									
Silk / Wool									
Total									
Payment Recd									
Waste Purchase									
RM In Transit									
Total Purchase									
Opening - Purchase + Frt.									
Wool /Top									
Viscose / Top									
Silk Sliver									
S/Wool									
Total									
Closing Balance									
W/Top									
Viscose/Top									
Silk Sliver									
Silk / Wool									
Total									
Closing Balance Total									
RM In Transit									
Closing Balance Total Net									
Raw Material Consumption									
Wool / Top									
Viscose / Top									
Silk Sliver									
Silk / Wool									
Total									
Net Consumption									

W-Wool Top-Ball of Sliver RM – Raw Material

RAW MATERIAL RECONCILIATION: DEPARTMENT WISE MONTHLY SUMMARY REPORT

Stock taking and raw material reconciliation exercises are important responsibilities of a department head and provide essential information for the management. The account department requires these for verification, stock valuation, accounting and preparing profit and loss accounts every month. Once the raw material is issued to production department, it keeps on moving from one department to another in different shape until the final output. Material measurement is also done within the department for its conversion in terms of weight or length. Hence, it is important for individual departments to keep record of inputs and outputs in terms of their respective measurements. At the end of the month, department heads have to take material stock, reconcile it, and submit the information to the accounts department. Accounts department may summarise these in the following formats. Individual departments also keep record of stock at different stages within their departments.

Example:

April – Quantity in Tons	Opening stock	Recvd	Total	Del.	Closing Stock		Equivalent Mtrs	
Dyeing	9.7	38.1	47.8	31.9	15.9			
Spinning	30.2	43.3	73.5	43	27.2	}	128657	mtrs
Yarn Room	26.8	43	70	43	26.5			
Weaving	65955	131170	197126	132596	64530			
Mending	29132	132596	161728	139660	22068	}	137776	mtrs
Finishing	56054	139660	195714	137903	51178			
						Total	266433	mtrs

May

	Opening stock	Recvd	Total	Del.	Closing Stock			
Dyeing Tons	15.9	23.9	39.8	34.8	5.0			
Spinning Tons	27.2	43.6	70.8	33.5	34.4	}	117612	mtrs
Yarn Room Tons	26.5	33.5	60	35	24.9			
Weaving	46580	113006	159587	115639	43948			
Mending	22075	116639	137714	124102	13612	}	110628	mtrs
Finishing	51178	124102	175280	115986	53068			
						Total	228240	mtrs

June

	Opening stock	Recvd	Total	Del.	Closing Stock			
Dyeing Tons	5.0	53.8	58.8	52.8	6.0			
Spinning Tons	34.4	55.5	89.9	51.3	35.2	}	122985	mtrs
Yarn Room Tons	24.9	51.3	76	53	23			
Weaving	40505	185287	225792	150060	75732			
Mending	13646	150060	163706	139748	23958	}	165655	mtrs
Finishing	53068	139748	192816	120516	65965			
						Total	288640	mtrs

July

	Opening stock	Recvd	Total	Del.	Closing Stock			
Dyeing	6.0	101.3	107.3	89.9	17.4	}	+17611 kg balance to lay blend	
Spinning Tons	35.2	92.8	128	70.4	52.2			26031 mtrs
Yarn Room Tons	23.0	70.4	93	60	33.2			

Weaving	64060	216494	280554	219967	60587	
Mending	23204	219967	243171	211938	31233	148256 mtrs
Finishing	65965	211938	277903	204971	56436	

Total   408587   mtrs

August

	Opening stock	Recvd	Total	Del.	Closing Stock	
Dyeing	17.3	61.1	78.4	76.7	1.7	+ 23312 kg balance to lay blend
Spinning	52.2	81.5	133.7	69.7	58.9	25044 mtrs
Yarn Room	33.2	69.7	103	60	42.5	
Weaving	54677	261783	316460	230663	85797	
Mending	31232	230663	261895	217678	44217	193562 mtrs
Finishing	56436	217678	274114	200273	63548	

Total   444046   mtrs

September

	Opening stock	Recvd	Total	Del.	Closing Stock	
Dyeing	1.7	57.8	59.5	48.7	10.8	+ 1205 kg balance to lay blend
Spinning	58.9	75.1	134	66.8	62.6	22271 mtrs
Yarn Room	42.5	66.8	109	59	50.5	
Weaving	81405	273092	354497	240989	113508	
Mending	44221	240989	285210	264760	20450	193825 mtrs
Finishing	63548	264760	328308	255197	59867	

Total   416526   mtrs

October

	Opening stock	Recvd	Total	Del.	Closing Stock	
Dyeing	10.8	96.4	107.2	95.7	11.5	+ 7118 kg balance to lay blend
Spinning	62.6	81.5	144.1	73.2	65.7	25165 mtrs
Yarn Room	50.5	73.2	124	78	45.4	
Weaving	106455	289120	395575	280432	115143	
Mending	20450	280432	300882	286020	14862	171327 mtrs
Finishing	59867	286020	345887	290224	41322	

Total   423022   mtrs

RAW MATERIAL (RM)

Date   11-Nov

Opening stock		Total	Received		Total OS + Received		Issued			Closing Stock		
Wool	Polyester		Wool	Polyester	Wool	Polyester	Wool	Polyester	Wool	Polyester	TOTAL
19.5	58.9	78.4	6.6	13.7	26.1	72.6	12.4	13.3	13.7	55	68.7

\# OS-Opening stock

WORK IN PROGRESS (WIP)

	Opening stock	Recvd	Total	Del.	Closing Stock	
Dyeing	11.4	28.7	40.1	29.1	11.0	+ 4033 kg balance to lay blend
Spinning	65.8	28.6	94.4	28.2	64.3	236815 mtrs

Yarn Room	45.4	28.2	74	31	42.8
Weaving	92130	112855	204985	108317	96668
Mending	14862	108317	123179	103573	19606
Finishing	41322	103573	144895	79103	61866

178140   mtrs

Total   414955   mts

MONTHLY WASTE AND DEFECTIVE PRODUCTION

PPC department needs to have individual quality and department wise records of waste generation and percentage of defective production. These figures provide the guidelines for assessing the RM requirement to produce the desired quantity of final product. In case of higher waste and defective generation compared to plan, PPC department sounds an alarm to the concerned department for taking corrective measures. In case this is not possible, it should be discussed with management for making provisions in budget accordingly.

Examples:

Fabric Production	Domestic Mtrs	Odd%	C&D%	Total %	Export Mtrs	Exp.%
Ave For Previous F Y	326521	2.22	6.12	8.33	62172	91.47
Current Year Apr	265804	2.37	6.43	8.80	53146	93.04
May	169817	2.90	7.93	10.83	53146	93.04
Jun	189042	2.52	6.35	8.87	22111	85.54
Jul	179076	2.39	5.72	8.11	24035	87.52
Aug	174860	1.95	5.56	7.51	10609	93.26
Sep	225728	2.02	4.95	6.97	4181	90.95
Oct	170943	2.52	6.16	8.67	8728	95.98
Nov	176388	2.02	5.30	7.32	17418	91.48
Dec	225917	2.17	6.14	8.31	63664	95.36
Jan	273791	2.15	5.86	8.01	39005	90.78
Feb	217444	2.45	7.20	9.65	3863	82.15
Mar	228665	2.50	7.11	9.61	9161	97.40
Average	208123	2.33	6.23	8.56	25756	91.38

*C&D- Cuts and defective, Exp-Export

WASTE RECONCILIATION APRIL - MARCH (FY - Kg)

Month	OS	Received from Deptt			* Received from QC/Dsg/Inp/Wh	Total Receipt	Issues (Sale)	CS	* WH	* QC	* Dsg	* Inp	* Total
		Spg	Wvg	Total									
Apr	17,285	3,325	4,642	7,967	1,266	9,233	3,899	22,619	423	191	280	372	1,266
May	22,619	4,168	3,476	7,644	1,657	9,301	13,078	18,842	923	191	251	292	1,657
Jun	18,842	5,691	3,922	9,613	957	10,570	9,445	19,967	400	170	201	186	957
Jul	19,967	7,097	3,744	10,841	1,088	11,929	11,035	20,861	437	159	195	297	1,088
Aug	20,861	9,137	5,970	15,107	1,346	16,453	8,773	28,541	-	711	224	411	1,346
Sep	28,541	8,080	5,826	13,906	1,558	15,464	14,122	29,883	547	209	391	411	1,558
Oct	29,883	8,962	6,264	15,226	1,938	17,164	14,230	32,817	580	232	206	920	1,938
Nov	32,817	7,079	5,744	12,823	1,657	14,480	13,032	34,265	376	264	218	799	1,657
Dec	34,265	6,721	6,210	12,931	1,280	14,211	13,271	35,205	468	256	152	404	1,280
Jan	35,205	4,857	4,375	9,232	2,381	11,613	17,408	29,410	1,590	97	182	512	2,381

Feb	29,410	4,565	2,930	7,495	1,145	8,640	1,602	36,448	-	338	330	477	1,145
Mar	36,448	5,005	4,692	9,697	2,383	12,080	9,599	38,929	745	281	841	516	2,383
Total		74,687	57,795	132,482	18,656	151,138	129,494		6,489	3,099	3,471	5,597	18,656
					12,167	(Excluding Warehouse waste)							

*OS- Opening stock, CS-closing stock, Spg-Spinning, Wvg- Weaving, FY-Financial year, QC- Quality Control, Deg-Designing, Inp-Inspection, WH-Warehouse

Note: Difference in Wvg Waste between Dept & Waste Godown is 1118 kgs in Apr to July, after verification, Godown figures were found correct

WASTE STOCK STATEMENT FROM APRIL - MARCH (FY)

										Total April to March (FY)		
S. N.	Material	Opening Stock 01-Apr 09	Receipts	Issue/Sales			Opg. Stock 01-May 09	Clg. Stock 31-Mar 10	Total Receipt	Issue	Value
		(Kgs.)	(Kgs.)	(Kgs.)	Rate Rs. / kg	Value in Rs.					
	COMBERS WASTE										
1	Merino Noils Waste	353.9	0.0	353.9	165.0	84960.0	0.0	606.25	25335	25083	5072935
	28.0 mic waste	720.5	0.0	720.5	118.0	58146.0	0.0	0.00	3434	4154	473421
2	Merino Wool Waste	0.0	0.0	0.0		0.0	0.0	0.00	29524	29524	858356
3	100% Polyester Waste	0.0	38.7	0.0		0.0	38.7	144.45	2632	2488	173801
	Sub Total	**1074.4**	**38.7**	**1074.4**	**283.0**	**143106.0**	**38.7**	**750.70**	**60925**	**61249**	**6578513**
	DEPARTMENTAL WASTE								0		
1	A/W Soft Waste	1261.5	157.0	0.0		0.0	1418.5	778.98	8387	8869	2250786
2	A/W Hard Cutting Waste	0.0	0.0	0.0		0.0	0.0	0.00	0	0	0
3	Recombing Noils Waste	741.5	994.0	0.0		0.0	1735.5	3822.10	23625	20544	977141
4	A/W Mixture Hard Waste	0.0	0.0	0.0		0.0	0.0	0.00	0	0	0
5	T/W Soft Waste	5358.0	768.0	0.0		0.0	6126.0	2009.00	11024	14373	1309583
6	T/W Hard Waste	3308.5	1245.1	0.0		0.0	4553.6	7534.00	34881	30655	779323
7	T/W Dornier Cutting Waste	2395.0	3437.5	3899.0	13.8	53611.3	1933.5	10857.80	41936	33473	475714
8	T/W Fine Sweeping Waste	64.0	404.0	0.0		0.0	468.0	4403.00	9360	5021	115525
9	T/W Hard Cutting Waste	4156.3	1266.0	0.0		0.0	5422.3	5886.35	16602	14872	129996
10	Oily Dornier Waste	0.0	961.0	0.0		0.0	961.0	834.00	2521	1687	11809
11	T/W Hard Cutting with cotton yarn							1497.00	1497	0	0
12	A/W Recombing Noils Waste O. G							932.00	932	0	0
13	A/W O.G Hard Waste							606.00	606	0	0
	Sub Total	**17284.8**	**9232.6**	**3899.0**	**13.8**	**53611.3**	**22618.4**	**39160.23**	**151370**	**129494**	**6049876**
	Grand Total	**18359.2**	**9271.3**	**4973.4**	**296.8**	**196717.3**	**22657.0**	**39910.93**	**212295**	**190743**	**12628389**

*AW-All wool, TW-Terry wool, FY-Financial Year

STOCK RECONCILIATION FROM RAW MATERIAL TO FINISHED PRODUCTS FOR A FINANCIAL YEAR

Stock of material is taken every month in each section and departments assess the quantity at different stages. The objective of stock taking is to reconcile the material and for its valuation for accounting purposes. In a textile-manufacturing factory, stock taking is an extensive exercise and for accurate stock taking process, operations activities are stopped. But it may not be possible to stop the operation every month to avoid production losses. Thus, the quantity of material on machines is taken on tentative basis and department wise stock reconciliation is submitted to the accounts department with explanations for any variations not within the norms. Many companies conduct physical stock taking every quarter but few companies conduct annual physical stock taking exercises by stopping operations and material transactions to achieve accuracy. Annual stock taking and its reconciliation is a critical exercise for a textile manufacturing company as it represents a good amount of data and gives a fair idea about material waste, fabric shrinkage, length and weight variations, etc.

Example:

STOCK RECONCILIATION FROM FIBER TO FINISHED FABRIC OF A COMPOSITE MILL

(Product-Worsted Suitings)

STOCK RECONCILIATION FROM RM TO FINISHED PRODUCTS FOR A FINANCIALS YEAR (Kgs on Net Weight)		
Dye House	**Kgs**	**Waste %**
Opening	13,447	
Received from RM Godown	8,96,909	
Total	9,10,356	
Delivered to Recombing	9,00,321	
Closing stock	1,630	
Waste Visible	3,052	0.34%
Waste Invisible	5,353	0.59%
Balance	**0**	
Dye house Waste %		**0.92%**
Recombing	**Kgs**	
Opening	11,468	
Received from Dye house	9,00,321	
Received from RM Godown	3,58,967	
Total	**12,70,756**	
Delivered to Spinning	**12,43,542**	
Closing stock	3,364	
Waste Visible	27,347	2.16%
Waste Invisible	(3,495)	-0.28%
Balance	-2	excess
Recombing Waste %		1.88%
Spinning	**Kgs**	
Opening	16,796	
Received from Recombing	12,43,542	
Total	**12,60,338**	
Delivered to Post Spinning	**11,95,420**	
Delivered to export yarn store	3,675	
Closing stock	11,965	
Waste Visible	42,488	3.40%
Waste Invisible	6,790	0.54%
Balance	**0**	
Spinning Waste %		**3.95%**
Post Spinning	**Kgs**	
Opening	9,807	
Received from Spinning	11,95,420	
Received from export yarn store	6,452	
Total	**12,11,679**	
Delivered to double yarn store	**11,97,009**	

Delivered to export yarn store	**6,517**	
Closing stock	6,466	
Waste Visible	1,799	0.15%
Waste Invisible	(30)	0.00%
Balance	(82)	short
Spinning Waste %		0.15%
Double Yarn Store	**Kgs**	
Opening	22,517	
Received from Spinning	11,97,009	
Total	**12,19,526**	
Delivered to Weaving	**11,20,953**	
Delivered to export yarn store	**9,838**	
Yarn sold	74,056	
Closing stock	14,684	
Waste Visible	-	0.00%
Waste Invisible	-	0.00%
Balance	**(5)**	**excess**
Spinning Waste %		**0.00%**
Weaving	**Kgs**	
Opening	35,269	
Received from double yarn Store	11,20,953	
Received from export yarn store	2,00,856	
Total	**13,57,078**	
Delivered to Grey Perch	**12,15,993**	
Returned to export yarn store	**51,010**	
Lead Cloth Delivered to Finishing for their usage in m/cs	**839**	
Closing stock	34,120	
Waste Visible	56,676	4.28%
Waste Invisible	6,134	0.46%
Balance	**(7,694)**	**excess**
Weaving Waste %		**4.75%**
Finishing (By weight)	**Kgs**	
Opening	35,738	
Received from Wvg/Grey Perch	12,15,993	
Total	**12,51,731**	
Delivered to warehouse-Fabric	**11,50,406**	
Delivered to warehouse-Remnant	**8,565**	
Closing stock	40,925	
Other Wastes (Quality control /Finishing/Inspection/Designing)	12,167	1.00%
Finishing Weight Loss	39,691	3.28%
Balance	**(23)**	
Finishing Loss %		**4.28%**
Finishing (By length)	**Mtrs**	
Opening	1,01,815	
Received from Wvg	34,50,381	
Total	**35,52,196**	
Delivered to warehouse-Fabric	**33,00,294**	**96.32%**

Delivered to warehouse-Remnant	**28,549**	**0.83%**
Closing stock	1,25,769	
Shrinkage	97204	2.84%
Balance	**380**	**Short**
Finishing Shrinkage %		**2.84%**
Total Waste & Waste % upto Weaving (Visible)	1,31,362	10.33%
Total Waste & Waste % upto Weaving (Invisible)	14,752	1.32%
Total Waste & % (both visible & invisible) upto Finishing	1,46,114	12.65%
Net Weight of Wool/cashmere/Flex Issued to Plant	7,49,069	
Conditioned Weight of Wool/cashmere/Flex Issued to Plant	7,73,448	
Difference between Net & Conditioned Weight (In case of wool)		3.15%
	OR	
Net Weight of Wool /Cash/Flex/ Polyester/Viscose Issued to Plant	12,55,876	
Conditioned Weight of Wool & Polyester Issued to Plant	12,81,290	
Difference between Net & Conditioned Weight (In case of actual blend)		1.98%
Finishing Weight Loss		
Total Weight of the Fabric transferred to warehouse (% of grey fabric used)	11,50,466	94.90%
Total Weight of the Rmnt transferred to warehouse	8,565	0.71%
Other Wastes (Quality control /Finishing/Inspection/Designing)	12,167	1.00%
Total	11,71,198	96.61%
Total Weight of Grey Fabric used for total weight of Fabric/Rmnt Transferred	12,12,356	3.39%
To warehouse	**OR**	
Finished Fabric & Shrinkage		
Fabric transferred to warehouse in Mtrs (% of grey fabric used)	33,00,294	96.33%
Fabric-Rmnt transferred to W/h in Mtrs	28,549	0.83%
Total	33,28,843	97.16%
Grey Fabric Mtrs used or Fabric transferred to ware house	34,26,047	
Shrinkage	97,204	2.84%

Note- Raw material godown stock reconciliation on condition weight

Kg	Opening Stock	Recd.	Issue	Return	Closing
Wool & Others (Cndt)	66,773	8,30,087	7,73,448	3,647	1,19,765
Polyester/Viscose	1,15,144	4,57,538	5,07,842	-	64,840
Yarn	21,371	1,69,856	1,49,565	1,292	40,370
Total	2,03,288	14,57,481	14,30,855	4,939	2,24,975

Note- Raw material godown stock reconciliation on net weight

Wool & Others (Net wt)	65,164	8,03,683	7,49,069	3,564	1,16,214
Polyester/Viscose	1,15,119	4,57,538	5,07,822	-	64,835
Yarn	21,371	1,69,856	1,49,565	1,292	40,370
Total	2,01,654	14,31,077	14,06,456	4,856	2,21,419

Note- Weight loss depends upon the type of qualities produced (fine, medium, and coarse) which happens due to fiber loss in Scouring, Singing, Shearing and Milling, etc. The difference in weight may also be on account of variation in moisture content between raw material and finished product.

FACTORS AFFECTING PRODUCTION AND LEAD TIME

Examples:

(1) DYEING: Excess load of production and no. of shades on small size machine.

Production program on small size dyeing machines increased by 17%, however it is decreased by 8% on bigger size dyeing machines.

Below is the comparative study of 2012 (Apr. to Sep.) with 2013 (Apr. to Sep.) & (Oct to Dec)

Dyeing Machine Capacity	400 kgs & above	100 to 400 kgs	Below 100 kgs	Total
Month	No Of Shades	No Of Shades	No Of Shades	
Apr.12 to Sep.12	407	976	908	2291
% age	17.77	42.60	39.63	
Apr.13 to Sep.13	237	731	1253	2221
% age	10.67	32.91	56.42	
Oct.13 to Dec.13	38	183	368	589
% age	6.45	31.07	62.48	

Dyeing program depends on market requirements. In the fashion industry, there is a continuous trend of small lot sizes.

Under-utilisation of dyeing machines, not only causes higher chemical and energy costs but also delays delivery as there will be queuing for dyeing in smaller size of machines.

(2) SPINNING: More no. of smaller lot size & higher quantity of finer count of yarn requirement.

No. of small lots increased as shown in above dyeing report, yarn count is also becoming finer by 17% in comparison to previous year. This will reduce total spinning production and will cause late delivery.

Example:

Month	Kgs	Avg. Count Nm
Apr.11 to Sep.11	444894	40.9
Apr.12 to Sep.12	450432	40.6
Apr.13 to Sep.13	328372	47.1
Oct.13 to Dec.13	124508	46.3

(3) Downward trend in smaller lot size of fabric sales order causes smaller beam length in weaving for domestic market.

	S- 12	W-12	S-13	W-13	S-14	W-14
Domestic	Mtrs.	Mtrs.	Mtrs.	Mtrs.	Mtrs.	Mtrs.
Order Recd	708400	1000325	517050	876470	307983	613762
Avg Warp length	543	559	548	436	448	353

*S-Summer season, W-Winter season

Small size of beams in warping and weaving means more stoppages, changeover time, and higher cost of production.

Sales team of different regions should improve coordination amongst themselves and try to issue the manufacturing order in bulk instead of piece meal so that the PPC team may club shades and qualities together before releasing programs for production. Close coordination between PPC and sales will help in increasing the lot size in dyeing and spinning. It will also help in increasing the warping and weaver's beam length. It will increase productivity and reduce the lead-time, waste generation and overall cost.

Looking at the present situation and trend of smaller lots' order especially in fancy materials, management may review and decide to add smaller size of dyeing machines so as to increase productivity & reduce the lead-time and cost of dyeing.

Following table shows examples of sales orders released to PPC on different dates. Quality nos. may be different but blend or mixing no. is same (same group). If these orders are clubbed together, lot size for dyeing, spinning and weaving will increase. At times, sales orders are not enough to fill the production capacity. In such situations, the PPC department works under pressure, not being in a position to hold the orders, club orders and release small orders for production instead of allowing the machines to be idle. It is important for PPC to have sufficient sales orders to plan in the best way possible, utilise optimum plant capacity, and minimise the cost of manufacturing.

Example: In the following table, different qualities having same blends are issued on different dates by the sales team to PPC. This does not represent a good example of coordination among sales representatives from different areas. The orders are received by different sales teams from different regions, may be for the same blend group or quality. It is good to have a Planning In-charge in the sales team who coordinates with all the sales representatives, receives their orders, clubs them according to blend/quality and issues to PPC for better planning.

Blends Group	Quality No	Order Date	Fabric Mtrs	Blends Group	Quality No	Order Date	Fabric Mtrs
A	105103	04-Mar	1500	D	555526	10-Mar	600
	105241	10-Mar	300		555527	16-Mar	600
		16-Mar	900		555528	17-Mar	600
	105244	04-Apr	1800		555529	20-Mar	600
	105250	10-Apr	1200		555530	16-Apr	600
	105251	15-Apr	900		555531	12-Mar	600
	105252	16-Apr	900		555532	14-Mar	600
	105253	20-Apr	900		555533	10-Mar	600
	105259	01-May	900		555534	18-Mar	600
	105260	25-Apr	900		555535	20-Mar	600
	105261	30-Apr	1200		555536	16-April	600
	105262	04-May	900	Total			12600
Total			12300	E	880084	20-Apr	900
B	203039	16-Mar	2000		880085	22-Apr	900
	206245	20-Mar	3300		880087	24-Apr	900
		27-Mar	600		880088	26-Apr	1200
		23-Apr	300		880091	04-May	900
	206254	16-Apr	2800		880092	14-May	900
		25-Apr	900		880094	16-may	1200
		28-Apr	1500	Total			6900
	206255	16-Mar	900	F	973187	10-Mar	400
		04-Apr	900		973210	16-March	200
	206320	04-May	900		973214	18-Mar	400
	206338	15-May	900		973227	20-Mar	1500
		26-May	600		973228	26-Mar	900
	206339	04-Apr	900			09-Apr	400
	206340	06-Apr	900		973229	10-Mar	900
	206341	14-Apr	900		973230	10-Mar	900
	206342	17-Apr	900		973231	10-Mar	900
	206343	04-May	1200			03-Apr	800
	206344	14-May	1200		973232	16-April	900
Total			21600		973233	10-May	1200
C	1900	25-Mar	2000		973234	15-May	1200
	1900	04-Apr	5200		973235	10-Mar	900
	1900	09-Apr	400		973236	16-Apr	900
	1900	17-Apr	800		973237	19-Mar	900
Total			8400	Total			13300

7

Management Information System (MIS)

An organisation is a group of different people working in different sections or functions, who follow the policies of the organisation to achieve its target. During each of these functions, various kinds of critical information is generated which is then used by other sections or the management. Management information system, or MIS, broadly refers to a computer-based system that provides managers with the tools to organise, evaluate and efficiently manage departments within an organisation.

It gathers all information, reports, or data from multiple online systems related to different functions, scrutinizes their accuracy, analyses them and provides a report to the concerned department heads and management in a proper format for appropriate actions and decision-making. Organisations use information systems at all levels of operation to collect, process, and store data. Management aggregates and disseminates this data in the form of information needed to carry out the daily operations of business.

In a manufacturing company, commonly, reports are generated daily, weekly, fortnightly, monthly, quarterly, and annually. Daily and weekly reports are essential to assess the performance and take timely corrective measures to achieve targets. Historical data stored by MIS is used to compare with present data, set the norms, and decide future targets.

MANAGEMENT REPORTS

Reports are generated depending on size of manufacturing organisations and requirement of management. Daily, monthly, quarterly and annual reports are prepared by MIS for use by concerned departments and management to compare with targets, identify reasons for any variance and take timely corrective actions to avoid slippages.

FINANCE:

- Bank position
- Cash Flow
- Profit and Loss (P&L) Statement
- Monthly performance report
- Department-wise expense details
- Trial Balance, Balance Sheet
- Audit Reports
- Related statutory compliance, etc.

COMPANY SECRETARY:

- Audit Committee Reports
- Nomination and Remuneration Committee (NRC)
- Stakeholders' Relationship Committee (SRC)

- Corporate Social Responsibility (CSR) Committee
- Risk Management Committee (RMC)
- Presentation for board meetings and AGM, etc.

SALES:

- Sales order booking – quality, customer, agent and area-wise details
- Customer wise sales and revenue collection, outstanding status
- Stock verification report at customers' end
- Agent-wise sales and revenue collection
- Area-wise sales and revenue collection/bad debt, if any
- Performance of area sales managers
- Booking details of area sales managers
- Quality wise sales detail
- Details of sales promotion, conferences, places, investment, booking during event, record of dealers attending conference, overall feedback of trades
- Details of discounted sales
- Details of customer complaints and claims sanctioned
- Details of goods return, reasons, value, its verification, etc., and analysis report
- Performance of dealers/wholesellers/retailors
- Performance reports of MBOs/EBOs, etc.

HRD AND ADMINISTRATION:

- New recruitment and turn out detail
- Training and development program, annual calendar and expenses
- Department wise manpower detail
- Daily attendance report, reconciliation of workers, department wise over time report, detail of disciplinary action, late comers and trends etc.
- Wages and salary detail, performance incentive, bonus, PF, gratuity monthly allocation, etc.
- Workforce and productivity ratio
- Statutory compliance
- Expenses details of company's transport, vehicles, etc.
- Cost savings, etc.

PRODUCTION REPORTS
RAW MATERIAL PROCUREMENT:

- Raw material booking status- in-transit, opening stock, received, issue and closing stock

- Budget *vs* actual stock, age wise inventory
- Tracking report of raw material booked
- Details of raw material received - Weight of individual bale, cartons, etc. and verification of the gross and net weight
- Records of moisture, oil content - actual *vs* declared by supplier and variance
- Raw material price trends and history
- Daily stock movement
- Cost saving/suggestions

DYEING:

- Daily production report
- Individual machine capacity, production, utilisation and reasons for stoppages
- Monthly dyes & chemicals consumption – quantity, price and total amount
- New recipe developed and cost saving
- Dyeing cost of dyes and chemicals and energy – Top/Fiber, fabric and yarn
- Details of redyeing – No. of lots, quantity and reasons
- Opening stock of material in department, received, delivered and closing stock.
- Waste percentage
- No. of new shade developed /strike rates
- No. of shade adjustments, topping, etc.
- Detail of break down, stoppages and spare parts consumption, etc.
- Cost saving, etc.

SPINNING:

- Individual machine utilisation, operating efficiency and installed efficiency
- Machine-wise, section-wise production and efficiency
- Ring Frame production, efficiency, and yarn delivery
- Section wise stock position – opening, received, delivered and closing stock
- Total number of lots run
- Details of lot size (0 – 25 kgs, 26-50 kgs, 51-100 kgs, 100-200 kgs and above 200 kgs)
- Total quantity and number of lots
- Machine wise, section wise waste detail
- Details of machine, preventive maintenance, break downs and spare parts consumption
- Opening stock, received, delivered and closing stock
- Cost saving, etc.

WEAVING:

- Weaving preparatory reports - individual machine utilisation, operating efficiency and installed efficiency
- Individual machines utilisation, operating efficiency and installed efficiency
- Reasons for machines stoppages and loss of efficiency or specific reason
- Weaving Machines reports – Machine wise, quality wise, operator wise production, efficiency and utilisation
- Machine-wise, Quality-wise breakages report
- Average beam length, Nos. of article change and stoppages details
- Average time for beam gaiting, knotting, Article changes
- Details of machine preventive maintenance, break downs and spare parts consumption
- Opening stock, received, delivered and closing stock
- Cost saving, etc.

FINISHING:

- Individual machine utilisation, operating efficiency and installed efficiency , reasons for specific stoppages and loss of efficiency
- Fabric delivery to folding, rework details – meters and percentage
- Stock position - opening, received, delivered and closing stock
- Details of preventive maintenance and break down details, spare parts and energy consumption
- Quality wise shrinkage
- Chemical consumption
- Detail of rework with reasons
- Cost saving, etc.

FOLDING:

- Daily production report
- Opening stock, received, delivered and closing stock
- Details of inspection report with department-wise, reason-wise cuts and defective percentage
- Outside production – total received and delivered to warehouse
- Details of packing cost
- Cost saving, etc.

WAREHOUSE:

- Quality wise opening stock, received, dispatch and closing stock
- Age wise stock
- Customer wise dispatch
- Order wise requirement *vs* dispatch date and variance

- Details of assorted and un-assorted dispatches
- Season-wise stock summary
- Details of sound and seconds materials dispatch
- Goods return summary
- Cost saving, etc.

PRODUCTION PLANNING AND CONTROL:

- Monthly production plan
- Raw material booking and consumption detail
- Raw material Stock movement – Godown opening stock, received, issued and closing stock
- Departments WIP status – Comparative reports of norms *vs* actual
- Department stock reconciliation – raw material to finished goods and waste
- Sales order status – Received, delivered to warehouse and balance to deliver
- Sales order, location wise movement status from RM godown to warehouse
- Department wise lead-time of completion of each assorted order
- Total lead time for individual orders
- Department wise rework
- Historical data of sales and production
- Cost saving, etc.

QUALITY ASSURANCE:

- Test reports of fibre, yarn, fabric, outsourced yarn and fabric and relative impact on the end products
- Test reports of dyes, chemicals, lubricants, packing materials, etc.
- Test report of moisture content, oil content, gross and net weight of raw materials received in godown
- Test report of moisture content of material at various stage and moisture difference between raw material and finished goods, if any
- Cuts & Defectives reports (department-wise, defect-wise analysis)
- Department wise audit reports of process controls
- Lot wise single yarn and double yarn quality confirmation report before delivery to yarn store.
- End breakages in winding, warping weaving and Ring frames
- Frequency of faults in weaving, loom shed defects – no. of times, looms – no. of times
- Count and blend wise defect details
- Historical data related to quality
- Department wise rework
- Cost saving, etc.

PROCUREMENT AND STORES:

- Department wise purchase and consumption details - Budget *vs* actual and variance
- Procurement of Capex, selection process, price and quality (viability) comparative charts etc
- Vendor-wise quality approval and rejection
- Vendor-wise purchase
- Price of individual item, history and cost saving measures
- Lead time of individual order
- New vendor development
- Development of import substitute
- Opening outstanding of suppliers, amount of purchase, payment made to vendors and closing outstanding amount
- Payment terms agreed as per policy (no. of days *vs* payment done (actual days
- Department-wise inventory details (opening stock, purchased, issued and closing stock
- Details of slow moving or non-moving inventory
- Indigenous development
- Cost savings, etc.

ENGINEERING:

- Consumption and stock reports of power and fuels, steam, air and water – Opening stock, received, consumption and closing stock
- Power factor, consumption of power, fuel, steam, air and water per unit of production
- Maintenance and breakdown reports of power house, transformers, panels, generator sets, air compressors, etc.
- Reports of plant breakdowns and scheduled maintenance – Electrical, Mechanical and Instrumentation
- Motor burning frequency (with serial no. of each motor in the plant, reasons and rewinding expenses
- Illumination in each section – Norms *vs* actual
- Noise level in each section – Norms *vs* actual
- Effluent treatment plant working report – Scheduled clearance certificate from state pollution control board
- Consumption of tube lights
- Preventive maintenance, breakdowns, spare parts consumption of boiler, generator sets, transformers panels, powerhouse, etc. and their approval certificates
- Fuel storage scheduled clearance certificates
- Scheduled energy audit report
- Cost saving, etc.

EXAMPLES OF FORMATS FOR COMMON REPORTS

MIS reports are important tools for each lower, middle, and higher management to study their daily performance, compare with pre-set targets, and visualise reasons for improvement. For example, a manufacturing department can see machine-wise and quality-wise daily production report, and various reasons for breakdowns and loss of production. Similarly, the sales department can see the quality wise, customer-wise, area-wise sales, collection trends and pass on the information to the production planning section to review the production plan. Account and finance department will prepare the profit and loss statement, cash flow, balance sheet based on the above data.

A few reports are provided below as guidelines:

DAILY PRODUCTION AND SALES REPORT

	Target			Actual		
	P/Month	P/day	To date	For day	To date	Avg/day
No. of Working days-Production						
No. of Working days-Sales & Collection						
Production Kg/Mtr						
Dyeing						
Fabric						
Fiber						
Yarn						
Wool Scouring						
Carding						
Combing						
Recombing						
Spinning						
Average RPM						
Average Count						
Average TPM						
Average efficiency						
Spinning Stock -WIP						
Yarn delivery Own						
Weaving						
Domestic						
Export						
Institutional						
Total						
Looms Average RPM/PPM						
Average Picks/Inch or /CM or / Per Meter						
Average Installed Efficiency						
No of Looms Run/Shift/day						
Mending Stock						
Finishing Production						
Domestic						
Export						
Institutional						
Total						
Finishing Stock						

SALES							
A- Quantity							
Domestic							
Export							
Institutional							
Total							
B- Value (Rs lacs)							
Domestic							
Export							
Institutional							
Total							
C- Average price							
Domestic							
Export							
Institutional							
Revenue collection (Rs lacs)							
Domestic							
Export							
Institutional							
Total Collections							

DAILY SALES, EXPENSES, AND TENTATIVE PROFIT & LOSS REPORT

	Sales Quantity Basis						Rs.in lacs	
	Budget			Actual				
PARTICULARS	For Month	Today	To date	To day	To date	Balance to achieve	To day	To date
SALES								
Domestic - A								
Domestic - B								
Export								
Other Income including DEPB, overdue charges								
TOTAL INCOME								
EXPENSES								
Raw material consumption								
Material cost								
MANUFACTURING EXPENSES								
Power Cost/ Utility cost								
Dyes & Chemicals								
Stores & spares								
Other Manufacturing Cost								
Wages								
Personnel Cost								
Administration cost								
Corporate Administration cost								
Selling Expenses								

Outside Weaving & Mending						
TOTAL OPERATING COST						
TOTAL COST						
GROSS MARGIN						
SALES VARIABLE COST						
Publicity Charges						
Bonus & Commission						
TOTAL SALES VARIABLE						
PBDIT						
Average Selling price						
Domestic - A						
Domestic- B						
Export-						
Qty in lacs mtrs						
Domestic - A						
Domestic - B						
Export-						
Total						

MONTHLY PERFORMANCE REPORT AND COMPARATIVE ANALYSIS OF A WEAVING & PROCESSING UNIT ALONG WITH A SMALL DIVISION OF PLASTIC ITEMS

A1	SALES QUANTITY		Month to date	Last Year	Year to date	% Sale Year	Last year	% Sale of
	A. TEXTILES	Quality Details	Mtrs	March		To date	April -March	Last year
		Ankara	804282	750143	7647323	54%	9499548	67%
		Crepeline	210		210	0%	58468	0%
		Base Cloth/Grey Fabric	18788	95	105050	1%	175463	1%
		Bed Sheet	5300		12931	0%		0%
		Foam Materials	203039	126909	1625463	11%	864178	6%
		Furnishing	30847	14560	248349	2%	293752	2%
		Shirting	311650	235005	3776876	27%	2625499	19%
		Suiting	25551	35905	786564	6%	665901	5%
		Total (Mtrs)	1399667	1162617	14202766	100%	14182809	100%
		Yarn			46151		3013	
		Rags & Fents(Bags)	1294	852	8180		6486	
		Job Work - Printing (Mtrs) (N.S.F)	20962	11683	116165		253462	
A2	B. SEWING THREAD/ PLASTICS							
		Sewing Thread (Bags)	3615		47253			
		Plastics (Dozs)	76469		125294			
		Vinyl Tube(V.T)(Cartons)	1418		4007			
		Vinyl Tube Rejection(Kgs)	27271		69883			
	SALES VALUE N=000							
B(1)	A. TEXTILES							
		Ankara	75830	71671	726035	51%	905930	69%

		Crepeline	13		13	0%	3201	0%
		Bed Sheet	435		1067	0%		0%
		Foam Materials	18633	11890	148255	10%	82349	6%
		Furnishing	2318	1121	19233	1%	22786	2%
		Grey Cloth	1211	13	7163	1%	17232	1%
		Shirting	18986	14691	234198	16%	165086	13%
		Suiting	2948	4302	93118	7%	76351	6%
		Rags & Fents (Begs)	6890	3992	41952	3%	30866	2%
		Job Work	363	1238	2961	0%	5691	0%
		Yarn			10586	1%	1143	0%
		Sub Total A (1)	127627	108918	1284581	90%	1310635	100%
B(2)	B. SEWING THREAD/ PLASTICS							
		Sewing Thread (Bags)	7024		100045	7%		0%
		Plastics (Dozs)	8668		20762	1%		0%
		Vinyl Tube (V.T) (Ctn)	8103		22897	2%		0%
		Vinyl Tube Rejection (Kgs)	1463		3438	0%		
		Sub Total B (1)	25258		147142	10%		0%
		GRAND TOTAL (A1+B1)	152885	108918	1431723	100%	1310635	100%
C	COLLECTIONS=N='000							
C(1)	Banks							
		Cotton loan/Term Loan			119675			
		B/A & import financ	21166		195366			
		Unutilised Forex			26111			
		Sub total	21166		341152			
C(2)	Group							
		Export proceeds						
		Group Company	7600		66738			
		Sub-total			66738			
C(3)	Textile							
		Debtors - Fabric	113315		1148568			
		Others (Cash & waste yarn)	7473		71668			
		Sub-total	120788		1220236			
C(4)	Sewing thread/Plastic							
		Debtors - Sewing thread	8101		94456			
		Debtors - Plastics	9723		21192			
		Debtors - Vinyl Tube	8266		21406			
		Sub-total	26090		137054			
	Insurance	Insurance claim received	21152		66860			
		Grand total of Collection	189196		1832040			
D	PAYMENTS=N='000							
		Forex/LC	113248		420837			
		Yarn Supplier	20011		88536			
		Other Group co	1422		80732			
		Grey/Yarn	35930		314469			
		Chemicals	10538		145445			
		Power & fuel	12513		124865			

		Salaries & wages	11037		113278			
		Finance cost	2466		32407			
		Loans and advances	4560		25177			
		Other Expenses	6171		64350			
		Insurance			18113			
		Engineering	3678		74321			
		Dividend			2962			
		Overseas Travelling	105		2869			
		Duties & Taxes	3009		37760			
		Prepayments	6145		95484			
		Gratuity	437		11856			
		Loan Repayment	1800		162380			
		TOTAL	233070		1815841			
F	GREY PURCHASED (MTRS)							
		Ankara	295377	207451	812139		1016746	
		Shirting			702243			
		Suiting			74645			
		Total mtrs	295377	207451	1589027			
E	PRODUCTION MTRS							
		Weaving : Sulzer	1100245	1263527	14211665		13859453	
		Processing	1462785	1262107	15657631		14715717	
		Printing	1140462	1070314	10880198		11304585	
		Job work -Group Co		20956	19412		261510	
		Folding		1197759	11358712		14091263	
		Plastics (kgs)	47722		166880			
		Sewing thread (Kgs)	23708		235980			
		Vinyl Tube (Ctns)	2345		3880			
		Vinyl Tube (Kgs)	75354		117850			
G	FUEL & OIL CONSUMED (LTRS)							
		LPFO	502584	429302	5246021			
		A.G.O	51465	19168	328520			
		Kerosene	51430	22640	397363			
					Qty.	**Rate**	**Amount**	**NO. OF**
H	STOCK: METERS							**MONTHS**
		Weaving: Grey (mtrs)			337716	54	18327847	
		Processing (mtrs)			419035	74	30912212	
		Folding (mtrs)			339472	78	26448301	
		Finished Goods (mtrs)			775871	77	59563579	
		Total			1872094		135251939	
		Yarn (Kg) - In godown			275516	238	65572808	
		In process			272143	240	65314320	
		Sewing thread			71449	315	22506435	
		Plastics(kgs)			251525			
		PP Kg			307875	110	27667750	
		PVC Kg			125400	95	29248125	
		DOP Kg			14420	122	15298800	
		Chemicals Kg				500	7210000	
		Total for plastic and sewing thread					101931110	

		TOTAL STOCK					368070177	
I	DEBTORS							
	I(1)	Textiles=N000						
		Fabrics-Current			75829			
		Non current			28367			
		Discount			6000			
		Yarn						
		Total			110196			
	I(2)	Bilplast/Sewing thread						
		Sewing thread (Kgs)			18987			
		Vinyl Tube (Ctns)						
		Vinyl Tube (Kgs)			1616			
		Total			20603			
J	SUNDRY CREDITORS O/S=N='000							
		Chemicals			40924			
		Grey/yarn			39214			
		Grey-1			42598			
		Power&Fuel			8456			
		Foreign			99095			
		Others						
		Total			230207			
J	INTRA GROUP O/S (N'000)							
		Total			-245610			
K	TOTAL BANK EXPOSURE (N'000)							
		Overdraft			-146225			
		Cotton Loan			-5400			
		Lease finance			-19632			
		Term Loan			-41444			
		Foreign Loan			-48420			
		Banker's Acceptance (Regular)			25000			
		Banker's Acceptance (Import Fin.)			-70268			
		Total			-356389			
L	FOREIGN BILLS O/S							
		Due			2380549			
		Not due			2723439			
		Total			5111988			
M	LETTER OF CREDIT							
		Confirmed						
		Unconfirmed						
		TOTAL						

CASH FLOW FORMAT OF A COMPOSITE TEXTILE MILL

	Total	Total	Due	Projected	Deferred	Net	Actual
	Outstanding	Overdue	In July-12	In July-12		Projection	Recvd As on
	(Rs. In Lakhs)					July-12	30.07.2012
Bank Balance							
INFLOW							

Sales Collections							
Domestic							
Export							
Refund/DEPB							
Waste/scrap sales and others							
Insurance claim							
Working Capital Infusion							
Total inflow							
OUTFLOW							
ERP							
Computers							
Outflow for Publicity/Capital expenses etc. Publicity Charges / Conference General Contribution Deposit/Normal Capex Management Charges /TDS/Service Tax etc. Entry Tax Payable							
Payment for Raw Materials Raw Materials Speciality Fibers Outsourced Grey Yarn Outsourced Dyed Yarn Payment for Polyester Specialty Yarn							
Manufacturing expenses Dyes/Chemicals, etc. Stores/spares, consumables, lubricants etc. Stores / spares under-CAD, Custom duty Electricity Charges Furnace Oil/Diesel /Lub oil Water charges Freight & Clg. charges Packing materials							
Administrative Expenses Salaries/ Statutory Liabilities Wages Ex-gratia/LTA Staff Settlement Transport hire charges House Keeping/ Others Drivers Salary Security & Canteen Legal/Professional fees Legal/Professional fees Corporate Audit fee Printing / stationery etc Telephones House rent Vehicle running /maintenance Traveling expenses, etc. Subscription etc. Rates, Local Taxes, etc. General Expenses							
Sales & Sale variable expenses Advertisements, Sign board, design books, sales promotion expenses, etc. Retail shop Rent							

Agency Commission for Domestic / TDS Directors Travelling/Others Agency Commission for Export/service tax/ texworld							
Insurance for Personal Accident/Workmen Compensation							
Outflow (a) Plant Operation							
Payment for documents and clearing charges							
Custom duty on fabric							
Yarn outsourced Godown Rent & Other Expenses Payment for Outsourced fabrics Weaving/Processing Charges for Sampling							
Total Outflow							
Closing Balance							

BUDGET ANALYSIS FOR A FINANCIAL YEAR (APRIL TO MARCH)

(A weaving unit, Processing on job work)

Financial Highlights:

Particulars	Budget Apr 95 to Mar 96		Estimated Actual Apr 94 to Mar 95		Growth Apr 95 to Mar 96	
	-N-M	% to sales	-N-M	% to sales	-N-M	% to Acts
A-TOTAL NET SALES	1,156.37	100	734.889	100	421.476	57.4
B-Raw Materials	578.56	50	461.219	62.8	117.341	25.4
C-Direct Expense	136.122	11.8	75.556	10.3	60.566	80.2
D-Stock Adjustment			25.29	3.4	-25.29	-100
E-Cost Of Sales (B+C+D)	714.682	61.8	562.065	76.5	152.617	27.2
F-Gross Profit (A-E)	441.683	38.2	172.824	23.5	268.859	155.6
G-Other Income	0	0	0	0	0	
H-Admin/Sales Expenses	172.801	14.9	137.033	18.6	35.768	26.1
I-PBDIT (A-E+H)	268.882	23.3	35.791	4.9	233.091	651.3
J-Finance charges Local	20.437	1.8	17.082	2.3	3.355	19.6
K-Finance charges Overseas		0	281.983	38.4	-281.983	-100
I-Total Finance Charges	20.437	1.8	299.065	40.7	-278.628	-93.2
M-Cash Profit (Before Depreciation) (I-L)	248.445	21.5	-263.274	-35.8	511.719	
N-Depreciation	111.431	9.6	101.878	13.9	9.553	9.4
O-PBT (M-N)	137.014	11.8	-365.152	-49.7	502.166	
Break Even Levels						
Break Even Sales	797.65					
Budgeted Sales	1156.365					
Planned Achievement	145.00%					
Quantitative Details						

	Budgeted	Estimated Actuals	Growth	%	
No. of Working Days	300	263	-37	14.1	
Production M Mtrs	10.83	6.892	-3.94	57.1	
SALES					

Fabric M Mtrs	8.959	7.802	-1.15		14.8
=N=	780.656	608.455	-172.20		28.3
Rate/Mtr	87.14	77.99	9.15		11.7
Fabric Govt. supply M Mtrs	0.732	0.314	0.42		133
=N=M	305.25	116.23	189.02		162.6
Rate	417.1	370	47.09		12.7
Fabric TOTAL M Mtrs	9.691	8.116	1.57		19.4
=N=M	1085.905	724.685	361.22		49.8
Rate	112.05	89.29	22.76		25.5

*N- Currency in Naira, M- Million

COMMENTS ON ACTUAL PERFORMANCE FOR LAST FINANCIAL YEAR (1994-95)

The performance of the unit during the year was poor. Against the Budgeted Turnover of N 1,344M the unit could achieve a turnover of only N 735M, an achievement of only 55% - that indeed is a poor performance.

The unit incurred a loss of (N 365M) against the budgeted profit of N 15M.

The unit also just managed to achieve only 70% of the budgeted production level.

COMMENTS ON ANNUAL BUDGET (1995-96)

1. The unit has budgeted a sale of N 1,156M as against the estimated actual of N 735M a growth of 57%, which seems to be too optimistic looking at the past performance of the unit.

2. The unit has budgeted a Gross Profit of N 442M i.e. 38.2% of sales as against the Actual Gross Profit of N 173M i.e. 23.5% of sales. The Gross Profit further analysed between Uniform and Government Orders and Others is as under:

Particulars	Uniform material	Govt. Orders	Others	TOTAL
Sales.	147.727	157.440	851.198	1156.365
RM+Direct Cost	69.013	50.346	595.323	714.682
Gross Profi	78.714	107.094	255.875	441.683
Gross Profit %	53.3%	68.0%	30.1%	38.2%

From the above it can be seen that 42% of the Gross Profit is from Uniform and Govt. Orders and as such the success of the unit depends on procurement of orders from them.

3. The unit has budgeted Cash Profit at N 248M as against the Estimated Actuals Cash Loss of (N 263M). This is mainly because of increased sales, change in the product mix & waiver of interest on overseas loan.

4. The unit has budgeted a Net Profit of N 137M as against the Estimated Loss of (N 365M).

5. The unit has budgeted a Capital Expenditure of N 58.750M.

6. The unit has projected a growth of N 36M (26.1%) in Overhead Expenses and a rise of N 60M i.e.53% in Direct Expenses. This rise is mainly in the case of Management Fees to F&C, Sales promotion expenses, Sales Commission & Discounts that are directly related to increase in Turnover.

   The increase in the Direct Expenses is mainly due to increase in Repairs & Maintenance expenses by N 26M, AGO consumption by N 20M, and Wages by N 8M.

7. The unit has not provided any Manpower Budget nor any planned reduction in work force.

8. The unit has budgeted Overseas Liability at 31.03.96 at N 2078M i.e. USD 24.738M. The unit has budgeted a repayment of N 180M i.e. USD 2.142M to Overseas Creditors.

9. The unit has not given the purchase budget.

LIQUIDITY ASSETS, NET WORKING CAPITAL AND NET WORTH

PARTICULARS	-N-M	LEVEL IN MONTH		COMMENTS
		ACTLS.	NORM	
Inventory: R.M.	80.328	1.67	1	
W.I.P.	33.92	0.7	0.5	High
F.G.	101.851	1.06	1	Okay
Others				
Sundry Debtors	313.532	3.25	1	Very High
Cash and Bank Balance	0.576			
Other Current Assets	4.7111			
Total Current Assets	534.918			
Less Current Liabilities :				
Trade Creditors	261.26			
Bank Borrowings	77.886			
Other Liabilities	23.412			
Total Current Liabilities	362.558			
Net working Capital	172.36			
NET FIXED ASSETS	1592.035			
TOTAL TENGIBLE ASSETS	2126.953			
Dues to Overseas Crs.	2078.134			
Other Outside Liabilities	362.558			Same as Current Liabilities
Total Outside Liabilities	2440.692			
NET WORTH(31.03.98)	-1242.451			
Share Capital	54.788			
Reserves & Surplus	1430.253			
P & L : Previous Year	-1935.794			
: Current Year	137.014			
NET WORTH (31.03.96)	-313.739			

Ratio Analysis

PARTICULARS	RATIO		DEFINITION OF RATIO	COMMENTS
	BUDGET	NORM		
Profitability Ratios :				
Gross Profit to Sales-%	38.2	40		Profitabilit
Cash Profit (before Depreciation) -%	21.5	20		Below
Net Profit to Sales-%	11.8	15		Norms
Liquid Ratios :				
Current Ratio	1.48	2	Measure of 'Margin of safety' to cover Current Liabilities.	Poor
(Current Assets/Current Liabilities)			Liquidity position of the unit to cover Current Liabilities.	Poor
Liquid Ratio	0.88	1	Assets coverage to Total Outside Liabilities (incl. Long Term Borrowing)	Poor
(Liquid Assets/Liquid Liabilities)				

Solvency Ratio	0.87	1.25		
(Tangible Assets/Total Liabilities)				
Efficiency Ratios :				
Inventory Turnover Ratio	5.35	4.8	No. of times the inventory (RM, WIP & FG) is turned over in a year.	Good
(Sales Turnover/Total Inventory)				
Working capital Turnover Ratio	2.16	3.4	No. of times the Current Assets is turned over in a year.	Poor
(Sales Turnover/Current Assets)				
Total Assets Turnover Ratio	0.54	2	No. of times the Total Assets is turned over in a year.	Very Poor
(Sales Turnover/Total Assets)				

ADMINISTRATIVE OVERHEADS

Annexure 1.

Particulars	Budgeted	Est. Actuals	Growth	%
Salaries	36.06	26.939	9.1121	33.9
Vehicle Maintenance	4.32	3.412	0.908	26.6
Communication	2.4	1.788	0.612	34.2
Printing & Stationery	2.1	1.709	0.391	22.9
Travel	4.044	3.416	0.628	18.4
Professional Expenses	18.372	19.971	-1.599	-8
Personal relations (PR)	18.6	20.055	-1.455	-7.3
Rents Rates & Taxes	1.536	1.138	0.398	35
Repairs & Maintenance	0.96	0.852	0.108	12.7
Insurance Sales Promotion	4.428	3.91	0.518	13.2
Sales Promotion	6	2.017	3.983	197.5
Management Fees	34.691	28.521	6.17	21.6
Turn over commission/Discounts	39.29	23.305	15.985	68.6
Total	172.801	137.033	35.768	26.1
Details of Direct Expenses				
Wages	30.984	22.703	8.281	36.5
Over time (OT)	8.72	7.114	1.606	22.6
Others	2.496	2.201	0.295	13.4
Terminal Benefit	6.864	5.645	1.219	21.6
Bonus	2.862	2.277	0.585	25.7
Utilities	6.18	5.281	0.899	17
Repair & Maintenance (R & M)	35.416	9.017	26.399	292.8
Carriage	2.16	1.838	0.322	17.5
Insurance	4.8	4.218	0.582	13.8
AGO	35.64	15.775	19.865	125.9
Total	136.122	76.069	60.053	78.9

ANNUAL PROFIT/LOSS AND BALANCE SHEET OF A COMPANY

(A medium size weaving and processing unit along with a section of plastic materials)

COMBINED EXPENSES STATEMENT FOR AND UPTO MARCH' 03

Value N, 000 (Naira)	FOR THE MONTH			TO DATE			Incr. from Prv.Yr.
PARTICULARS	Actual	Budget	Prv. Year	Actual	Budget	Prv. Year	
A. MANUFACTURING							%
FACTORY, WAGES & GRATUITY	14598	14598	11479	150502	150504	136309	10%
MAINT. MACHINE &LOCAL SPARE	9804	8810	10715	129060	95310	82618	56%
IMPORTED SPARES	2455	2985	2116	27540	28385	34324	-20%
SUB TOTAL	3255	3255	2927	33149	35805	22304	49%
B. ADMIN EXPENSES	30112	29648	27237	340251	310004	275555	23%
DIRECTOR'S REMUNERATION	15	15	15	180	165	180	0%
RENT & RATES	371	136	125	4113	1496	1604	156%
PRINT & STATIONERY	366	210	470	3010	2310	2548	18%
POST & TEL.	379	100	96	2514	1100	1254	100%
OVERSEAS TRAVELLING	347	347	347	4928	3817	4164	18%
LOCAL TRAVEL/LEAVE TRAVEL	130	130	150	1145	1430	2801	-59%
INSURANCE	2000	1500	1200	20300	16500	14400	41%
AUDIT	50	45	45	882	495	618	43%
MEDICAL	291	250	206	2839	2750	3579	-21%
HOUSE MAINTENANCE	117	60	179	1074	660	1162	-8%
LEGAL & PROF. FEES	225	24	0	1222	264	1053	16%
UPSTAMPING FEES	0	0	0	0	0	0	
STAFF WELFARE	564	300	455	4884	3300	4427	10%
RECREATION AND ENTERTAINMENT	738	138	141	353	1518	2353	-85%
SECURITY	225	180	366	2721	1980	2721	0%
TRANSPORTATION	0	0	0	0	0	0	
MOTOR MAINT.	226	100	149	2365	1100	1984	19%
MOTOR RUNNING	121	100	69	1024	1100	838	22%
GENERAL EXPENSES	74	150	284	2317	1650	2038	14%
CORPORATE EXPENSES	1290	600	1137	18355	6600	15076	22%
SECRETARIAL SERVICES	95	9	44	1372	99	564	
SEWING THREAD EXP	0	0	0	0	0	0	
SUB TOTAL	7624	4394	5478	75598	48334	63364	19%
C.FINANCE	6786	11730	3057	68879	109620	41923	64%
INTEREST& BANK CHARGES	6786	11730	3057	68879	109620	41923	64%
D. MARKETING EXPENSES							
CARRIAGE OUTWARDS		60	6	84	660	760	
ADVERT & SALES PROMOTION & EXPORT	1009	20	0	3287	220	0	
SUB TOTAL	1009	80	6	3371	880	760	
E.DEPRECIATION							
WORKS	4046	4046	2500	37784	38624	27493	37%
ADMINISTRATION	0	0	0	0	0	0	
SUB TOTAL	4046	4046	2500	37784	38624	27493	37%
F.SEWING THREAD	0	0	357	0	0	2350	
GRAND TOTAL	49577	49898	38635	525883	507462	411445	28%

*Incr-Increase, Prv-Previous, Yr-Year

COMBINED OPERATING STATEMENT FOR AND UPTO MARCH'03

PARTICULARS	FOR THE MONTH			TO DATE			Incr.fr.Prv.Yr.
	Actual	Budget	Pre. Year	Actual	Budget	Pre. Year	
A. PRODUCTION							
WEAVING ('OOO MTR)	1100	1492	1264	14212	16814	13859	3%
PROCESSING ('OOO MTR)	1463	0	1262	15658	0	14715	6%
PRINTING ('OOO MTR)	1140	0	1091	10899	0	11566	-6%
SEWING THREAD (KGS)	23708			235980			
VT -Cartons (Vinyal tubes)	2345			6122			
PLASTIC (KGS)	47722			166880			
B.SALES (QTY'000MTRS)							
FABRIC	1400	1800	1163	14203	22860	14183	0%
SEWING THREAD - Cartons	3615	9700	0	47253	321100	0	
PLASTIC - Dozen	76469	147680	0	125294	1033760	0	
VT Cartons	1418	6000		4007	42000		
C.SALES (VALUE) N= 000							
FABRIC	120053	153992	104412	1223922	1804662	1277060	-4%
RAGS & FENTS	6890	1200	3771	40497	13200	30607	32%
SEWING THREAD	7024	17130	9044	100045	197114	99370	1%
PLASTIC SALES	8334	20349	0	20425	142443	0	0%
VT SALES	8994	36000	0	25764	252000	0	0%
YARN/CHEMICALS SALES	0	0	1074	12046	0	2217	0%
EXPORT SALES	0	0	0	0	0	0	
TOTAL	151295	228671	118301	1422699	2409419	1409254	1%
D. REALISATION PER MTR (FABRIC) N=	85.75	85.55	89.78	86.17	78.94	90.04	-4%
E. GROSS CONTRIBUTION	62973	100776	57510	660795	994477	637612	4%
F. CONTRIBUTION OF SALES	43.61%	44.30%	50.21%	47.81%	41.50%	46.25%	3%
G. ANKARA DISCOUNT/T.O COMMN	7679	4620	2522	90781	54140	40979	122%
H.NET SALES	143616	224051	115779	1331918	2355279	1368275	-3%
I(a). NET CONTRIBUTION	55294	96156	54988	570014	940337	596633	-4%
% OF SALES - ON NET	38.50%	42.92%	47.49%	42.80%	39.92%	43.60%	-2%
ADD:							
I(b) OTHER INCOME							
OTHERS (INCL. JOB WORKS)	463	300	109	20364	3000	5517	269%
MISC. INCOME	0	0	1500	36522	0	11250	225%
I(c) NET CONTRIBUTION PLUS	55757	96456	56597	626900	943337	613400	2%
OTHER INCOME							
J.EXPENSES							
MANUFACTURING	30112	29648	27237	340251	310004	275555	23%
ADMINISTRATION	7624	4394	5478	75598	48334	63364	19%
FINANCE	6786	11730	3057	68879	109620	41923	64%
MARKETING	1009	80	6	3371	880	760	
DEPRECIATION	4046	4046	2500	37784	38624	27493	37%
SEWING THREAD	0	0	357	0	0	2350	0%
TOTAL	49577	49898	38635	525883	507462	411445	28%
K.OP. PROFIT/LOSS	6180	46558	17962	101017	435875	201955	-50%
L.LESS: GLOBE AGRO	-100	0	-100	-1200	0	-1200	

PARTICULARS						
MANAGEMENT EXPENSES (not considered above)	-3500	-3500	-2500	-42000	-31500	-30000
M.PROFIT/LOSS	2580	43058	15362	57817	404375	170755
Add: S. THREAD NET INCOME		0				
TOTAL PROFIT/LOSS	2580	43058	15362	57817	404375	170755

COMBINED CONTRIBUTION STATEMENT FOR AND UPTO MARCH'03

N 000	FOR THE MONTH			TO DATE			Incr.from Prv.Yr.
PARTICULARS	Actual	Budget	Actual (Prv)	Actual	Budget	Actual (Prv)	
A. SALES							
FABRIC	120053	153992	104412	1223922	1804662	1277060	-4%
RAGS & FENTS	6890	1200	3771	40497	13200	30607	32%
SEWING THREAD	7024	17130	9044	100045	197114	99370	1%
PLASTIC	8334	20349		20425	142443		
VT	8994	36000		25764	252000		
YARN/CHEMICAL SALES	0	0	1074	12046	0	2217	
OTHERS	0	0	0	0	0	0	
TOTAL SALES	151295	228671	118301	1422699	2409419	1409254	1%
LESS: ANKARA DISCOUNT	0		0	0	0	0	
LESS: DISCOUNT - OTHERS	239		8	1314	0	126	943%
LESS: DISC ON TO (PROV)	0		0	0	0	0	
LESS: DISC (ACTUAL)	7440	4620	2514	89467	54140	40853	
NET SALES	143616	224051	115779	1331918	2355279	1368275	-3%
B. COST OF SALES							
FIN. GOODS-OPEN STOCK	94310	0	28393	35842	0	19326	85%
ADD. COST OF PRODN.	86070	0	70911	818120	0	790829	3%
TOTAL SALES	180380	0	99304	853963	0	810155	5%
FIN. GOODS-CLOS. STOCK	92059	0	38513	92059	0	38513	139%
COST OF SALES	88322	127895	60791	761904	1414942	771642	-1%
C. CONTRIBUTION	55294	96156	54988	570014	940337	596633	-4%
D. COST OF PRODN.							
i) OPENING STOCK							
- RAW MATERIAL	257844	0	193141	210014	0	119597	76%
- WORK IN PROGRESS	96978	0	54761	67567	0	58395	16%
- SUB TOTAL	354822	0	247902	277581	0	177992	56%
ii) ADD:PURCHASES							
- IMPORTED YARN	23804		42867	266624	0	85410	212%
- IMPORTED CHEM	693		0	121309	0	4500	2596%
- LOCAL YARN	20323		32975	201449	0	595356	-66%
- LOCAL CHEM	5946		12915	166858	0	137555	21%
- LOCAL GREY CLOTH	22153		11825	109967	0	58282	89%
- LOCAL OTHERS	2298		933	18301	0	10240	79%
- SUB TOTAL	75217	0	101515	884508	0	891343	-1%
TOTAL	430039	0	349417	1162089	0	1069335	9%
iii) LESS CLOSING STOCK							
- RAW MATERIAL	251902	0	211056	251902	0	211056	19%
- WORK IN PROGRESS	92067	0	67450	92067	0	67450	36%
- SUB TOTAL	343969	0	278506	343969	0	278506	24%
iv) COST OF PRODN.	86070	0	70911	818120	0	790829	3%
v) COST OF PRODN.	86070	0	70911	818120	0	790829	3%

BALANCE SHEET

PARTICULARS– N, 000	31.3.03
A. FIXED ASSESS	
GROSS ASSETS	412162
LESS: LAST YR. DEPRECIATION	139816
OPENING BALANCE	272346
ADD/SOLD DURING THE YEAR	180705
TOTAL ASSETS	453051
LESS: CURRENT YR. DEPREC.	37784
NET FIXED ASSETS	415267
B. CURRENT ASSETS	
I) STOCKS - RAW MATERIAL	251902
- WORK IN PROGRESS	92067
- FINISHED GOODS	92059
- GOODS IN TRANSIT	0
- FUEL	7613
- LOOSE TOOLS/SPARES	33131
- SUB TOTAL	476772
II) DEBTORS (NET OF PROVISIONS FOR BAD-DEBTS)	90651
AMOUNT OWED BY GROUP CO -1	16909
AMOUNT OWED BY GROUP CO -2	0
AMOUNT OWED BY GROUP CO- 3	33143
AMOUNT OWED BY GROUP CO- 4	-152
PREPAYMENTS	12122
FEM L/C DEPOSITS	861
OTHER IMPORT DEPOSITS	279
ADVANCES & STAFF LOAN	24511
OTHERS	47672
CASH IN HAND - CHEQUE IN HAND	17372
WITH BANK	-1994
SUB TOTAL	241374
TOTAL CURRENT ASSETS	718146
TOTAL ASSETS (A+B)	**1133413**

BALANCE SHEET AS AT

PARTICULARS	31.3.03
LIABILITIES	
A. SHAREHOLDERS FUNDS	
SHARE CAPITAL	48686
CAPITAL RESERVE	7698
REVENUE RESERVE	97324
PROFIT/(LOSS) DURING THE YEAR	101017
SUB TOTAL	254725

B. LOANS & ADVANCES	
I) LOAN FROM JUPITO ASSETS MANAGEMENT	48420
II) BANK LOANS - TERM LOAN A/C	0
RM LOAN	0
TERM LOAN	40018
COTTON LOAN	5400
BANKERS ACCEPTANCE	95268
IMPORT FINANCE	0
BANK LEASE A/C	26819
III) BANK OVERDRAFT	167990
SUB TOTAL	383915
C. CURRENT LIABILITIES	
CREDITORS (TRADE&SUNDRY)	194233
CREDITORS (GROUP CO)	-1112
GROUP CO	178068
CREDITORS (GROUP CO)	111
CREDITORS (GROUP CO)	16412
OTHER CURRENT LIABILITIES	-4488
PROVISIONS (PREVIOUS+CURRENT)	99385
LIABILITY TOWARDS TAXATION	-2149
DIVIDEND PAYABLE	14312
OTHER CREDITORS	0
SUB TOTAL	494772
TOTAL LIABILITIES (A+B+C)	**1133413**

STOCK SUMMARY

	STOCK SUMMARY AS ON					01-03-03	01-04-02	
S.NO.	PARTICULARS	MTR/KG	QUANTITY	AMOUNT		QUANTITY	AMOUNT	
1	RAW MATERIAL			N=				
A)	YARN	KG	196529.05	117231643.9	236.1	118328278	177087212	224.74
B)	CHEMICAL & DYES	KG		130738310.3		137319358	26528253.78	
	TEXTILES			34437447.72		43472731.6		
	PLASTIC			47598116.54		39428230.6		
	VT			46408930.16		53061852.9		
	PACKING MATERIAL			2293815.91		1356543.46		
C)	YARN IN SEWING THREAD	KG	11566	3392440	340	2196400	6398778.04	290.4
			508095.05	251902394.2		257844036	210014244	
2	FINISHED GOODS							
A)	FABRICS	METRE	801786.21	58647217.39	73.15	59871959.6	26708166.12	77.64
B)	SEWING THREAD	Cartons	12362	20397300	1650	21605400	9133916.4	
C)	PLASTIC	KGS	35904	4667520	130	6747150		
D)	VT	Cartons	2264	7924000	3500	5260000		

	VT WASTAGE	KGS	8450	422500		825800		
			814148.21	92058537.39		94310309.6	35842082.52	
3	WORK IN PROGRESS							
A)	GREY CLOTH IN STORE	METRE	627025	34929303.88	55.71	25882724	15098539.61	58.97
B)	PROCESSING DEPTT	METRE	437083	27858072.54	63.74	40783230.4	41010087.62	83.36
C)	FOLDING DEPTT	METRE	364576.4	29199218.39	80.09	30231962.4	11458605.27	85.06
D)	PLASTIC							
E)	VT			80000		80000		
	TOTAL		1428684.4	92066594.81		96977916.8	67567232.5	
	GRAND TOTAL			436027526.4		449132263	313423559	
	WIP			92066594.81		96977916.8	67567233	
	FG			92058537.39		94310309.6	35842083	
	RAW MATERIAL			251902394.2		257844036	210014244	
				436027526.4		449132263	313423559	
				7613141.85		5524473.25	5230916.42	
				443640668.2		454656736	318654475.3	

SALES REGISTER

1-Apr-01 to 31-Mar-02

Particulars	Inwards Qty.	Rate	Amount	Outwards Qty.	Rate	Amount
FINISHED GOODS STORE				14184816.51 MTRS	89.75	1273134270
AFRICAN PRINTS				9500601.98 MTRS	95.36	906022706
BEDSHEET				4058.70 MTRS	86.06	349303.81
CRAPELINE				58467.25 MTRS	54.77	3202006.07
FOAM MATERIALS				860119.61 MTRS	95.33	81999178.5
FURNISHING				295771.92 MTRS	77.83	23019225.9
GREY CLOTH				3556 MTRS	109.82	390535
SHIRTINGS				2720138.23 MTRS	63.26	172069253
SUITINGS				742102.82 MTRS	116	86082062
JOB WORKS						4762496.03
BRIGHT YARN (N.S.F)				733.17 KGS	253.67	185983.5
NM (NSF MATERIAL)				250091.29 MTRS	18.3	4576512.53
RAGS & FENTS				6485 KGS	4757.81	30854366.2
ASPR (ASSORTED PIECES OF RAGS)				210 KGS	500	105000
BAPF (AFRICAN PRINT FENTS)				1134 KGS	3703.13	4199353.95
BAPR (AFRICAN PRINT RAGS)				2875 KGS	6908.38	19861586.2
BSR (BED SHEET RAGS)				3 KGS	3000	9000
CDF (COTTON DRILL FENT)				26 KGS	1642.69	42710
CDR (COTTON DRILL RAG)				323 KGS	2231.89	720900
CFF (COTTON FURN. FENT)				10 KGS	2650	26500
CFR (COTTON FURN. RAGS)				108 KGS	2841.2	306850
CLR (CLEAN RAGS)				317 KGS	4390.54	1391800
CRAPR (CRAPE LINE RAG)				37 KGS	2636.76	97560

Particulars				Outwards Qty.	Rate	Amount
DR (DIRTY RAGS)				867 KGS	2844.23	2465950
FMF (FOAM MATERIALS FENTS)				1 KGS	2400	2400
FMR (FOAM MATERIALS RAGS)				217 KGS	3153	684200
GCR (GREY CLOTH RAGS)				74 KGS	2750	203500
GYSF (GRINDLE YARN SUIT FENTSA0)				2 KGS	2858	5716
GYSR (GRINDLE YARN RAGS)				3 KGS	3000	9000
PAPF - PREMIUM AFR PRINT FENTS				5 KGS	3700	18500
PAPR - PREMIUM AFR PRINT RAGS				4 KGS	5600.01	22400.03
STF (SUITING FENTS)				44 KGS	2887	127028
STR (SUITING RAGS)				225 KGS	2464.05	554412
SEWING THREAD DIVISION						99297254.9
SEWING THRD-IRREGULAR				45 BGS	3382.22	152500
SEWING THRD-SST				208 BGS	2617.58	544457
SEWING THRD-ST				254 CTN	3257.63	827438
SEWING THRD-TK				48180 CONE	84.12	4052899.8
SEWING THRD-TK 30				1939 BGS	2538.83	4922800.14
SEWING THRD-TK 48				26144 CTN	3396.48	88797459.9
YARN STORE				5842.31 KGS	354.29	2069860.75
Yarn in Yarn Stores				5842.31 KGS	354.29	2069860.75
Grand Total :						1410118248

SALES REGISTER (Fabric and Plastic Items)

1-Apr-02 to 31-Mar-03

Particulars	Inwards Qty.	Rate	Amount	Outwards Qty.	Rate	Amount
CHEMICAL STORE				10195 KGS	163.4	1666075.00
FINISHED GOODS STORE				14105651.8 MTRS	86.68	1222740972.30
AFRICAN PRINTS				7647251.11 MTRS	94.94	726028598.50
BASE CLOTH				988 MTRS	60	59280.00
BEDSHEET				12930.59 MTRS	82.5	1066729.32
CRAPELINE				210 MTRS	62	13020.00
FOAM MATERIALS				1625463.10 MTRS	91.21	148255127.58
FURNISHING				248348.59 MTRS	77.44	19233310.42
GREY CLOTH				7019.96 MTRS	109.5	768508.40
SHIRTINGS				3776876.49 MTRS	62.01	234198001.37
SUITINGS				786563.96 MTRS	118.4	93118388.71
GREY STORE				97042 MTRS	65.28	6334670.00
GRE-40401 (N.S.F. SAMPLE)				12123 MTRS	101	1224250.00
GRE-40402 (SAMPLE BASE CLOTH GROUP Co)				382 MTRS	100	38200.00
GRE-5760 (BASE CLOTH)				84537 MTRS	60	5072220.00
JOB WORKS						2961283.06
150/2 BRIGHT GOLD				492.82 KGS	150	73923.00
150/2 BRIGHT LILAC				556.09 KGS	150	83413.00
CDS36A (COTTON DRILL SUITING)				18385.51 MTRS	40	735420.40
NM (NSF MATERIAL)				116165.20 MTRS	17.81	2068526.16

						20507252.98
PLASTIC SALES						
001/1 LT CONTAINER BOT.				12237.5 DOZ	105.8	1294166.90
002/1 LT CONTAINER TOP.				30318.5 DOZ	64.75	1963129.73
003/1 LTR BOTTOM TR				2280 DOZ	120	273617.60
004/1 LTR COVER TR				90 DOZ		
005/ KEEP FRESH 1 LTR. BOT.				5362 DOZ	145.5	779997.40
006/ KEEP FRESH 1 LTR. COVER				1145 DOZ	0.83	952.40
007/ KEEP FRESH MOIN-MOIN BOT.				4337 DOZ	63.72	276340.31
008/ KEEP FRESH MOIN-MOIN COVER				945 DOZ		
009/ MOIN-MOIN BOTTOM				8360 DOZ	58.47	488777.60
010/ MOIN-MOIN COVER				7720 DOZ		
011/ KEEP FRESH 1/2 LTR. BOT.				240 DOZ	85.71	20570.70
012/ KEEP FRESH 1/2 LTR. COVER				240 DOZ		
013/ DUST PAN				3137 DOZ	146.7	460091.95
014/ 1 LTR MUG				973 DOZ	104.2	101413.78
015/ HANGER				1748 DOZ	111.2	194284.86
016/ MIX - BOWL				2676 DOZ	380.3	1017743.10
018/ TUB BASIN				710 DOZ	753.9	535272.51
019/ UTILITY BASIN				822 DOZ	375.6	308735.48
020/ 7 LTR. BUCKET				3289 DOZ	547.5	1800788.64
021/ 7 LTR. BUCKET HANDLE				3273 DOZ		
023/ 10 LTR. BUCKET				1075 DOZ	709.5	762735.98
024/ 10 LTR. BUCKET HANDLE				1055 DOZ		
026/ 15 LTR. BUCKET				552 DOZ	841.8	464674.41
027/ 15 LTR. BUCKET HANDLE				542 DOZ	24.64	13357.20
029/ REC. CONTAINER				4507 DOZ	255.9	1153162.16
030/ REC. CONT.COVER				4517 DOZ	1.14	5143.00

SALES REGISTER: 1-Apr-2002 to 31-Mar-2003

Particulars	Inwards Qty.	Rate	Amount	Outwards Qty.	Rate	Amount
031/ SHOPPING BASKET				600 DOZ	588.19	352915.94
033/ 7 LT MARBLE BUCKET BOT				41 DOZ	571.43	23428.62
034/ 7 LT MARBLE BUCKET HANDLE				41 DOZ		
18" TUBE				4 DOZ	857.14	3428.56
1 L CONTAINER NEW				110 DOZ	115.16	12667.1
1 LTRS SET				1290 DOZ	105.53	136140
BIG BUCKET				1005 DOZ	1049.1	1054382.33
CRUNCH BOX				2030 DOZ	256.33	520341.7
HANGER				100 DOZ	95.24	9524
MEDIUM BUCKET				870 DOZ	729.43	634607.27
MEDIUM KEEP FRESH				385 DOZ	95.28	36681.3
MOIN - MOIN SET				1502 DOZ	56.4	84719.6
MUG DEEP				30 DOZ	104.75	3142.5

MUG TRANSPARENT				10 DOZ	119.05	1190.5
PERFORATED CONED				5000 DOZ	8.5	42500
SMALL BUCKET				10504 DOZ	540.43	5676627.85
RAGS & FENTS				8180 KGS	5128.7	41952307.9
ASPF (ASSORTED PIECES OF FENTS)				10 KGS	6030	60300
ASPR (ASSORTED PIECES OF RAGS)				441 KGS	706.12	311400
BAPF (AFRICAN PRINT FENTS)				1488 KGS	3687.6	5487165.72
BAPR (AFRICAN PRINT RAGS)				4239 KGS	7034.5	29819289.9
BSF (BED SHEET FENTS)				2 KGS	2200	4400
BSR (BED SHEET RAGS)				8 KGS	3075	24600
CDF (COTTON DRILL FENT)				2 KGS	1460	2920
CDR (COTTON DRILL RAG)				134 KGS	2411.9	323200
CFF (COTTON FURN. FENT)				3 KGS	3033.3	9100
CFR (COTTON FURN. RAGS)				137 KGS	2973.7	407400
CLF (CLEAN FENT)				6 KGS	3000	18000
CLR (CLEAN RAGS)				302 KGS	4396.7	1327800
CRAPF (CRAPELINE FENT)				1 KGS	2000	2000
CRAPR (CRAPELINE RAG)				3 KGS	2600	7800
CYFF (COLOURED YARN FURN FENTS)				15 KGS	2733.3	41000
DR (DIRTY RAGS)				723 KGS	2981.1	2155350
EXFF (EXPORT FURNISHING FENTS)				17 KGS	2300	39100
FMF (FOAM MATERIALS FENTS)				3 KGS	2118.7	6356
FMR (FOAM MATERIALS RAGS)				399 KGS	3257.9	1299900
GCR (GREY CLOTH RAGS)				45 KGS	2695.2	121285
OFFC (ASSORTED OFF CUT RAGS)				19 KGS	763.16	14500
PAPF - PREMIUM AFR PRINT FENTS				2 KGS	6230	12460
PRSHF (PRINTED SHIRTING FENTS)				1 KGS	2314.3	2314.28
PRSHR (PRINTED SHIRTING RAGS)				1 KGS	3667	3667
RDFF (RAIN DROP FURNISHING FENTS)				9 KGS	2300	20700
SAPF (SOSO AFRICAN PRINT RAGS & FENTS)				1 KGS	3700	3700
SAPR (SOSO AFR PRINT RAGS)				4 KGS	7000	28000
STR (SUITING RAGS)				165 KGS	2415.8	398600
SEWING THREAD DIVISION				1417590 CONE	70.57	100044916
SEWING THRD-TK				1417590 CONE	70.57	100044916

Sales Register: 1-Apr-02 to 31-Mar-03

Particulars	Inwards Qty.	Rate	Amount	Outwards Qty.	Rate	Amount
VINYL- TUBE SALES						25763812.03
VINYL- TUBE				3907 CTN	5714.29	22325728.65
VINYL- TUBE REJECTION HARD				69882.79 KGS.	49.2	3438083.38
YARN STORE				46150.98 KGS	229.37	10585519.45
YARN IN YARN STORE				46150.98 KGS	229.37	10585519.45
GRAND TOTAL						1432556808

DEPARTMENTS' COMBINED PRODUCTION REPORT

Departments' Combined Production Reports

							Month- April 2019
					Date	18-Apr	
S No	Department	Units	Target April 18	To day 18/04/19	Actual	Working Day	Remarks
1	Dye House		(Month to Date)		(Month to Date)		
	Tops/Fiber	Kgs	26,000	1,996	21,709		
	Yarn	Kgs	-	57	1,523		
	Fabric	Kgs	19,800	2,029	26,518		
	Wool Tops Printing	Kgs	-	270	1,952		
	Total Dye house	Kgs	45,800	4,352	51,702	13	
2	Spinning						
	Own	Kgs	35,300	3,395	19,421		
	Job Commn	Kgs	28,700	225	14,873		
	Total Spinning	Kgs	64,000	3,620	34,294	13	
3	Weaving	Mtrs					
	Civil	Mtrs	175,000	8,980	115,380		
	Govt/Institutional	Mtrs	21,000	929	14,128		
	Export	Mtrs	14,000	100	12,677		
	Total Weaving	Mtrs	210,000	10,009	142,185	13	
4	Finishing	Mtrs					
	Domestic	Mtrs	174,000	6,294	88,306		
	Govt /Institutional	Mtrs	34,000	301	22,920		
	Export	Mtrs	27,000	650	23,532		
	Total	Mtrs	235,000	7,245	134,758	13	
	Outsource Fabric	Mtrs		964	964		
	Total Finishing Prod	Mtrs	235,000	8,209	135,722		
5	Defective						
	Domestic /Institutional	%	5.5	4.0	4.5		
	Govt	%	0.7	-	-		
	Export	%	4.3	-	1.2		
	Total Avg	%	4.7	3.6	3.9		
6	No of Hands Engaged /Day	Nos	1,076	910	974		

DYEING DEPARTMENT MONTHLY PERFORMANCE AND TENTATIVE EXPENSES

Particulars	Unit	April	May	Jun-12	Jul-12	Aug	Sept
Working Days	Nos	23	26	26	26	30	30
Production (Tops)							
Wool	Kg	25,829	31,453	54,833	64,106	77,498	72,722
Poly	Kg	12,020	20,967	22,277	23,644	34,447	32,221
Total (Tops)	Kg	37,849	52,420	77,110	87,750	1,11,945	1,04,943
Yarn Dyeing	Kg	3,098	4,358	4,890	5,362	7,510	7,007
Beam Dyeing - T/V	Kg	4,719	8,382	9,056	8,518	8,625	2,855
Beam Dyeing - TW/AW	Kg	-	-	-	-	1,203	735
Jet Dyeing - TW / AW	Kg	11,608	13,848	18,392	16,998	26,332	19,436

Wool Tops Printing	Kg	370	1,282	2,480	3,270	2,380	1,960
Total	Kg	57,644	80,290	1,11,928	1,21,898	1,57,995	1,36,936
Tops Delivered to Spg	Kg						
General Stores	Lac Rs	1.00	0.52	0.75	1.69	1.45	2.56
Per Kgs	Rs/kg	1.73	0.64	0.67	1.39	0.92	1.87
Repair & Maintenance	Lac Rs	0.50	-	0.01	0.01	0.38	0.04
Per Kgs	Rs/kg	0.87	-	0.01	0.01	0.24	0.03
Dyes/Chem	Lac Rs	12.61	20.04	26.90	26.63	33.48	32.84
Per Kgs	Rs/kg	21.87	24.96	24.03	21.84	21.19	23.98
Power	Kwh	57,293	72,939	1,00,547	1,09,434	1,37,093	1,23,097
Per Kgs	Kwh/kg	0.99	0.91	0.90	0.90	0.87	0.90
Water	KL	7503	12196	12275	14289	13140	16,703
Steam	MT	538	915	920	840	821	916
Prodn/Day/Person		84	86	102	104	112	101
Hands Dyehouse/day	Nos	30	36	42	45	47	45
No of Lots dye							
Wool	No	285	426	585	760	827	812
Poly	No	140	279	275	290	426	401
Yarn	No	117	135	156	184	245	235
Total	No	542	840	1016	1234	1498	1448
Redyeing							
Wool	Kgs	714	2,231	1,015	1,067	765	2,041
Poly	Kgs	724	2,050	1,457	1,782	1,439	2,366
Yarn	Kgs	3	39	139	164	148	105
Total	Kgs	1,441	4,320	2,611	3,013	2,352	4,512
Redyeing %	%	3.5%	7.6%	3.2%	3.2%	2.0%	4.0%

Same may be followed daily

MONTHLY PERFORMANCE REPORT (Worsted Dyeing Department)

PARTICULARS	April	May	June	JULY	AUGUST	SEPT	OCT
No.of Shades Delivered To Spinning	182	313	346	423	434	468	414
No. of Lots Dyed (wool Tops)	285	426	585	760	827	812	824
No.of Lots Dyed (Ploy Tops)	140	279	275	290	426	401	440
No. of Lots Dyed(Yarn)	117	135	156	184	245	235	221
No.of Lots Dyed P/W Fabric	109	152	103	100	182	129	122
No. of lots							16(WR)
No.of Lots Dyed P/V Fabric	24	49	47	36	41	17	30
Total Production (KG)	57644	80290	111928	121898	157995	136936	158079
Under capacity lot Dyed %	43%	39%	44%	31.40%	22%	18%	24%
Average cost per kg (Rs)	21.82	24.96	24.83	21.84	21.19	23.98	21.46

CATEGORY WISE DYEING PRODUCTION

	April	May	June	JULY	AUGUST	SEPT	OCT
Light/Medium 0-1.5%	52%	37%	38%	51.50%	48.50%	39%	42.95%
Dark/Extra dark 1.51& above	45%	58%	60%	46%	49%	56.50%	53.60%
Special Bright	3%	5%	2%	2.50%	2.50%	4.50%	3.45%
Avg. Shade Depth	2.58%	2.85%	3.17%	3.02%	2.84%	3.12%	2.46%
Avg. Cost/kg	21.82	24.96	24.86	21.83	21.84	23.98	21.46
Avg. Blend %	P/W 32/68	P/W 39/61	P/W 28/72	P/W 26/74	P/W 30/70	P/W 30/70	P/W 25/75

CONSUMPTION OF DYES AND COST PER UNIT

Sr. No.	Dyes	Supplier	Rate/kg	April		May		June	
				Quantity	Value	Quantity	Value	Quantity	Value
	A. Wool Dyes								
1	---BLACK FB M-SRGL		579.35	74.7	42428.11	502.98	285682.58	546.98	310673.7
2	----DARK BLUE FB M-TR		457.88	159.24	71482.84	195.96	87966.44	322.18	144626.6
3	----RED FB M-BR		950.21	2.19	2080.96	11.51	10936.92	86.44	82136.15
4	---- SCARLET FB M-L		898.04	26.61	23896.84	5.89	5289.46	148	132909.92
5	---BLACK GB		266.52			242	64497.84	752.34	200513.66
	--- RED DM		321.5			8.62	2771.33	31.34	10075.81
6	---- BLUE E2RN		1825.60	0		0		0	
	---- BORDEAUX 2S-B		722.8	18.66	13223.04	24.98	17701.58	123.17	87281.96
7	---- GREY 2SBL		747.19	4.07	3041.06	23.19	17327.34	15.11	11290.04
8	---- BROWN G01		1426.63						
9	---- BROWN MDBL		608.00	70.8	43046.4	75.59	45958.72	147.31	89564.48
10	---- OLIVE SG		1112.25	7.37	8187.92	7.37	8187.92	15.11	16786.91
11	---- TURQUOISE M5G		367.32	10.13	3506.09	13.51	4675.95	18.93	6551.86
12	---- YELLOW M2GLN250		509.36	82.36	41950.89	76.77	39103.57	110.87	56472.74
13	---- BLUE 2R		1335.00	11.25	15018.75	41.17	54961.95	26.05	34776.75
14	---- BROWN G		637.78	10.77	6868.89	19.28	12296.4	19.95	12723.71
	---- BLACK 3G/DM		264.65	400.56	106008.3	0		75	19848.75
15	---- RED 3BN/COLOMIL RED		381.08	2.03	597.88	0.18	29.45	45.27	13332.92
16	---- RED RS/COLOMIL RED		325.53	0.4	125.34	0.26	81.47	0.01	3.13
	B. Polyester Dyes								
1	---- BLUE F2RL		402.87	20.01	8079.24	25.48	10287.8	31.12	12565.01
2	---- PINK REL 200%		1256.74	30.51	34657.83	68.77	78119.28	41.85	47539.5
3	---- RUBINE GFL		308.22	24.13	7437.35	19.64	6053.44	14.37	4429.12
4	---- SCARLET 2RLS		190.68	2.23	425.22	34.98	6669.99	29.59	5642.22
5	---- YELLOW BROWN RELT		266.48	74.38	18461.12	128.45	31821.29	135.1	33531.82
6	---- BLACK CCR		311.31	68.88	19465.49	159.09	44958.83	268.96	76008.1
7	---- YELLOW BROWN S2RI		307.6	36.6	11037.83	16.08	4849.41	12.61	3802.92
8	---- BROWN F3RL 150%		275.04	51.49	15536.59	78.94	23819.36	92.64	27953.19
9	---- CLASSIC NAVY S2G		291.64	149.59	35508.18	240.09	56990.16	320.66	76115.06

#	Name								
10	---- BLUE FBL		555.12	86.26	39037.83	86.97	39359.14	121.95	55189.69
11	---- RED F2BL		239.78	0.9	214.69	1.93	460.38	37.65	8981.03
12	---- TURQUOISE BLUE F2GL		797.98	0.48	383.03	16.5	13166.67	5.27	4205.35
13	---- YELLOW F4GL 200 %		244.82	39.1	9522.81	57.44	13989.51	51.32	1298.99
	C. Viscose Dyes								
1	---- BLUE CA		1680.09	6.64	10940.86	6.85	11286.88	4.83	7958.49
2	---- FAST RED CA		1583.93	2.09	3116.42	3.5	5218.89	3.42	5099.6
3	---- YELLOW CA		837.33	8.48	6961.15	11.31	9284.27	7.29	5984.29
4	---- BLACK NN		234.52	24.61	5771.54	23.34	5473.7	18.61	4364.42
5	---- BRILL RED 3BSA		329.71	6.54	2113.92	10.7	3458.56	2.31	746.66
		Avg. Rate/ kg INR	640.32						
		Value of Dyes (INR)			610134.4		1022736.48		1610984.6
		Total Quantity in kg		1514.06		2239.32		3683.61	
		Value of Chemicals (INR)			650700		981281.12		1167729.3
		Total Value INR			1260834		2004017.12		2778714.3
		AVG Shade Depth		2.63%		2.80%		3.29%	
		Production in kg		57644		80290		111928	
		AVG Dyeing cost/kg		21.82		24.96		24.83	

DAILY PERFORMANCE REPORT (Worsted Spinning Department)

COMBING/RECOMBING CAPACITY VS ACHIEVED PRODUCTION AND EFFICIENCY

							Actual
							Prod /Eff %
WOOL SCOURING			@100%- 5500Kg				2519 Kg
							50%
CARDING	Wrapping	Speed	Production				Actual/Day
	Gms	Mtr/Minute	@ 100%	@80%	@85%	@90%	Prod /Eff
CARD 1	24	40	1382.4	1105.92	1175.04	1244.16	
CARD 2	24	40	1382.4	1105.92	1175.04	1244.16	
CARD 3	24	40	1382.4	1105.92	1175.04	1244.16	
CARD 4	20	40	1152	921.6	979.2	1036.8	
TOTAL			5299.2	4239.36	4504.32	4769.28	2519 Kg
							47.54%
GREY COMBING							
GILLING				@60%	@65%	@70%	
I	30	130	5616	3369.6	3650.4	3931.2	
II (Two coiler)	34	130	6364.8	3818.88	4137.12	4455.36	
III (Twin coiler)	34	125	6120	3672	3978	4284	2519 Kg
							41.16%
	Feed weight Gms		Production				
COMBER NO		Nips/Minute	@100%	@80%	@ 85%	@90%	
1	340	160	454.35	363.48	386.20	408.91	
2	340	160	454.35	363.48	386.20	408.91	
3	340	160	454.35	363.48	386.20	408.91	

			Prod	Prod	Prod	Prod	
4	340	160	454.35	363.48	386.20	408.91	
5	340	160	454.35	363.48	386.20	408.91	
6	340	160	454.35	363.48	386.20	408.91	
7	340	160	454.35	363.48	386.20	408.91	
8	340	160	454.35	363.48	386.20	408.91	
9	340	160	454.35	363.48	386.20	408.91	
10	340	160	454.35	363.48	386.20	408.91	
11	340	160	454.35	363.48	386.20	408.91	
12	340	160	454.35	363.48	386.20	408.91	
TOTAL			5452.19	4361.75	4634.36	4906.97	2519 Kg
				60%	65%	70%	46.20%
PREFINISHER	22	120	3801.6	2280.96	2471.04	2661.12	
FINISHER	22	120	3801.6	2280.96	2471.04	2661.12	2519 Kg
WINDING							
			Prod	Prod	Prod	Prod	
CONVERTER			@ 100%	@80%	AT 85%	AT 90%	
(polyester)	23	244	8081.28	6465.02	6869.09	7273.15	3666 Kg
			Prod	Prod	Prod	Prod	45.36%
GILLING			@100%	@60%	@65%	@70%	
1	25	244	8784	5270	5710	6149	
2	25	244	8784	5270	5710	6149	
3	25	244	8784	5270	5710	6149	3666 Kg
							41.73%
RECOMBING							
LINE A							
DEFELTER1							
(wool opener)							
DEFELTER2							
(blending)							
LINE C							
GILLING 1							
(wool opener)							
GILLING 2							
(blending)							
GILLING			Prod	Prod	Prod	Prod	
LINE A GC 13			@100%	@60%	@ 65%	@ 70%	
GILLING I	25	200	7200	4320	4680	5040	
GILLING II	25	200	7200	4320	4680	5040	
GILLING III(16+16)	32	200	9216	5529.6	5990.4	6451.2	
LINE C GN 6							
GILLING I	25	140	5040	3024	3276	3528	
GILLING II	25	140	5040	3024	3276	3528	
GILLING III (16+16)	32	140	6451.2	3870.72	4193.28	4515.84	
TOTAL			15667.2	9400.32	10183.68	10967.04	5702 Kg
							36.39%

			Prod	Prod	Prod	Prod	
RECOMBER NO			@100%	@ 80%	@ 85%	@ 90%	
1	320	165	441	353	375	397	
2	320	165	441	353	375	397	
3	320	165	441	353	375	397	
4	320	165	441	353	375	397	
5	320	165	441	353	375	397	
6	320	165	441	353	375	397	
7	320	165	441	353	375	397	
8	320	165	441	353	375	397	
9	320	165	441	353	375	397	
10	320	165	441	353	375	397	
11	320	165	441	353	375	397	
12	320	165	441	353	375	397	
			5292	4233	4498	4763	
GILLING			Prod	Prod	Prod	Prod	
LINE 1			@100%	@ 60%	@ 65%	@ 70%	
I	20	180	5184	3110	3370	3629	
II	20	180	5184	3110	3370	3629	
			Prod	Prod	Prod	Prod	
LINE 2			@ 100%	@ 60%	@ 65%	@ 70%	
I	20	150	4320	2592	2808	3024	
II	20	150	4320	2592	2808	3024	
TOTAL			9504	5702.4	6177.6	6652.8	
NEW NSC							
OPENING							
BLENDING							
GILLING							
I							
II							
III							
RECOMBING	Feed weight		Prod	Prod	Prod	Prod	
		nNps/m	@ 100%	@ 80%	@ 85%	@ 90%	
1	320	170	454.35	363.48	386.20	408.91	
2	320	170	454.35	363.48	386.20	408.91	
3	320	170	454.35	363.48	386.20	408.91	
4	320	170	454.35	363.48	386.20	408.91	
5	320	170	454.35	363.48	386.20	408.91	
6	320	170	454.35	363.48	386.20	408.91	
7	320	170	454.35	363.48	386.20	408.91	
TOTAL			3180	2544	2703	2862	
Total No Combers-19			8472	6778	7201	7625	5702 Kg
							67.30%
			Prod	Prod	Prod	Prod	
			@ 100%	@ 60%	@ 65%	@ 70%	

PREFINISHER	20	180	5184	3110	3370	3629	
FINISHER	20	150	4320	2592	2808	3024	
TOTAL GILLING PROD			13824	8294.4	8985.6	9676.8	5702 Kg
							41.25%

SPINNING DAILY PERFORMANCE REPORT

		April	01-May	02-May	03-May	04-May	31-May	To date	Ave
A: PERFORMANCE	Target								
No. of Working Days		25						24	
Utilisation%	94.29	83.23	91.98	95.13	93.51	86.37	54.42	72.29	
Operating Efficiency %		93.1%	90.43%	93.69%	94.68%	96.07%	90.34%	93.77%	
Installed efficiency %		77.6%	83.16%	89.06%	88.43%	82.57%	49.40%	67.91%	
Average Count Nm		64.0	81.3	80.8	79.6	75.1	46.9	65.9	
Speed Index		0.96	0.97	0.97	0.98	0.98	0.91	0.98	
Avg gms/spndl/day@48Nm	552	514.1	499.2	517.2	522.6	530.3	498.7	517.6	
Avg gms/spndl/day@60Nm	335	311.6	302.5	313.4	316.7	321.4	302.2	313.7	
B: PRODUCTION KG.									
Carding		4629	3363.0	3909.0	1679.0	3612.0	2346.0	93292	3887
Top Prod (Net Wt)		4344	3183.0	2631.0	3428.0	1222.0	3249.0	87129	3630
Roving Prod		5688	5191.0	4541.0	3549.0	899.0	6288.0	117440	4893
RF Prod Actual	6678	5603	3749.4	4220.9	4395.0	4638.7	5886.3	112894	4704
RF Prod @48 Nm	11221	9241	9907.1	10608.7	10534.5	9835.8	5884.9	194151	8090
RF Prod @60 Nm	6700	5790	6222.0	6658.1	6616.1	6175.0	3659.3	122167	5090
Autoconer Prod. @Actual		5636	3886.0	4304.0	4842.0	2836.0	5461.0	114191	4758
TFO Prod. @Actual		3917	3180.0	2400.0	3832.0	3480.0	3732.0	84064	3503
Yarn Delivery									
Packing (Actual)	6611	4821	1799.1	5305.6	2052.5	2734.4	12156.3	111835	4660
Packing @60 Nm	7186	5119	1512.5	6283.6	2351.1	3210.9	18470.6	124331	5180
Dispatch									
C: YIELD JCS Yield%		65.6%				62.6%	60.99%	64.2%	
Greasy to Scoured		66.3%				63.3%	63.53%	65.1%	
Scoured to Top								91.6%	
Top to Yarn	94.95	95.0%		93.8%		95.5%	92.1%	95.7%	
Fiber to Yarn	86.8								
D: STOPPAGES % age									
Electrical		0.59%	0.17%	0.00%	0.22%	0.08%	0.08%	0.17%	
Electronics Works		0.41%							
Lot Changes		0.91%	1.23%	0.88%	1.09%	0.81%	0.38%	0.56%	
Maintenance Work		0.43%	0.83%	1.33%	1.28%	0.92%	0.06%	0.41%	
Hand Shortage		0.45%	0.92%	0.00%	0.00%	0.00%	0.00%	2.14%	
Power Failure		1.91%					0.00%	0.05%	
Remnants Change		0.04%	0.00%	0.06%	0.06%	0.14%	0.07%	0.03%	

Run Ot/Idle Time		3.58%	0.00%	2.60%	2.51%	5.76%	0.84%	7.34%	
RM Or Program Short		8.62%	3.00%	3.15%	1.33%	5.92%	44.15%	17.01%	
Total stoppages		16.9%	6.15%	8.02%	6.49%	13.63%	45.58%	27.71%	
E: RF ENDS BREAKAGES %	5%								
Ends Down % (Snap Study)	1%		0.67	0.83	0.90	0.71	0.58	0.77	
F: STOCK									
Scoured Wool		12798	13892	13892	7142	0	0	10716	
Top		12397	31441	36156	36404	30923	22906	23830	
Roving		0	0	0	0	0	0	0	
Total WIP	85 MT	51417	38182	32843	33785	38746	40530	47580	
Finished Goods (Kg)									
Moving/Current Stock									
Non-Moving/Non-Current									
G: HANDS ENGAGED									
Permanent	254		197	201	204	211	213	5126	
Badli/Contract	80		139	140	135	134		1871	
Contract							107.5	889	
OT			8					19	
Total Hands Engaged	334		344	341	339	345	320.5	7904	
Production /Hand @48 Nm			28.8	31.1	31.1	28.5	18.4	24.6	
Production /Hand @60 Nm			18.1	19.5	19.5	17.9	11.4	15.5	
H: POWER CONSUMPTION		58633	55202	57620	54604	55620	49830	1343000	55958
Power Units/Kg Actual		10.46	14.72	13.65	12.42	11.99	8.47	11.90	
Power Units/KG @48 Nm		6.34	5.57	5.43	5.18	5.65	8.47	6.92	
Power units /Kg @60 Nm		10.13	8.87	8.65	8.25	9.01	13.62	10.99	
Store Consumption Rs.									
Consumption/Spindle									
Rewinding(Kg) System		524		1025	150	860	705	18657	777
Rewinding (Kg) Quality issue	1.50%								
I: WASTE (%)									
Noil	1.5	1.25	1.71	0.43	0.28	1.57	1.26	1.22	
Soft	0.4	0.28	0.60	0.32	0.42	0.19	0.27	0.28	
Hard Waste	1	1.27	1.60	1.43	1.46	1.34	1.10	1.38	
Sweeping Waste	0.4	0.41	0.47	0.39	0.34	0.35	0.27	0.36	
Pneumafil/Bonda Waste	1.75	1.68	2.35	1.68	1.80	1.34	1.00	1.49	
TOTAL WASTE	5.05	4.89	6.73	4.25	4.29	4.80	3.89	4.73	

MONTHLY PERFORMANCE AND TENTATIVE EXPENSESOF A SPINNING DEPT (worsted)

Particulars	Unit	April	May	June	July	Aug	Sept
Working Days	Nos	23	26	26	27	29	29
Spindles	Nos	11,724	14,410	16,211	13,978	18,176	15,435
No of Shift Worked	Nos	52	65	65	61	83	83
Production							

Prodn(Own)	Kg	58,377	68,475	90,143	1,16,109	1,52,412	1,36,936
Prodn(Job)	Kg	34,319	48,927	22,278	23,705	42,842	33,316
Total	Kg	92,696	1,17,402	1,12,421	1,39,814	1,95,254	1,70,252
Avg Count Nm		35.6	34.3	42.5	40.5	40.8	40.2
Running Effcy%	%	82.4	75.8	76.9	77.6	78.2	80.8
Waste%							
Own	%	5.5	5.4	5.9	5.7	5.3	5.5
Job	%	5.0	5.6	5.0	6.4	6.2	6.1
Yarn Delivered to Post. Spinning	Kg						
TFO Production	Kg						
General Stores	Lac Rs	4.84	7.78	5.50	8.86	11.98	10.46
Per Kgs	Rs/kg	5.22	6.63	4.90	6.34	6.13	6.15
Repair & Maintenance	Lac Rs	2.06	0.98	1.85	0.21	1.02	1.31
Per Kgs	Rs/kg	2.22	0.84	1.65	0.15	0.52	0.77
Dyes/Chem	Lac Rs	1.51	1.06	1.49	0.79	0.24	0.19
Per Kgs	Rs/kg	1.63	0.91	1.33	0.57	0.12	0.11
Power	Kwh	4,91,202	6,16,289	7,37,602	8,72,695	11,95,817	9,58,246
Per Kgs	Kwh/kg	5.30	5.25	6.56	6.24	6.12	5.63
Water	Klt	2604	2870	2840	3090	2640	1,820
Steam	MT	139	175	135	174	292	205
Per kg Prodn/Person		23	24	21	23	25	23
Hands Engaged /day							
Spinning	Nos	173	188	206	228	273	260
Defelter	Nos	10	13	15	16	17	18
Post Spinning	Nos	72	71	67	74	91	79

ENDS BREAKAGES STUDY IN SPINNING DEPARTMENT

Date :

Report No :

Ends Breakage Study in Ring Frame

RH % : 63.9   Temperature : 28.4 C

Time (From 11.30 am To 12.30 pm)				
Ring Frame No.	RF No (28)	RF No (29)	RF No (30)	
No.of Spindles	432	432	432	
No.of Idle Spindles	–	–	–	
Spindle Speed (rpm)	8000	8400	9000	
Doff Position	4/8	4/8	6/8	
Count (Nm)	2/90	2/90	2/90	
Blend	65:35	65:35	50:50	
Lot No.	29980	29899	29860	
Shade No.	Dark (Black)	Dark (Blue)	White	
Traveller No.	27	27	28	
Anti static %	0.61	0.49	0.53	
Moisture %	4.3	4.4	6.5	

Rubbing Frame	RF2B	FRC	SSR2	
	Reason of Breakages			
Creel Breakages				
Shifting of Apron				
Apron Lapping				
Roving guide chocked				
Roller Lapping				
Undrafted	5	8	9	
Spinners Double				
Fluff in Travellers				
Others	35	35	25	
Total Breakages	40	43	34	
Breaks/100 Spindle/Hours	9.25%	9.95%	7.87%	
Remarks	Ends Breakages rateis onhigher side			

DEPARTMENT WISE WASTE GENERATION REPORT (worsted)

	Dept wise waste %					
Month				Annual average Kg		
Dept	Type	Norms		Prodn	Waste	%
Converter	Poly	0.2		720383	1856	0.26
Recombing	Soft	1.36		1084674	745	0.07
	Noil			1084674	15583	1.42
	Total	1.36		1084674	16328	1.48
Spinning	Sweep			980259	3802	0.39
	Soft	4.5		980259	27360	2.72
	Hard			980259	13627	1.37
	Total	4.5		980259	44787	4.37
Total (Re + Sp)		5.86		980259	61115	5.87
Weaving	Hard			1258435	6362	0.50
	Catch Selv			1258435	49760	3.80
	Chindi			1258435	1697	0.13
	Total	4		1258435	57819	4.39
Folding	Chindi	0.19		1134576	2250	0.20
Dyeing	Soft	0.01		1044061	100	0.01
	Hard			1044061	99	0.01
	Total	0.01		1044061	199	0.02
Grand Total		10.26			0	10.74

COTTON SPINNING AND WEAVING MILL DAILY WASTE GENERATION REPORT

Items	Opening Balance	Receipt	Sale	Sale Value	Sale Rate	Closing Balance
	(Kg)	(Kg)	(Kg)	0	Rs./Kg	(Kg)
Spinning Deptt.						
Dropping Unwillow	9703	2644	12187	137713	11.3	160
Card Fly	4547	7763	9481	326525	34.4	2829

Top Roller	1537	1449				2986
Rottary Filter	1663	1463	2652	31824	12	474
kiti dust	4081	1475	5467	31162	5.7	89
Spg hard waste	55	1462				1516
Doubling Hardwaste	28	283				311
Hard waste Mech Corner	50	600				650
Hard waste RJK	27	500				527
Hard waste Auto Corner	24	200				224
Hard waste PS Mettler	44	300				344
Total Spg.A	21759	18138	29787	527224	17.7	10110
Weaving Deptt.						
Hard Waste Wvg	39	237				276
Size long cotton	190	307				497
Unsize long cotton	205	75				280
Size short cotton	60	200				260
Size long P/C	33	245				278
PV long dyed	480	271				751
P/V /P/C Leno dyed	1548	1041	2280	19950	8.75	309
Grey Chindi Cotton	3	186	60	939	15.65	129
Grey Chindi P/V	218	170	240	840	3.5	148
Total Wvg B	2776	2732	2580	21729	8.42	2928
Total (A+B)	24535	20870	32367	548953	16.96	13038

COTTON SPINNING MONTHLY WASTE RECORD

MONTH	Apr	May	Jun	Jul	Aug	Sep
Cotton Consumption	230510	207992	222467	178214	170701	156855
Ring Frame Production	201797	181012	194064	157302	148581	135845
Salable waste %						
Dropping (Kg)	6761	7328	6905	4944	4021	3377
Dropping %	2.93	3.52	3.1	2.77	2.36	2.15
Card Fly (Kg)	9651	9313	10095	7952	7933	8213
Card Fly %	4.19	4.48	4.54	4.46	4.65	5.24
Top Roller (Kg)	1887	2388	1925	1816	1836	1404
Top Roller %	0.82	1.15	0.87	1.02	1.08	0.9
Top Roller Frame & Lakdi Mixed (Kg)	1702	1563	2154	1676	1648	1364
Top Roller Frame & Lakdi Mixed %	0.74	0.75	0.97	0.94	0.97	0.87
Kiti Dust (Kg)	939	654	406	449	855	643
Kiti Dust %	0.41	0.31	0.18	0.25	0.5	0.41
Total Saleable Waste (Kg)	20940	21246	21485	16837	16293	15001
Total Saleable Waste %	9.08	10.21	9.66	9.45	9.54	9.56
Sweeping Dust (Kg)	1559	844	1292	2306	1803	1699
Sweeping Dust %	0.68	0.41	0.58	1.29	1.06	1.08
Useable Waste %						

Card Waste (Kg)	418.4	458.3	439.7	326.8	337.5	340
Card Waste %	0.21	0.25	0.23	0.21	0.23	0.25
Frame Silver (Kg)	1329.2	1350.6	1384.2	1069	1216	1126
Frame Silver %	0.66	0.75	0.7	0.68	0.82	0.83
Frame Roving (Kg)	85.5	119.5	146.5	118.8	114.5	111.5
Frame Roving %	0.04	0.07	0.08	0.08	0.08	0.08
Ring Pneumafil (Kg)	6374	6652	7775	5312.5	5020	5845
Ring Pneumafil %	3.16	3.67	4.01	3.38	3.38	4.3
Ring Roving (Kg)	314.4	334.3	301.8	249	263.7	320.7
Ring Roving %	0.16	0.18	0.16	0.16	0.18	0.24
Ring Bonda (Kg)	496.5	571.2	508.8	431.9	423.3	509.8
Ring Bonda %	0.25	0.32	0.26	0.27	0.28	0.38
Total Useable Waste (Kg)	9018	9485.9	10556	7508	7375	8253
Total Useable Waste %	4.47	5.24	5.44	4.77	4.96	6.08
Hard Waste %						
Ring Frame (Kg)	1287	1903	895	1457	955	1041
Ring Frame %	0.64	1.05	0.46	0.93	0.64	0.77
Doubling (Kg)	496	232	498	500	385	319
Doubling %	0.25	0.13	0.26	0.32	0.26	0.23
Sale Yarn (Kg)	2450	2977	2845	2450	1708	1711
Sale Yarn %	1.21	1.64	1.47	1.56	1.15	1.26
Total Hard Waste (Kg)	4233	5112	4238	4407	3048	3071
Total Hard Waste %	2.1	2.82	2.18	2.8	2.05	2.26

WEAVING DEPARTMENT MONTHLY PERFORMANCE AND TENTATIVE EXPENSES

Particulars	Unit	April	May	June	July	Aug	Sept
Working Days	Nos	24	26	26	28	30	30
No. of Shifts Worked	Nos	72	80	78	82	87	90
No. of Looms	Nos	117.0	103.5	59.7	60.8	84.6	78.1
Production							
Civil	Mtr	1,76,640	1,50,205	1,86,300	2,04,443	3,10,603	2,94,796
Govt/Institutional	Mtr	47,580	21,117	15,093	18,584	30,090	29,264
Export	Mtr	21,043	15,135	22,831	35,790	69,751	52,161
Total Mtr		2,45,263	1,86,457	2,24,231	2,58,817	4,10,444	3,76,221
Should be Production	Mtr				2,48,216	3,99,692	3,54,570
Avg Speed	RPM				330	321	318
Total Picks(Lac)	Nos	247	142	177	207	292	277
Avg Picks		57.4	58.4	52.8	58.9	57.8	61.4
Effcy%	%	73.3	72.6	70.2	73.0	77.5	77.4
No. of Beam Gaiting Started					10	12	13
No. of Beam Knotting Started					11	15	16
Waste	%	4.3	4.4	4.6	4.0	4.0	4.1
General Stores	Lac Rs	6.24	12.73	2.24	5.34	6.87	14.51

Per Mtr	Rs/mtr	2.55	6.83	1.00	2.06	1.67	3.86
Repair & Maintenance	Lac Rs	0.02	0.15	0.09	0.69	0.35	0.04
Per Mtr	Rs/mtr	0.01	0.08	0.04	0.27	0.09	0.01
Chemical	Lac Rs	0.39	-	0.78	-	0.01	-
Per Mtr	Rs/mtr	0.16	-	0.35	-	0.00	-
Power	Kwh	2,26,305	2,07,731	2,43,465	2,91,631	3,44,507	3,19,578
Per mtr	Kwh/mtr	0.92	1.11	1.09	1.13	0.84	0.85
Water	KL	1960	2012	2020	2690	2420	1,720
Prodn/Day/Person		46	38	46	46	62	60
Hands /day	Nos						
Weaving	Nos	223	186	188	203	222	210
G.Perch	Nos	14	13	12	13	15	14
Mending	Nos	9	10	10	11	13	11

WEAVING DEPARTMENT DAILY/MONTHLY PERFORMANCE REPORT

A. PERFORMANCE	April	May	Jun	July	Aug	Sep	Oct	Nov	Dec	Avg
Utilisation %	68.51	69.98	69.32	61.11	61.27	62.49	65.27	54.03	57.16	63.24
Operation efficiency %	70.57	70.73	71.34	74.34	77.40	75.93	76.68	74.31	75.23	74.06
Installed efficiency %	48.35	49.5	49.45	45.43	47.42	47.45	50.05	40.15	43	46.76
(Results are Poor,wide scope of Improvements)										
B.Reason wise stoppages-loss in percentage										
Beam gaiting & Checking/ Awaiting	5.99	5.41	6.65	5.18	6.38	5.65	6.45	9.21	8.5	6.60
Repair & Overhaul	6.3	1.6	2.74	9.19	5.56	9.97	5.23	3.77	3.98	5.37
No Beam/ No Weft	0.79	0.58	0.75	0.67	0.95	0.90	0.93	1.37	0.46	0.82
Hand shortage	0	0	0	1.45	3.08	0.05	1.46	2.45	0.79	1.03
Power failure	0.08	0.1	0.36	2.86	2.78	0.75	0.32	0.49	0.85	0.95
No Materials etc.	18.33	18.33	20.18	19.54	19.98	20.19	20.34	28.68	28.26	21.54
Total Stoppages	31.49	30.02	30.68	38.89	38.73	37.51	34.73	45.97	42.84	36.76
C. OTHER INFORMATION										
Prodn. Per day, Calculated mtrs	9571.82	6462.09	8197.2	8799.88	12469.4	11761.2	11905.03	10079.1	10792.68	10004.27
Avg. Gaiting time + Avg. Awaiting time Hrs	7.35	7.07	11.73	7.71	8.14	6.49	6.45	9.98	10.3	8.36
Avg. Knot Pull time + Avg. Await time Hrs.	4.66	3.26	4.78	4.72	4.9	4.95	6.31	8.49	7	5.45
Avg Warp Prodn Length mtrs/ day	10389.71	7116.2	9966.5	11063.21	15317.5	14439.07	12922.26	11495.6	12908.34	11735.38
Avg. Beam Length on loom mtrs	399	358	456	497	584	528.2	453	415	526	468.47
Avg Loom speed	256	298	304	304	304	304	304	304	304	298.00
No. of Looms	94.67	60.23	64.19	72.64	92.06	92.47	93	78.7	78.4	80.71
Avg pick per calculated mtrs	63.22	58.7	54.77	60.2	57.48	59.77	62.44	64.11	63.33	60.45
Total No. of Quality	58	63	43	56	56	63	67	72	55	59.22
No.Of Samples	165	175	12	5	24	33	46	235	69	84.89
Prod (mtrs) Domestic	154230	134344.75	178714.3	199340.25	306550	291064	314002.5	202511	206735	220832.39

Prod (mtrs) Export	21042	15135.25	22830.5	35789.5	69540.8	51661.5	8093.25	75995	38962.75	37672.28
Prod (mtrs) Institutional	7793.25	20529	15092.25	18586.25	30089	9119.75	29717.25	8632.75	12143.75	16855.92
Prod (mtrs) Sample	10684.25	6488	2035.5	1277	1887.25	3553.5	2366.25	7447.25	3424.75	4351.53
Govt.	39787.25	587.75	0	0	0	20144.5	28439.5	12276.3	60882.75	18013.11
Stock lots	11726.25	9371.5	5559.25	3795.5	2377.25	677.5	2158.25	4789	8970.5	5492.56
Total Production	245263.00	186456.25	224231.8	258788.5	410445	376220.8	384777	311651	331127.5	303217.78

WEAVING MACHINE BREAKAGE STUDY AND ANALYSIS

DATE								TOTAL	% age
	NO OF STUDIED	178	64	94	18	7	9		%
		ANKARA	45"	BED	POLY	SCH	BASE		
	QUALITY		SHIRTING	SHEET	SUIT	CHECH	CLOTH		
	PICKS PER HOUR								
	EFFICIENCY								
	LOOM'S SPEED RPM								
	LOST TIME DUE TO STOPAGES								
	BEAM'S SIZE								
	HUMIDITY & TEMP. IN DEPT								
	CAUSES OF ENDS BREAKAGES								
PREP.FAULT	WARP COUNTS Ne	20/1C	40/1P	12/1C	150/2TW	1	40/1P		
	MISSING ENDS	11	2	7	2	5	1	24	
	CROSSING ENDS	14	33	6	2		4	74	
	STICKING ENDS	10	7	8				25	
	SLACKING ENDS	12	5					17	
	EXTRA/SPARE ENDS			2				2	
TOTAL								142	37.0%
SPIN.FAULT	YARN'S SLUBS (UNDRAFTED YARN)	20		18				38	
	FOREIGN FIBRE	1		5				6	
	BAD PIECING	4		3				7	
	WEAK YARN	9	4	4				17	
	THICK & THIN PLACES	10	6	3			2	21	
	BAD/LOOSE KNOT	3		1				4	
	WASTE YARN	2		1				3	
TOTAL								96	25.0%
WVG.FAULT	COTTON/FLYING FLUFF	1						1	
	DOUBLE ENDS	2						2	
	CROSS DENTING OR DRAWING					4		4	
	ENDS BREAK AT SELVEDGES AREA	7	1		1			9	
	ENDS BREAK BETWEEN BEAM & DROP WIRE		5					5	
	ENDS BREAK BETWEEN BEAM & DROP SHAFT		1					1	
	ENDS BREAK BETWEEN READ & GUIDE TEETH								
	ENDS BREAK BETWEEN HEAD & REED	7	13	2	1			23	
	ENDS BREAK IN FRONT OF REED								
	MISCELLANEOUS: (OTHER FAULTS) NEP	1	7					8	
TOTAL	TOTAL WARP BREAKS⟹	124	84	60	6	10	7	53	14.00%
	AVG. WARP BRKS/LOOM/HR	0.7	1.31	0.64	0.33	1.43	0.77		
	CAUSES OF WEFT BREAKAGES								

SPIN.FAULT	WEFT COUNTS	20/1C	150/1 text	150/1 TEX	150/2 TEX	150/1 Tex	150/1 Tex		
	SLUB (UNDRAFTED YARN)	21						21	
	BAD PIECING	4						4	
	WEAK YARN	7						7	
	THICK & THIN PLACES	3						3	
	SLOUGHING OFF								
	ENTANGLED CONES	2		7				9	
	FOREIGN FIBRE								
	HOOKING NEPS								
TOTAL WVG. FAULT								44	11.00%
	TRANSFER FAILURE		1	1			1	3	
	PICK DROPS IN THE SHED								
	CONE EXHAUTED & BOTTOM CONE CHANGED	30	2	8	4	1		45	
	MISCELLANEOUS: (ACCUMMULATOR)								
	TOTAL WERT ⟹	67	3	16	4	1	1	48	13.00%
MECHANICAL STOPS	AVG. WEFT BRKS/LOOM/HR	0.38	0.05	0.17	0.22	0.14	0.11		
	IDLE STOPPAGES								
	BROKEN GUIDE TEETH								
	BROKEN HEALD WIRE								
	BAD REED								
	TEMPLE AREA								
	TOTAL MECHANICAL ⟹								
	AVG. MECH BREAKS/LOOM/HT								
	G TOTAL							383	

SIZING MATERIAL CONSUMPTION IN WEAVING PREPARATORY

										CUMULATIVE		AVERAGE			
DESCRIPTION	O S IN M.STORE 1-Mar-03 KG	RECVD IN M.STORE Mar-03 KG	OS IN DEPT. 1-Mar-03 KG	ISSUE TO DEPT Mar-03KG	CONS FOR MONTH Mar-03 KG	CB IN M.STORE 31-Mar-03	CS IN DEPAT 31-Mar-03	RATE IN N=	TOTAL CONS VALUE IN N=	CONS FOR THE YEAR	VALUE FOR THE YEAR N=	RATE N=/KG 2003	RATE=N=/KG 2002	CONSUMP. PER MONTH 2003	CONSUMP. PER MONTH 2002
P.V.A. KURRAY	19440		80	820	780	18620	120	260	202800	18960	4929600	259.4	259.4	740	1748
SIZE SUPRA								71				69.96	69.96		
QUALLEX								100				99.92	99.92		
NYCIL TEXSIZE 304								157		250	39250	156.88	156.88		83.33
NYCIL TEXSIZE 506								76				75.44	75.44		
NYCIL A5 776								72				71.52	71.52		
PURETEX AT								70		1480	103600	69.92	69.92		246.67
SIZO SOFT	800		140		40	800	100	68	2720	1080	73440	67.92	67.92	26.67	100
LUB JPC	175	200		50	40	325	10	295	11800	968	18665	214.8	214.8	33.33	90.3
WAX RC	720	900	60	240	240	1380	60	120	28800	4275	513000	119.88	119.88	240	379.5
PUREWET	360	270				630		200		180	36000	200	200		60
NYCIL DEFORMER								215				214.8	214.8		
FORMMASTER			40		10		30	350	3500	132.4	46329.5	349.28	349.28	15	44.13
LANCOLVM 2000										25					83.3
PLYSTRAN CR			150				150								33.3
PURESIZE C-4	10775	1000	600	6600	6875	5175	325	105	721875	92225	10397900	115.5	115.5	6500	7960
RAN 1000								535				535	535		
R SIZE P 100								665				665	665		
MONOETHYLENE GLYCOL								199.5		56	11172	199.5	199.5		18.67

RSL 40								460				460	460		
TOTAL	32270	2370	1070	7710	7985	26930	795		971495	119631	16336957			7555	10847.2

*OS-Opening Stock, CS-Closing Stock, CONS-Consumption, DEPT-Department, N= Naira (Currency)

FINISHING DEPARTMENT MONTHLY PERFORMANCE AND TENTATIVE EXPENSES

Particulars	Unit	April	May	June	July	Aug	Sept
Working Days	Nos	22	23	24	26	30	30
No. of Shifts Worked	Nos	44	46	48	52	65	65.5
Transfer to Inspection(Grey Basis)	Mtr	2,66,756	1,86,056	2,08,574	2,32,889	4,09,003	3,82,186
Production							
Civil	Mtr	1,71,547	1,51,383	1,44,708	1,80,510	2,79,651	2,78,232
Govt/Instt	Mtr	47,075	23,654	15,736	18,726	27,723	24,935
Export	Mtr	26,364	4,788	27,010	20,201	72,661	42,760
Total	Mtr	2,44,986	1,79,825	1,87,454	2,19,437	3,80,035	3,45,927
Outsourced Fabric	Mtr	-	-	-	-	3,038	43,089
Grand Total		2,44,986	1,79,825	1,87,454	2,19,437	3,83,073	3,89,016
Defective%							
Domestic/Instt	%	5.3	5.0	4.2	5.7	5.3	5.7
Govt	%	0.1	0.8	-	-	-	-
Export	%	2.0	2.3	3.4	3.4	2.1	2.4
Total	%	4.1	4.7	4.1	5.5	4.7	5.2
General Stores	Lac Rs	5.77	2.65	7.13	4.90	4.81	10.70
Per Mtr	Rs/mtr	2.35	1.47	3.80	2.23	1.27	3.09
Repair & Maintenance	Lac Rs	0.18	-	0.03	0.03	0.18	0.49
Per Mtr	Rs/mtr	0.07	-	0.02	0.01	0.05	0.14
Chemical	Lac Rs	8.15	5.16	4.50	4.80	4.75	3.27
Per Mtr	Rs/mtr	3.33	2.87	2.40	2.19	1.25	0.94
Packing (Primary)	Lac Rs	1.40	4.92	2.24	2.50	6.53	5.42
Per Mtr	Rs/mtr	0.57	2.73	1.19	1.14	1.72	1.57
Power	Kwh	1,11,270	1,00,956	1,07,142	1,16,484	1,63,322	1,55,972
Per Mtr	Kwh/mtr	0.45	0.56	0.57	0.53	0.43	0.45
Water	Klt	21083	18923	21524	20482	30242	30,725
Steam	MT	1592	1172	1185	1226	1980	1,798
Prodn/day/Person (Inspection)		186	122	124	130	165	153
Hands / day - Finishing	Nos	92	86	84	85	100	96
Inspection	Nos	60	64	63	65	77	75
Rejection							
-Dyehouse	Mtr				2,267	-	
Spinning	Mtr				-		
Mending	Mtr				26,236	37,645	27,669
Finishing	Mtr				9,530	8,929	6,434
Rejection Total	Mtr	14,933	18,302	23,571	38,033	46,574	34,103
Rejection - %age	%	5.60	9.84	11.30	16.33	11.39	8.92

PROCESSING DEPARTMENT DAILY AND MONTHLY MACHINE STOPPAGE REPORT (Cotton)

PROCESSING DEPARTMENT DAILY AND MONTHLY MACHINE STOPPAGE REPORT

MACHINES	TOTAL TIME OF WORKING HOURS	TOTAL STOPPAGES HOURS DUE TO MECH. FAULT	TOTAL STOPPAGES HOURS DUE TO ELECT. FAULT	TOTAL STOPPAGES HOURS DUE TO MECH & ELECT. FAULT	STOPPAGES DUE TO MECH & ELECT. FAULT %	OTHER STOPPAGES HOURS DUE TO NO STEAM, AIR , WATER & MAINTENANCE	OTHER STOPPAGES %	STOPPAGES HOURS DUE TO NO MATERIALS. NO WORKERS & SHORTAGE OF ORDERS & OVERHAULING OF MACHINES	STOPPAGES DUE TO NO MATERIAL SHORTAGE %	TOTAL STOPPAGES %
OLD YARN DYEING M/C	684	30	151.3	181.3	26.51	7.45	1.09	95	13.89	41.48
NEW YARN DYEING M/C	684		125.4	125.4	18.3	7.45	1.09	100.3	14.66	34.09
YARN DRIER M/C	684		125.4	125.4	18.3	6.35	0.93	100	14.62	33.88
JET DYEING M/C OLD	684		75.78	75.78	11.08	6.3	0.92	294.1	43	55
JET DYEING M/C NEW	684		70.95	70.95	10.37	6.3	0.92	213.3	31.18	42.48
J-T-10 NO 1 M/C	684	2.15	9.7	11.85	1.73	2.8	0.41	233.2	34.09	36.24
J-T-10 NO 2 M/C	684	12.8	85.38	98.18	14.35	2.8	0.41	126	18.42	33.18
J-T-10 NO 3 M/C	684	28.15	38.48	66.63	9.74	0.55	0.08	184	26.9	36.72
J-T-10 NO 4 M/C	684	8.55	6.9	15.45	2.26	4.55	0.67	240.5	35.16	38.08
J-T-10 NO 5 M/C	684	41.25	17.28	58.53	8.56	4.75	0.69	117.2	17.13	26.39
J-T-10 NO 6 M/C	684	36	14.13	50.13	7.33	5	0.73	90	13.16	21.22
J-T-10 NO 7 M/C	684	25.8	11.88	37.68	5.51	2.8	0.41	141	20.61	26.53
J-T-10 NO 8 M/C	684	25.8	60.33	86.13	12.59	2.8	0.41	161.4	23.6	36.6
M & P DRYER	684	0.5	10.61	11.11	1.62	2.8	0.41	161.4		3.21
FINISHING DRYER	684	10	4.03	14.03	2.05			62	9.06	11.11
MERCERIZER M/C (OLD)	684	55.8	10.03	65.83	9.62	2.2	0.32	108.55	15.87	25.82
MERCERIZER M/C (NEW)	684		0.25	0.25	0.04			568	83.04	83.08
CALENDAR M/C (OLD)	684		6.4	6.4	0.94			143	20.91	21.85
CALENDAR M/C (NEW)	684	33	7.33	40.33	5.9	2.2	0.32	127	18.57	24.79
STENTER 1 M/C	684	16.8	54.93	71.73	10.49	42.65		14	2.05	18.76
STENTER 2 M/C	684	11	32.2	43.2	6.32	50.9	7.44			14.14
STENTER 3 M/C	684	9.12	58.87	67.99	9.94	71.1	10.39	26	3.8	24.09
SCOURING JIGGER 1 M/C	684	35.53	3.13	38.66	5.65			132	19.3	25
SCOURING JIGGER 2 M/C	684	12	3.33	15.33	2.24	1.25	0.18	104	15.2	17.63

PROCESSING DEPARTMENT MACHINE WISE DAILY AND MONTHLY PRODUCTION REPORT (Cotton)

MACHINE WISE PRODUCTION						MARCH 2003
No. of working days = 28.5						
MACHINE	Prod. @ 100%		Avg. Prod. Achieved Per Day	Total Prod. In Month	Efficiency %	Reasons for Low Production
	Per day	Per month				
YARN DYEING M/C - 1 (Old) (Kg)	1500	42750	438.6	12500	29.24	
YARN DYEING M/C - 2 (New) (Kg)	1500	42750	596.46	17000	39.77	
YARN DRYER M/C	4000	114000	1078.95	30750	26.97	
JET DYEING M/C STAR (OLD)	17000	484500	2821.65	80417	16.6	

JET DYEING DEVREKHA (NEW)	26000	741000	6462.56	184183	24.86	
STENTER 1 M/C	65000	1852500	41623.05	1186257	64.04	
STENTER II M/C	72000	2052000	35130.46	1001218	48.79	
STENTER III M/C	50000	1425000	26424.28	753092	52.85	
M&P DRIER M/C	58000	1653000	50519.72	1439812	87.1	
FINISHING DRYER	43000	1225500	15332.28	15332.28	1.25	
CALENDAR M/C - (Old)	36000	1026000	28437.26	810462	78.99	
CALENDAR M/C - (New)	57600	1641600	34416.39	980867	59.75	
MERCERIZING M/C (OLD)	57500	1638750	30512.95	869619	53.07	
MERCERIZING M/C (NEW)	57500	1638750	3402.11	96960	5.92	
JT 10 NO. I M/C	10000	285000	5579.93	159028	55.8	
JT 10 NO. II M/C	10000	285000	5175.33	147497	51.8	
JT 10 NO. III M/C	10000	285000	4399.12	125375	43.99	
JT 10 NO. IV M/C	10000	285000	4201.02	119729	42.01	
JT 10 NO. V M/C	10000	285000	5710.91	162761	57.11	
JI 10 NO. VI M/C	10000	285000	5761.54	164204	57.62	
JT 10 NO. VII M/C	10000	285000	5433.16	154845	54.33	
JT 10 NO. VIII M/C	10000	285000	5150.49	146789	51.5	
SCOURING JIGGER M/C - I	3400	96900	1135.19	32353	33.39	
SCOURING JIGGER M/C - II	3400	96900	1109.54	31622	32.63	

FABRIC STOCK MOVEMENT AND RECONCILIATION IN PROCESSING DEPARTMENT

QUALITY NO	OPENING STOCK	RECEIVED FROM GREY STORE	TOTAL	ISSUED TO FINISHED FOLDING	CLOSING STOCK
	Mtr	Mtr	Mtr	Mtr	Mtr
3519	199298	748460	947758	692502	255256
2096	121145	144701	265846	241891	50955
49225	24096	148940	173036	163167	9869
34011	0	0	0	0	0
5560	7071	0	7071	798	6273
73229	12109	40666	52775	52775	0
90110	200	0	200	200	0
3707	13281	22646	35927	23814	12113
3710	0	4124	4124	2679	1445
1362	52107	36512	88619	70029	18590
1604	40193	20160	60353	56273	4080
3085	5198	0	5198	0	5198
School check	1916	12242	14158	14158	0
Sample	3043	0	3043	0	3043
3071	20384	0	20384	20384	0
3544	0	8909	8909	8909	0
24003	3390	0	3390	3390	0

81019	7222	4599	11821	6709	5112
3080	0	0	0	0	0
3086	120000	107165	227165	162000	65165
3612	0	0	0	0	0
85304	666	712	1378	1378	0
			0		
TOTAL ⟹	631319	1299836	193155	1494056	437099

FABRIC SHRINKAGE MONTHLY REPORT (worsted suitings)

ITEMS		April	May	June	July	Aug	Sept	Nov	Dec
GREY LENGTH	Mtrs	145210.40	121992.20	126988.31	215429.20	210729.90	268439.18	304585.30	273151.23
FINISH LENGTH	Mtrs	137903.05	115986.30	120515.95	204970.53	200273.18	255196.17	290223.70	259928.55
CHINDIES / RAGS	Kgs	217.30	255.27	236.70	286.47	358.93	511.27	446.93	390.61
CHINDIES / RAGS	Mtrs	708.88	764.85	668.70	953.35	1141.47	1393.76	1378.63	1210.30
DIFFERENCE	Mtrs	6598.47	5241.05	5803.66	9505.32	9315.25	11849.25	12982.97	12012.38
SHRINKAGE	%	4.54	4.30	4.57	4.41	4.42	4.41	4.26	4.40
CHINDIS / RAGS	%	0.49	0.63	0.53	0.44	0.54	0.52	0.45	0.44
TOTALSHRINKAGE	%	5.03	4.92	5.10	4.85	4.96	4.93	4.72	4.84
FROM W/H									
CHINDIES / RAGS	Kgs	406.65	220.5	347.5	332.0	325.2	711.9	988.8	716
CHINDIES / RAGS	Mtrs	1219.95	661.5	1042.5	996.0	975.6	2135.7	2966.4	2148

FINISHING DEPARTMENT MACHINE WISE DAILY PERFORMANCE REPORT (worsted)

Machine Wise Finishing Efficiency Daily Report									18-May-07
Machine	**Unit**	**DAILY**							
		Actual Prodn	% Eff	Stoppage ... Hrs					
				Hands Shortage	No Material	Mech Repairs	Elec/ Inst Repairs	No Steam / Less Temp	Total stop
Washing Mc No 1	Lots	2	33.3		16				16
2	Lots					24			24
3	Lots				24				24
4	Lots	5	68.8						
5	Lots	6	82.5	8					8
6	Lots	7	96.3						
Stentor 1	Mtrs	20729	80.0				2		2
Stentor 2	Mtrs	22740	87.7						
Stentor 3	Mtrs	5609	21.6	16					16
Stentor 4	Mtrs	20179	77.9						
Shearing Mc No 1	Mtrs	20862	48.3	8					8
2	Mtrs			16					16
3	Mtrs	13770	31.9	8					8
Jigger 1	Lots			8					8
Jigger 2	Mtrs	14071	55.2	8					8
Contisol	Mtrs	6150	32.0	8					8

Contifi	Mtrs				8				8
Relaxing	Mtrs			24					24
KD 1	Lots				24				24
KD 2	Lots	13	32.5	8					8
KD 3	Lots	34	85.0						
N Press 1	Mtrs	7308	72.5						
N Press 2	Mtrs				24				24
Super Finish	Mtrs	30624	96.7						

CUMULATIVE

Actual Prodn	% Eff	Stoppage ... Hrs					
		Hands Shortage	No Material	Mech Repairs	Elec/Inst Repairs	No Steam / Less Temp	Total stop
22	20.4	280	56				336
				408			408
46	35.1	200	40	13			253
82	62.6	112	32				144
97	74.1	88					88
97	74.1	64	8				72
320855	68.8			33	2	3	37
259043	55.5	96	16	16	2		130
101058	21.7	256	24	6			286
302890	64.9	8		7	1		16
330372	42.5	24	8				32
11160	2.2	224	8	16			248
145620	18.7	200	32		2		234
12	27.5	72					72
185076	40.3	80	96	3		7	186
59111	17.1	176	8	17			201
46096	24.3		72				72
27912	8.1	400					400
			408				408
268	37.2	16	83				99
512	71.1			5	4		9
39684	21.9			244			244
14070	7.8	16	336				352
492888	86.4			5			5

ENGINEERING DEPARTMENT- DAILY / MONTHLY AND ANNUAL ENERGY CONSUMPTION (example of worsted unit)

Example 1

Date	Power Units Consumption																		
	Spg M/c's	Spg H-Plants	Spg H plant units/kg	Spg. Units / kgs	Fsg	Fsg units / mtr	Wevg M/c's	Wevg H-plants	Wevg H plant units/mtr	Wevg units / mtr	Dyg	Dyg units / kg	Ltg	Comp	Utiy	DG Aux	Tsfo losses	Total	Total Units /Fsg mtr
01-Jul-13	12460	6331	2.26	6.70	2106	0.26	6030	5985	0.57	1.14	2020	1.06	2281	6838	3570	770	2009	50400	6.18
02-Jan-13	17215	5497	1.91	7.89	2788	0.28	6816	4861	0.45	1.08	2248	0.62	2294	6866	3146	745	624	53100	5.25
03-Jan-13	17293	5207	1.84	7.94	2379	0.30	5576	3680	0.35	0.89	88	0.00	2012	6578	2567	722	1598	47700	6.08

Date																		
04-Jan-13	17054	5884	1.95	7.62	2411	0.33	6463	3628	0.39	1.07	2628	0.53	2233	6738	3331	713	2017	53100	7.28
05-Jan-13	19899	2747	0.99	8.17	2662	0.43	6600	3980	0.36	0.96	3065	0.61	2298	6669	3431	747	1452	53550	8.70
Total	83921	25666	8.95	38.32	12346	1.60	31485	22134	2.12	5.14	10049	2.81	11118	33689	16045	3697	7700	257850	33.50
Avg/Day	16784	5133	1.79	7.66	2469	0.32	6297	4427	0.42	1.03	2512.25	0.70	2223.60	6737.80	3209.00	739.40	1540.00	51570.00	6.70
Avg/this year	15728	5950.33	2.53	9.18	2215	0.32	5388	4049.78	0.51	1.18	2081	0.78	2148	6283	2943	793	1214	47798	6.64
Dec-12	17922	6866	2.52	9.11	2603	0.30	6166	5293	0.58	1.25	2584.43	0.77	2321.26	6646.32	3260.77	744.32	1312	55469	6.44
Nov-12	17165	7196	2.57	8.71	2748	0.31	6439	4747	0.47	1.11	2144	0.70	2316	6742	3219	812	1277	54510	6.08
Oct-12	16966	7475	2.68	8.75	2669	0.30	6048	4828	0.53	1.19	2419	0.71	2310	6600	3214	811	1331	54511	5.84
Sep-12	17435	7387	2.63	8.83	2502	0.24	6444	4878	0.48	1.10	1970	0.79	2221	6557	3529	790	1297	51341	5.44
Aug-12	16755	6232	2.38	8.79	2403	0.39	5795	4414	0.47	1.09	2264	0.81	2138	6179	2668	765	1164	47898	7.41
Jul-12	16329	5806	2.20	8.72	1914	0.27	5336	4403	0.55	1.20	2507	0.75	2058	6281	2571	770	1226	48048	7.26
Jun-12	12576	3886	2.59	10.54	1656	0.39	4008	2623	0.47	1.16	1957	0.80	1892	6016	2534	794	931	38201	8.87
May-12	12179	4105	2.09	8.15	1747	0.34	4096	2343	0.43	1.15	1447	0.82	2021	5510	2938	870	1149	38404	7.28
Apr-12	14221	4600	3.08	11.02	1697	0.34	4164	2919	0.57	1.39	1436	0.91	2051	6019	2550	784	1236	41798	7.88
2011-12 Av/D	12832	3690	1.89	8.46	1867	0.30	4501	2366	0.39	1.12	1628	0.75	1918	5338	2490	731	959	38232	6.57
2010-11 Av/D	23142	5700	2.25	11.40	2207	0.30	8841	3254	0.42	1.55	1936	0.80	2724	5918	2400	767	1295	49234	6.55

Date	Steam Cons Tons										FO Cons in Ltrs				Water cons. In KL	
	Dyeing	Dyeing Prodn kgs	Dyg Steam consKg / Kg	Spg	Spg prodn kgs	Spg Steam consKg / Kg	Fsg	Fsg Prodn mtrs	Fsg Steam consKg / mtr	Total Steam cons. In Tons	FO consboiler ltr	FO cons TFH ltr	Total FO cons in ltr	FO lts/ Fsg mtr	Total Water cons In KL	Water ltr / Fsg mtr
01-Jul-13	16.45	1911.9	8.60	5.17	2804	1.84	21.62	8152.2	2.65	46.99	3480	2350	5830	0.72	852	104.51
02-Jan-13	15.79	3645.8	4.33	4.96	2878	1.72	20.75	10107.4	2.05	45.1	3340	2050	5390	0.53	631	62.43
03-Jan-13	0.00	0	0.00	3.32	2833	1.17	13.88	7842.9	1.77	30.18	2235	2085	4320	0.55	402	51.26
04-Jan-13	18.88	4991	3.78	5.93	3010	1.97	24.81	7291.8	3.40	53.94	3995	2285	6280	0.86	745	102.17
05-Jan-13	18.19	5050.4	3.60	5.72	2773	2.06	23.91	6154.8	3.88	51.98	3850	2130	5980	0.97	791	128.52
Total	69.30	15599.10	20.32	25.10	14298.00	8.77	104.97	39549.10	13.76	228.19	16900.00	10900.00	27800.00	3.63	3421.00	448.88
Avg/Day	17.33	3899.78	5.08	5.02	2859.60	1.75	20.99	7909.82	2.75	45.64	3380.00	2180.00	5560.00	0.73	684.20	89.78
Avg/this year	12.87	2841.48	5.08	4.05	2456.25	1.76	16.93	7914.53	2.49	37.09	2723.25	1902.74	4582.25	0.61	595.89	91.84
Dec-12	16.49	3840.41	4.78	5.04	2767.06	1.84	20.86	10268.67	2.38	45.36	3359.19	2287.74	5646.94	0.65	691.94	79.23
Nov-12	15.12	3151.11	5.51	4.65	2806.86	1.66	19.45	8965.62	2.20	43.36	3220.36	2272.50	5492.86	0.59	622.36	66.84
Oct-12	14.94	3536.65	4.40	4.47	2811.73	1.61	18.71	9327.72	2.15	40.68	3016.30	2068.70	5085.00	0.54	618.73	68.98
Sep-12	12.77	2526.20	5.38	4.27	2923.38	1.51	17.60	11250.85	1.78	38.84	2900.43	2317.08	5318.26	0.47	608.08	63.05
Aug-12	13.69	2960.31	5.06	4.41	2628.03	1.68	18.00	8010.93	3.05	40.11	2970.00	2088.80	4864.23	0.63	568.84	110.02
Jul-12	13.55	3486.04	4.18	4.32	2738.48	1.99	17.81	7597.33	2.41	40.28	2597.26	1900.56	4252.58	0.64	604.52	88.35
Jun-12	11.23	2563.50	4.90	3.34	1750.32	2.13	14.47	5021.50	3.60	31.46	2329.81	1474.62	3749.81	0.84	547.07	137.21
May-12	9.51	1760.94	5.56	2.91	1933.45	1.46	12.87	5272.07	2.46	26.35	2088.27	1410.68	3498.95	0.66	526.55	102.54
Apr-12	8.50	1748.12	5.97	3.01	1746.92	1.99	12.59	5516.12	2.35	27.37	2027.60	1304.00	3331.60	0.60	574.88	110.34
2011-12 Ave/D	11.04	2174.84	5.08	3.40	1952.03	1.74	15.36	6288.67	2.44	30.41	2284.14	1333.65	3638.73	0.59	532.15	86.94
2010-11 Ave/D	11.46	2407.32	4.76	3.6	2529.45	1.42	17.68	7445.78	2.37	32.75	2460.15	1456.69	3916.84	0.52	581.16	78.05

*Dyg- Dyeing, Cons-Consumption, Spg-Spinning, Wvg-weaving, Fsg-Finishing, Prodn-Production, Ltg-Lighting, Comp-Compressor,Utiy-Utility, DG Aux-Generator set Auxiliary, Tsfo-Transformer

DAILY UTILITY REPORT

Example 2

Fabric Production (Mtr) /Day	10148	11992	11102	9521	9466	153909
Date	1-Jan	2-Jan	3-Jan	4-Jan	15-Jan	To date
Fuel						
Coal opening stock(MT)	85.94	74.78	168.85	89.45	75.25	
Coal consumption (Kg)	11160	13500	15640	14200	11900	170140
Coal Receipts (MT)	0	107.57	0	0	27.38	
Coal Closing Stock (MT)	74.78	168.85	153.21	75.25	90.73	
HSD Opening Stock (Ltr)	28975	27965	26435	13905	12745	
HSD Consumption (Ltr)	1010	1530	1320	1160	930	17160
HSD Receipt (Ltr)	0	0	0	0	0	
HSD Closing Stock (Ltr)	27965	26435	25115	12745	11815	
Steam Generation (MT)	89	107	124	98	94	1334
Steam Gen. vs Coal Consumption (Kg/Kg)	7.97	7.93	7.93	6.90	7.90	7.84
Steam Consumption/Mtr	8.77	8.92	11.17	10.29	9.93	8.67
HSD Consumption/Mtr	0.10	0.13	0.12	0.12	0.10	0.11
Power						
MD (KVA)	3459.6	3459.6	3459.6	3345.4	3345.4	3501.4
PF (Power Factor)	0.99	0.99	0.99	0.99	0.99	0.99
Power Failure (Time-Minutes)						40
Consumption-Fsg	3716	4842	4688	4082	3834	58752
Consumption-Dye House	3681	3185	5520	2865	3801	49811
Consumption-Weaving	7870	7564	8014	8540	8487	109998
Consumption-Post Spinning	10169	9569	10498	9737	9584	139009
Consumption-New Spinning	22154	23455	24853	21244	22593	324509
Consumption-Compressor	8136	8566	8854	8124	8192	115894
Consumption-Miscellaneous	6274	6119	7073	6508	5909	85047
Total Consumption (KWH)	62000	63300	69500	61100	62400	883020
DG Consumption (KWH)	0	0	0	0	0	920
Power Consumption/ Mtr of Fabric	6.11	5.28	6.26	6.42	6.59	5.74
Water						
Total Raw Water Discharge (KL)	1910	2473	2876	2220	2125	31409
Soft Water Consumption-Dye House (KL)	410	410	540	360	390	4970
Soft Water Consumption-Fabric section (KL)	150	111	151	129	162	1728
Soft Water Cons-Finishing (KL)	387	973	1111	825	700	11397
Total Soft Water Consumption (KL)	947	1494	1802	1314	1252	18095
Soft Water Consumption/Mtr	0.09	0.12	0.16	0.14	0.13	0.12

STORES AND PURCHASE DEPARTMENT DAILY /MONTHLY SUMMARY REPORT

LOCAL

Status as on 14th Feb, 2021

Orders Pending	Value in Rs Lacs	Pending Order in The Month	Value in Rs Lacs	Total Pending Orders	Value in Rs Lacs	No. of Invoices Received	Value in Rs Lacs	Payments for No. Of Suppliers	Value in Rs Lacs
63	46.54	11	3.46	74	50.00	72	20.15	0.00	0.00

IMPORT

Status as on 14ᵗʰ Feb, 2021	No. of Orders	Value in Rs.(Lacs)
Pending Order (Already Issued)	30	109.23
Advance Payment Made (In Transit)	4	14.90
COD Material in Transit (Documents Required)	0	0.00
Total Pending Import Orders	34	124.13
COD Material Recd. but not paid	0	0.00
Advance Payment made this month	1	3.98
Material Received During Feb-21		

MONTHLY CONSUMPTION

Description	Budget	Arp-20	May-20	Jun-20	Jul-20	Aug-20	Sept-20	Oct-20	Nov-20	Dec-20	Jan-20	Feb-21	Mar-21	Avg.
Production	40	31.22	31.47	32.75	72.82	7.97	7.15	50.14	19.60	17.26	9.69			28.87
Textile Lube	5	6.87	6.82	3.72	4.24	1.92	2.85	2.53	3.79	2.78	2.80			3.83
Engg.	7.5	14.53	16.08	6.19	5.80	5.86	3.62	7.02	4.58	7.52	6.08			7.73
Total	52.5	52.62	54.36	42.66	82.87	15.75	13.62	59.69	27.98	27.56	18.57			39.57
Fuel	Actual	4.50	5.13	4.52	5.66	4.97	3.97	5.63	3.10	4.90	1.77			4.42
Packing	Actual	15.40	17.35	11.35	12.00	11.80	9.30	6.85	11.40	9.91	9.68			11.51
Others	Actual	2.31	2.71	3.57	1.03	1.15	1.24	0.53	0.69	0.92	0.75			1.49
Capital	Actual	0.38	0.80	22.09	9.91	0.00	0.00	0.00	5.24	6.07	1.14			4.56

PAYMENT, CONSUMPTION, RECEIPTS & ADVANCE PAYMENT STATUS (Rs., In Lacs)

	Apr-20	May-20	Jun-20	Jul-20	Aug-20	Sep-20	Oc-20	Nov-20	Dec-20	Jan-21	Total	
Payment	115.25	95.57	87.12	15.48	5.58	68.92	94.00	43.45	38.54		563.91	563.91
Consumption (Without Coal)	71.00	76.82	82.45	105.91	28.79	24.22	66.97	45.92	45.66	31.03	585.39	
GST Payment	5.85	8.38	4.58	5.83	2.71	3.00	5.37	5.55	7.20	3.12	48.47	633.86
Receipts (Without Coal)	75.51	82.40	61.23	102.59	20.56	18.33	67.86	47.75	43.33	21.32	519.56	
GST Payment	5.85	8.38	4.58	5.83	2.71	3.00	5.37	5.55	7.20	3.12	48.47	
Advance with Local Supplier										2.00	2.00	
Advance with Import Supplier			2.57							12.33	14.90	571.79

SAVING WITHIN FOUR MONTHS FROM LAST SUPPLY (Rs., In Lacs)

In Local Supplies	5.99
In Import Supplies	1.89
Total	**7.88**

ONE TIME SPARE PARTS STATUS

Under One Time Budget of 200 Lacs for Production Department(Rs., In Lacs)

Priority	Indent Value	Order Placed	Payment Made	Material Received
1	77.02	77.02	77.02	77.02
2	77.04	40.29	0	0

Under One Time Budget of 50 lacs for Utility (Rs., In Lacs)

Priority	Indent Value	Order Placed	Payment Made	Material Received
1	25.79	23.60	23.60	23.60

URGENT ITEMS STATUS (Rs., In Lacs)

Supplier	Value	Today's Status	Expected in Plant
GV-20 Spares	9.14	Payment Made on 17.01.20	1 Month After Payment
Leather Apron	3.47	Payment Made on 17.01.20	1 Month After Payment
Circular Brush	2.68	Pending for Payment	1 Month After Payment
Carding Wire	3.39		Received at Stores

STORE INVENTORY DETAILS SUMMARY JAN-2020

Department	Regular Spares (A)	Slow Moving (B)	INSURANCE (C)	DEAD (D)	TOTAL (A+B+C+D)
Wool Preparatory/ Top Making	8	3	3	2	15
Wool Spinning	12	1	5	0	18
Wool Post Spinning	0	1	0	0	1
Instrumentation	12	1	12	1	26
Bearing	0	12	0	0	12
Others Spares (P. Return)	0	2	0	0	2
Packing	39	0	0	0	39
Textile Lubricants	4	0	0	0	4
DYEING	11	0	0	0	11
JOB WORK SPARES	1	0	0	0	1
JOB WORK PACKING	10	0	0	0	10
General Stores	6	0	0	0	6
TOTAL	104	18	20	4	146

LOCAL OUTSTANDING (Rs. In Lacs)

Description	30-60	61-120	121-180	Above 181 Days	TOTAL
DYES-CHEM	0	0	0	23	23
Stores & Spares	35	18	4	18	76
SALT	0	0	0	0	0
COAL	3	0	0	11	14
PACKING	8	11	9	6	33
TOTAL	47	29	13	57	146

ANNUAL SPARES AND STORES CONSUMPTION HISTORY

Details of Consumption in Rs., Lacs					
Department	FY 1	FY 2	FY 3	FY 4	FY 5
GENERAL/SPARES					
Combing	66	37.72	40.97	33.57	17.3
RM Godown	10	1.95	2.42	3.66	0.72
Dyeing	17	15.73	24.69	22.03	15.98
Re Combing	41	32.74	39.46	33.52	32.71
Spinning	102	74.95	89.8	93.28	55.86

Sulzer	55	56.94	40.21	26.63	29.37
Warping/ Yarn Store	10	9.79	14.6	13.33	7.1
Finishing/ Perching	115	95.64	86.15	82.04	75.98
Engineering/R&D	114	104.43	92.05	95.09	89.03
ETP	8	18.83	19.52	9.35	11.37
Others					
Total	537	448.72	449.87	412.5	335.42
OIL & FUEL					
Coal & Lignite	406	324.74	347.41	313.13	283.97
HSD/Lubricant	8	6.35	5	3.14	3.24
Total	414	331.09	352.41	316.27	287.21
DYES/CHEMICALS					
Top Dyeing	187	139.3	182.65	214.6	176.29
Fabric Dyeing	37	50.77	75.02	78.39	58.68
Finishing	102	89.96	81.02	125.98	84.23
Total	326	280.03	338.69	418.97	319.2
Packing Materials	218	164.2	158.54	138.36	119.08
Building Maintenance	38	35.18	40.36	33.59	28.01
G TOTAL	**1533**	**1259.22**	**1339.87**	**1319.69**	**1088.92**

VENDOR WISE DYES AND CHEMICALS PURCHASE AND CONSUMPTION

Property	Item	Brand Name	Average Rate/Kg	Lead Time	Avg. Monthly Cons.	Amount	Avg Daily Cons.	Value
			Rs	Days	Qty.Kg	Rs	Qty.Kg	Rs
Reduction Cleaner Agent			155.16	3	763	1,18,434	25	3,948
Acid			51.59	3	5,264	2,71,544	175	9,051
Anti Creasing Agent			77.94	3	60	4,676	2	156
Metal Complex Dyes(1:2)Wool Dyes			950.21	3	17	16,093	1	536
Metal Complex Dyes(1:2)Wool Dyes			448.90	3	300	1,34,670	24	10,669
Metal Complex Dyes(1:2)Wool Dyes			747.20	3	33	24,797	1	827
Metal Complex Dyes (1:2) Wool Dyes			579.35	3	200	1,15,870	15	8,690
Metal Complex Dyes (1:2) Wool Dyes			708.63	3	58	41,351	2	1,378
Metal Complex Dyes (1:2) Wool Dyes			898.04	3	43	38,920	1	1,297
Disperse Dyes (Polyester Dyes)			307.04	3	101	30,931	3	1,031
Disperse Dyes (Polyester Dyes)			309.70	3	253	78,240	8	2,608
Disperse Dyes (Polyester Dyes)			237.37	3	317	75,306	11	2,510
Reactive Cotton Dyes			475.38	3	37	17,727	1	591
Reactive Cotton Dyes			820.92	3	10.00	8,209	0.33	274
Reactive Cotton Dyes			1,552.88	3	4.00	6,212	0.13	207
Reactive Cotton Dyes			1,647.46	3	6.00	9,885	0.20	329
Reactive Cotton Dyes			323.24	3	6.00	1,939	0.20	65
Levelling Agent (Printing)			192.44	15	30	5,773	1	192

Levelling Agent& DISPERSING(Terene TOPS)			161.68	10	510	82,457	17	2,749
Scouring & Detergent Agent			75.43	10	600	45,258	20	1,509
Metal Complex Dyes (1:2) Acid Dyes			509.36	10	161	81,779	5	2,726
Metal Complex Dyes (1:2) Acid Dyes			608.00	10	147	89,601	5	2,987
Metal Complex Dyes (1:2) Acid Dyes			1,113.50	10	68	76,163	2	2,539
Metal Complex Dyes (1:2) Acid Dyes			346.03	10	32	10,900	1	363
Softner			341.91	15	175	59,800	6	1,993
Scouring & Detergent Agent			115.15	15	958	1,10,352	32	3,678
Acid Buffer			11.63	5	2,727	31,718	91	1,057
Levelling Agent (Fine Wool, Cotton)			5.00	5	3,099	15,497	103	517
Acid Buffer			15.01	5	1,814	27,225	60	908
Alkali (Cotton Dyes)			22.03	5	480	10,574	16	352
Reducing Agent			109.72	3	300	32,949	10	1,098
Washing for Wool Reactive			12.66	5	500	6,327	17	211
Bleaching Chemicals			42.50	10	120	5,100	4	170
Bleaching Chemicals			152.98	5	120.00	18,357.60	4	612
Bleaching Chemicals			19.79	5	120.00	2,374.80	4	79
Bleaching Chemicals			42.50	5	150.00	6,375.00	5	213
Blcaching Chcmicals			38.38	5	30.00	1,151.40	1	38
						1,57,650		5,255
Antistatic Agent (Polyester)			219.34	20	1,050	2,30,307	44	9,561
Wool Procter Agent			217.28	10	400	86,912	24	5,215
Deairting Agent/Wetting			178.29	10	700	1,24,803	25	4,534
Dispersing Agent (Cotton)			185.62	10	35	6,426	1	214
Levelling Agent (Wool)			165.28	10	800	1,32,224	27	4,463
Skin Peach Effect			356.66	15	350	1,24,831	12	4,161
Skin Peach Effect			455.02	10	233	1,06,171	8	3,539
Disperse Dyes (Polyester)			552.24	7	145	79,854	5	2,662
Disperse Dyes (Polyester Dyes)			243.55	7	83	20,283	3	676
Disperse Dyes (Polyester Dyes)			273.61	7	117	32,099	4	1,070
Disperse Dyes (Polyester Dyes)			863.94	7	30	25,918	1	864
Carrier Agent			163.66	10	240	39,278	8	1,309
Anti-Bleeding Agent (FNG)			167.49	15	120	20,099	4	670
Anti-Creasing Agent			57.90	15	400	23,160	13	772
						2,40,691		8,023
Softener			79.51	15	600	47,706	20	1,590
Foaming Agent			103.06	15	210	21,643	7	721
Body Builder Agent			142.12	10	140	19,897	5	663
Waxing Agent			80.75					

						89,245		2,975
Softner			131.30	15	67	8,753	2	292
Skin Peach Effect			410.46	15	150	61,569	5	2,052
Wrinkle Free Agent			354.62	15	33	11,821	1	394
Skin Peach Effect			377.36	15	67	25,157	2	839
						1,07,300		3,577
Waxing Agent			154.36	75	360	55,492	12	1,850
Tinting Agent (Colour)			516.98	75	16	8,384	1	279
Tinting Agent (Colour)			516.98	75	21	10,807	1	360
Tinting Agent (Colour)			516.98	75	19	9,581	1	319
Alkali (Cotton Dyes)			41.38	5	877	36,286	29	1,210
Water & Oil Repellent			1,143.08	15	25	28,577	1	953
Antistatic Agent(Wool)			107.61	10	600	64,566	20	2,152
Disperse Dyes (For Polyester)			266.48	12	231	61,666	8	2,056
Disperse Dyes (For Polyester)			1,255.55	12	95	1,19,091	3	3,970
Disperse Dyes (For Polyester)			402.87	12	60	24,172	2	806
Disperse Dyes (For Polyester)			183.17	12	60	10,990	2	366
Disperse Dyes (For Polyester)			365.05	12	60	21,903	2	730
Disperse Dyes (For Polyester)			321.50	12	25	8,038	1	268
Reactive Dyes (For Wool)			252.74	12	500	1,26,370	25	6,319
						3,72,230		14,514
GRAND TOTAL						33,03,511		1,27,861

ANNUAL COMPARATIVE SALES, EXPENSES, RM CONSUMPTION, AND SALES ANALYSIS

		2008-09	2009-10	2010-11	2011-12
1	Total Sales	12,45,181	12,70,804	14,34,406	18,80,435
2	Fabric Sales '000 Mtrs.				
3	Domestic	2,745	3,071	2,729	3,080
4	Govt/Institutional	695	699	750	708
5	Export	684	546	969	1,192
7	Total	4,124	4,316	4,448	4,980
8	Fabric Sales in 000 Rs				
9	Domestic	8,43,727	9,36,427	8,94,422	11,17,174
10	Govt/Instt	1,67,614	1,66,869	2,30,610	2,80,856
11	Export	2,16,854	1,58,084	2,74,462	4,37,929
12	Total Fabric sales	12,28,195	12,61,380	13,99,493	18,35,959
13	Yarn sales	369	286	2,011	2,068
14	Waste Sales	8,521	5,411	9,990	20,179
15	Export Benefi	8,096	3,727	22,911	22,229
16	Total Income	12,45,181	12,70,804	14,34,406	18,80,435
16 a	Realisation/mtr Rs (Sales Value/sales Mtrs)				
17	Domestic	307	305	328	363

18	Govt/Instt	241	239	307	397
19	Export	317	290	283	367
20	Fabric sales Ave Realisation/Mtr	298	292	315	369
21	Other Revenues	4	2	8	9
22	Total Realisation Per meter	302	294	322	378
	Raw material in 000 Rs				
23	RM Consumed	4,93,900	3,92,402	6,57,177	9,81,791
24	Cloth Purchases	22,701	86,484	50,933	55,346
25	(Inc)/Dcs in FG	(54,747)	(842)	(1,07,991)	(1,55,512)
26	(Inc)/Dcs in WIP	2,400	(25,297)	(44,424)	34,766
27	(Inc)Dcs Waste and scrap	601	134	(303)	7
28	Total RM Consumption (A)	4,64,855	4,52,881	5,55,392	9,16,397
29	RM Consumed Per metre of fabric sold (23/7)	120	91	148	197
30	Total RM Consumption Per metre of fabric sold	113	105	125	184
31	RM Consumption/(Fabric Sale)	37.8%	35.9%	39.7%	49.9%
32	Dyes & Chemical Consumed	51,676	45,957	60,124	58,720
33	Processing Charges	8,380	6,593	18,027	7,125
34	Stores, Consumables & Packing	59,478	63,498	72,931	84,304
35	Power	1,19,579	98,183	1,25,015	1,45,870
36	Fuel	58,347	61,791	59,941	79,426
37	Salary, Wages & benefit	1,74,021	1,86,002	2,12,330	2,20,661
38	Total Factory Cost (B) (Sum32-38)	4,71,481	4,62,025	5,48,369	5,96,105
39	Total Factory Cost Per metre of fabric sold (38/7)	114	107	123	120
40	Total Cost of Sales (A+B)	9,36,336	9,14,905	11,03,761	15,12,502
41	Total Cost of sales Per metre of fabric sold	227	212	248	304
42	Gross Profit (16-40)	3,08,845	3,55,899	3,30,645	3,67,933
43	Gross Profit Margin (42/16)%	24.8%	28.0%	23.1%	19.6%
44	Gross Profit per meter of fabric sold	75	82	74	74
45	WIP Quantity (KGs)	161	205	223	223
46	WIP Quantity (Mtrs)	123	112	204	204
47	FG Quantity (Mtrs)	742	685	1,269	1,657
48	% of RM Consumed to Sales	39.7%	30.9%	45.8%	52.2%
49	% of Cloth Purchases to Sales	1.8%	6.8%	3.6%	2.9%
50	% of Incs/Dcs in FG to Sales	-4.4%	-0.1%	-7.5%	-8.3%
51	% of Incs/Dcs in WIP to Sales	0.2%	-2.0%	-3.1%	1.8%
52	% of Incs/Dcs in Waste to Sales	0.0%	0.0%	0.0%	0.0%
53	% of Total RM consumed to Sales	37.3%	35.6%	38.7%	48.7%
54	Quantity of FG Mtr	742	685	1,269	1,657
55	Inc/Dec in FG Mtrs		(57)	585	388
56	Value of FG (Incld Pcs)	1,58,205	1,59,048	2,68,601	4,24,113
57	Per Mtr. FG	213	232	212	256
58	Inc/Dec in FG Value		842	1,09,553	1,55,512
59	FG Cycle (on sales)	47	46	70	84
60	FG Cycle (on quantity)	18%	16%	29%	33%
61	Qty of WIP (Kgs)	161	205	223	223
62	Value of WIP	46,726	73,441	1,00,098	1,10,894
63	Qty of WIP (Finishing -Mtr)	123	112	204	204

64	Value of WIP (Finishing)		21,497	20,080	37,847	46,877
65	Total Value (WIP)		68,224	93,521	1,37,945	1,57,771
66	Volume Mix					
67	Civil		68.7%	74.2%	63.9%	60.8%
68	Govt/Instt		13.6%	13.2%	16.5%	15.3%
69	Export		17.7%	12.5%	19.6%	23.9%
70	Total Fabric sales (%)		100.0%	100.0%	100.0%	100.0%
71	Growth Rate in Realisations					
72	Civil			-0.8%	7.5%	10.7%
73	Govt/Institutional			-1.0%	28.8%	29.0%
74	Export			-8.7%	-2.2%	29.7%
75	Total			-1.9%	7.7%	17.2%
76	Growth Rate in Volumes					
77	Civil			11.9%	-11.1%	12.9%
78	Govt/Institutional			0.6%	7.3%	-5.6%
79	Export			-20.2%	77.5%	23.0%
80	Total			4.7%	3.1%	12.0%
81	Growth Rate in Value					
82	Civil			11.0%	-4.5%	24.9%
83	Govt/Institutional			-0.4%	38.2%	21.8%
84	Export			-27.1%	73.6%	59.6%
85	Total Fabric Sales Revenue			2.7%	10.9%	31.2%
86	Other Income			-44.5%	270.5%	27.4%
87	Total Income			2.1%	12.9%	31.1%
88	Growth Rate in Closing Stock					
89	WIP Quantity (KGs)			27.4%	8.9%	0.2%
90	WIP Quantity (Mtrs)			-8.7%	81.4%	0.1%
91	FG Quantity (Mtrs)			-7.7%	85.4%	30.5%

*Incs -Increase, Dcs-Decrease, FG-Finished goods, WIP-Work in progress, RM-Raw material

MONTHLY SALES, RM CONSUMPTION, STOCK VALUATION

Sales (Mtrs)	Apr	May	Jun	Jul	Aug	Sep
Domestic	2,56,000	2,06,000	1,79,000	1,19,000	2,27,000	3,42,000
RMG/Govt	98,000	55,000	73,000	1,08,000	60,000	90,000
Export	1,30,000	1,60,000	92,000	1,96,000	2,27,000	56,000
Total Sales Mtrs	4,84,000	4,21,000	3,44,000	4,23,000	5,14,000	4,88,000
Realisation/Mtr						
Domestic Sales	341	285	304	299	328	393
Govt Sales	384	430	441	426	376	379
Export Sales	340	336	380	328	355	338
Average	349	324	354	345	346	384
Sales Value (Rs)						
Civil Sales Value	8,72,83,000	5,88,12,000	5,44,62,000	3,55,40,000	7,44,07,000	13,43,28,000
Govt Sales Value	3,76,20,000	2,36,28,000	3,22,02,000	4,59,99,000	2,25,54,000	3,41,31,000
Export Sales Value	4,42,00,000	5,37,91,000	3,49,71,000	6,41,97,000	8,06,35,000	1,89,31,000
Total Sales Value	16,91,03,000	13,62,31,000	12,16,35,000	14,57,36,000	17,75,96,000	18,73,90,000
Raw Material Consumption (Kg)						

Wool						
Opening Stock	54,725	66,072	50,130	45,408	1,08,465	1,42,726
Purchases	97,182	96,628	88,897	1,60,227	1,87,200	1,07,168
Issued to the Plant	85,835	1,12,570	93,618	97,170	1,52,939	1,28,608
Sale	-	-	-	-	-	-
Loss	-	-	-	-	-	-
Closing Stock	66,072	50,130	45,408	1,08,465	1,42,726	1,21,286
RM Consumption (A)	85,835	1,12,570	93,618	97,170	1,52,939	1,28,608
Polyester						
Opening Stock	1,49,429	2,15,190	2,18,564	2,22,845	2,16,226	1,81,432
Purchases	1,10,417	54,230	46,895	63,331	35,967	55,573
Issued to the Plant	44,657	50,856	42,438	69,951	70,762	78,713
Sale	-	-	176	-	-	-
Loss	-	-	-	-	-	-
Closing Stock	2,15,190	2,18,564	2,22,845	2,16,226	1,81,432	1,58,291
RM Consumption (B)	44,657	50,856	42,438	69,951	70,762	78,713
Yarn						
Opening Stock	1,49,429	2,15,190	2,18,564	2,22,845	2,16,226	1,81,432
Purchases	1,10,417	54,230	46,895	63,331	35,967	55,573
Issued to the Plant	44,657	50,856	42,438	69,951	70,762	78,713
Sale	-	-	176	-	-	-
Loss	-	-	-	-	-	-
Closing Stock	2,15,190	2,18,564	2,22,845	2,16,226	1,81,432	1,58,291
RM Consumption (C)	44,657	50,856	42,438	69,951	70,762	78,713
RM Consumption (A+B+C) Kg	1,75,148	2,14,282	1,78,493	2,37,072	2,94,463	2,86,035
RM Consumption Value Rs						
Wool						
Opening Stock	332	62,08,827	93,58,634	44,11,585	5,97,90,039	7,84,11,758
Purchases	6,66,83,142	7,78,91,172	7,72,62,731	13,85,65,843	12,44,33,339	8,35,69,923
Issued to the Plant	6,03,84,647	7,48,31,365	8,22,09,780	8,31,87,389	10,58,11,620	10,36,87,904
Sale	-	-	-	-	-	-
Loss	-	-	-	-	-	-
Closing Stock	62,98,827	93,58,634	44,11,585	5,97,90,039	7,84,11,758	5,82,93,777
RM Consumption (A)	6,03,84,647	7,48,31,365	8,22,09,780	8,31,87,389	10,58,11,620	10,36,87,904
Polyester						
Opening Stock	2,00,41,911	3,13,63,225	3,23,59,505	3,21,39,195	3,08,37,394	2,61,17,531
Purchases	1,82,58,971	89,05,740	60,22,788	86,01,983	52,54,831	74,97,691
Issued to the Plant	69,37,657	79,09,460	62,17,872	99,03,784	99,74,694	1,14,24,331
Sale	-	-	25,225	-	-	-
Loss	-	-	-	-	-	-
Closing Stock	3,13,63,225	3,23,59,505	3,21,39,195	3,08,37,394	2,61,17,531	2,21,90,891
RM Consumption (B)	69,37,657	79,09,460	62,17,872	99,03,784	99,74,694	1,14,24,331
Yarn						

Opening Stock	2,00,41,911	3,13,63,225	3,23,59,505	3,21,39,195	3,08,37,394	2,61,17,531
Purchases	1,82,58,971	89,05,740	60,22,788	86,01,983	52,54,831	74,97,691
Issued to the Plant	69,37,657	79,09,460	62,17,872	99,03,784	99,74,694	1,14,24,331
Sale	-	-	25,225	-	-	-
Loss	-	-	-	-	-	-
Closing Stock	3,13,63,225	3,23,59,505	3,21,39,195	3,08,37,394	2,61,17,531	2,21,90,891
RM Consumption (C)	69,37,657	79,09,460	62,17,872	99,03,784	99,74,694	1,14,24,331
Total RM Consumption (A+B+C) Rs	7,42,59,962	9,06,50,285	9,46,45,525	10,29,94,956	12,57,61,009	12,65,36,566
WIP Quantity						
Godown (KGs)	-	-	-	-	-	-
Shop Floor (KGs)	1,95,003.0	2,30,117.0	2,27,360.0	2,28,137.0	2,76,312.0	2,77,359.0
Finishing (Mtrs)	2,66,136.0	1,81,914.0	1,90,306.0	1,81,325.0	1,81,464.0	1,98,680.0
WIP Value						
Godown	-	-	-	-	-	-
Shop Floor	8,46,27,000.0	9,88,59,000.0	9,76,96,000.0	9,40,98,000.0	11,46,15,000.0	11,29,71,000.0
Finishing	5,53,04,000.0	3,94,16,000.0	4,18,00,000.0	3,79,04,000.0	3,68,12,000.0	4,05,28,000.0
Total WIP Value Rs	13,99,31,000.0	13,82,75,000.0	13,94,96,000.0	13,20,02,000.0	15,14,27,000.0	15,34,99,000.0
Opening stock	13,79,45,341.8	13,99,31,000.0	13,82,75,000.0	13,94,96,000.0	13,20,02,000.0	15,14,27,000.0
Adjustment (change in definition						
Closing Stock	13,99,31,000.0	3,82,75,000.0	13,94,96,000.0	13,20,02,000.0	15,14,27,000.0	15,34,99,000.0
Change in WIP Stock (Rs in '000)	(1,985.7)	1,656.0	(1,221.0)	7,494.0	(19,425.0)	(2,072.0)
FG Quantity						
Fabric (Mtrs)	11,09,501.8	11,03,789.9	11,20,337.6	11,82,382.0	11,56,000.0	11,79,311.0
Others (Pcs)						
FG Value						
Fabric	22,76,57,000	22,61,49,000	20,08,90,000	23,57,47,000	24,37,37,000	24,33,42,000
Others						
Total FG Value Rs	22,76,57,000.0	22,61,49,000.0	20,08,90,000.0	23,57,47,000.0	24,37,37,000.0	24,33,42,000.0
FG Value/Sales Value	1.3	1.7	1.7	1.6	1.4	1.3
Opening stock	26,86,00,563.7	22,76,57,000.0	22,61,49,000.0	20,08,90,000.0	23,57,47,000.0	24,37,37,000.0
Closing Stock	22,76,57,000.0	22,61,49,000.0	20,08,90,000.0	23,57,47,000.0	24,37,37,000.0	24,33,42,000.0
Change in FG Stock (Rs in '000)	40,943.6	1,508.0	25,259.0	(34,857.0)	(7,990.0)	395.0
Change in Waste/Scrap (Rs in '000)	(565.0)	(246.0)	(833.0)	(637.0)	(235.0)	1,231.0
Cloth Purchase (Rs in '000)	1,002.0	(23.0)	2,136.0	1,288.0	-	6,454.0
Total RM Consumption	7,42,99,356.4	9,06,53,179.7	9,46,70,865.9	10,29,68,244.1	12,57,33,358.6	12,65,42,574.5
Total RM Consumption/Sales Value	43.9%	66.5%	77.8%	70.7%	70.8%	67.5%

MONTHLY PROFIT AND LOSS SUMMARY

Profitability Statement													
All figures in 000													
Particulars	**Apr**	**May**	**Jun**	**Jul**	**Aug**	**Sep**	**Oct**	**Nov**	**Dec**	**Jan**	**Feb**	**Mar**	**Total**
Production Mtrs	324	469	505	511	548	460	448	422	495	491	401	326	5400
O/S Purchases	25	25	30	50	50	60	60	60	40	85	65	50	600
	349	494	535	561	598	520	508	482	535	576	466	376	6000
Sales Mtrs													
Domestic	137	152	186	286	365	424	420	380	247	422	338	343	3700
Govt/Institutional	18	70	118	138	140	143	62	71	56	82	80	22	1000

Export -EU	0	15	30	35	30	35	35	35	30	30	30	45	350
Export -US	30	30	40	45	45	40	40	45	40	45	60	40	500
Export -Others	18	9	31	27	36	50	43	47	46	48	61	34	450
Total	203	276	405	531	616	692	600	578	419	627	569	484	6000
Job Spg. Ton	40	40	40	50	50	50	50	50	50	50	50	50	570
000 Rs													
Sales Value													
Civil - Domestic	46458	53110	63011	96546	123055	156746	166715	133915	83974	137059	110368	110645	1281602
Less: TOI	3020	3452	4096	6275	7999	10189	10836	8704	5458	8909	7174	7192	83304
Net Sales	43438	49658	58915	90271	115056	146558	155879	125211	78516	128150	103194	103453	1198298
Govt./ Institutional	4887	18579	38042	40442	43474	39325	15483	18489	13688	19847	20559	4403	277217
Export -EU	0	4253	9696	11110	8713	9491	10166	10583	8445	9032	9032	12459	102980
Export -US	10089	10089	13198	14981	14308	12690	12690	15349	13739	15133	19631	13451	165346
Export -Others	4373	2014	7385	6382	6831	12130	9824	9801	11838	11000	15029	8002	104609
	62786	84593	127236	163186	188381	220194	204043	179432	126226	183162	167445	141768	1848451
Job. Spinning earning	3150	3150	3150	3938	3938	3938	3938	3938	3938	3938	3938	3938	44888
Waste Sale	274	392	454	463	484	411	390	362	425	425	316	258	4654
Other Income (Interest)	742	742	742	942	1092	1002	1092	942	992	1142	1242	1142	11900
Export Benefi	723	818	1514	1624	1493	1716	1634	1787	1701	1758	2185	1696	18647
Total Income	67675	89694	133095	170151	195387	227350	211096	186460	133281	190424	175125	148800	1928538

Manufacturing Exps													
(Incr.) Decr. in FG	(34070)	(50872)	(30337)	(7001)	4200	40138	21469	22402	(27070)	11901	24036	25203	0
(Incr.) Decr. in WIP	0	0	0	0	0	0	0	0	0	0	0	0	0
Raw Material	37274	56069	65622	66832	72047	64267	61798	53029	60369	63378	43642	33027	677352
O/S Purchases	3675	3675	4250	7750	7350	8500	8500	8500	5800	11775	9475	5750	85000
Dyes & Chemicals	3175	4602	5357	5480	5725	4878	4615	4255	5027	5006	3646	2915	54680
Packing	1292	1732	1777	2502	3226	4081	4317	3350	1816	3346	2746	2702	32887
Stores & Consumables	3180	4268	4538	4583	4860	4200	4110	3915	4463	4433	3758	3195	49500
Power & Fuel	14114	19646	20560	21126	22333	19282	18977	18241	21025	20873	17439	14579	228194
Total	28640	39118	71767	101271	119741	145345	123786	113692	71430	120711	104713	87370	1127613
Admin and Other Expo	3843	3843	3843	3843	3843	3843	3843	3843	3843	3843	3843	3843	46120
Publicity	2000	2000	8000	7000	7000	7000	10000	10000	10000	10000	10000	7000	90000
Selling Expenses	3209	4292	6378	8568	9989	12625	13707	10123	7059	10402	9326	8060	103737
Employees Cost	19241	19631	19799	20258	20073	19877	19757	19434	20112	19202	19422	19902	236708
Total Operating Exps.	56933	68885	109707	140040	160646	188691	171094	157092	112444	164159	147334	126175	1604179
EBITDA	10742	20809	23308	29212	34741	38659	40002	29369	20837	26265	27791	22625	324360
% of total income	15.9%	23.2%	17.5%	17.2%	17.8%	17.0%	18.9%	15.8%	15.6%	13.8%	15.9%	15.2%	16.8%
Corporate Expenses													
Fees / Commission	1433	1433	1619	1726	1686	1589	1589	1748	1652	1735	2005	1634	19848
Expenses.	0	0	500	0	0	0	0	0	500	0	0	0	1000
Net EBITDA	9310	19377	21189	27486	33055	37070	38413	27620	18686	24530	25785	20991	303512
% of Total income	13.8%	21.6%	15.9%	16.2%	16.9%	16.3%	18.2%	14.8%	14.0%	12.9%	14.7%	14.1%	15.7%
Interest	3533	3399	3425	3931	4264	4218	4552	4190	3856	3609	3324	3064	45366
PBDT	5776	15977	17764	23554	28791	32853	33861	23431	14830	20921	22461	17927	258146
Depreciation	9330	9343	9347	9367	9367	9370	9452	9644	9843	9895	9895	9885	114736
Net PBT	(3554)	6634	8418	14187	19424	23483	24409	13787	4987	11026	12567	8042	143410
Income Tax	0	0	0	2344	0	0	11642	0	0	8805	0	6450	29241
Net Income / (Loss)	(3554)	6634	8418	11843	19424	23483	12768	13787	4987	2221	12567	1591	114169

*TOI-Turn over incentive, Exps-Expenses, Incr-Increase, Decr-Decrease

DAILY CASH FLOW OF A WORSTED SPINNING UNIT

Particulars	Date	01-Dec	02-Dec	03-Dec	04-Dec	05-Dec	06-Dec	07-Dec	08-Dec	09-Dec	10-Dec	To date
Opening Balance Cash Flow	92.45	38.23	23.48	25.24	116.42	131.89	10.99	10.24	33.74	38.54	60.03	38.23
INFLOW (Collection)												
Domestic sale Area 1	145.98	3.56	2.23	1.49	16.57	0.63	5.23	11.59	2.71	6.55	5.58	56.15
Domestic sale Area 2	9.96			3.37						4.32		7.69
Total	155.94	3.56	2.23	4.86	16.57	0.63	5.23	11.59	2.71	10.87	5.58	63.84
Export sales Area 1	56.36	70.23										70.23
Export sales Area 2	988.31			10.31			104.24			68.15	8.88	191.58
Waste Sales (Export)	25.60			19.06						19.65		38.71
Waste sales (Domestic)	17.46									3.50		3.50
Total	1,244.33	73.83	2.32	34.29	16.57	0.63	5.24	115.88	2.71	102.23	14.50	368.21
Duty Draw Back	31.26		0.78	56.79						13.04		70.61
Scrap sale	0.65	0.05	0.08	0.07			0.01	0.05		0.06	0.04	0.37
Other Income	29.13						19.33					19.33
Total Inflo	1,304.71	73.83	3.10	91.08	16.57	0.63	24.57	115.88	2.71	115.27	14.50	458.15
OUTFLOW (Payment)												
R.M. Party Supplier 1	131.14							5.72		1.73		7.45
R.M. Party Supplier 2	29.91		-0.50	-0.36				32.67				31.81
R.M. Party Supplier 3	163.80											
R.M. Party Supplier 4												
R.M. Party Supplier 5												
R.M. Party Supplier 6	460.48	72.93										72.93
R.M. Party Supplier 7	325.00					120.00		40.00	40.00	10.00	5.50	215.50
Custom Duty/GST												
Raw Material Fright	16.95						0.29	0.25			0.20	0.74
Excise Duty/CST/VAT/ Entry Tax/Service Tax	1.45	0.06	0.00	0.04	0.00	0.00	0.14	0.12	0.05	0.05	0.01	0.48
Net R.M. Payment	1,128.72	72.99	-0.50	-0.31	0.00	120.00	0.43	78.75	40.05	11.78	5.71	328.91
Stores & Spares	28.08	10.84			0.60		0.50	0.09		0.12	0.27	12.43
Special Repairs Buldg & Others	0.54									0.15		0.15
Packing Materials	10.78											
Stores & Spares	39.40	10.84			0.60		0.50	0.09		0.27	0.27	12.58
Capex												
Net Stores Payment	39.40	10.84			0.60		0.50	0.09		0.27	0.27	12.58
Sales Commission/Claim										4.45		4.45
Logistic - Export / Import	21.08											
Electricity Bill	69.63											
Coal	10.56		0.61		0.49							1.09
Salary / Wages etc	166.68		0.97	0.08		0.20	21.23	11.18		0.52		34.17
Overheads	27.72	1.48	0.27	0.09		1.31	3.14	2.12		2.03	0.01	10.45
Legal expenses	5.83									3.96		3.96
Insurance & Bank Charges	8.92	1.27	0.01	0.04	0.01	0.01	0.01	0.24	-0.00	0.33	0.01	1.93
-Net Overheads	42.48	2.75	0.27	0.13	0.01	1.32	3.16	2.36	-0.00	6.33	0.02	16.34

Account	1	2	3	4	5	6	7	8	9	10	11	12
Corporate expenses	15.10	2.00										2.00
Special expenses									-42.13	70.43		28.30
Overseas Payment	-134.72											
TOTAL OUTFLOW	1,358.93	88.58	1.34	-0.11	1.10	121.53	25.32	92.38	-2.08	93.78	5.99	427.85
Closing Bank Balance 1	38.23	23.48	25.24	116.42	131.89	10.99	10.24	33.74	38.54	60.03	68.53	68.53
Bank 2	64.44	64.44	64.44	64.44	64.44	64.44	64.44	64.44	64.44	64.44	64.44	64.44
- Gross Closing Bank Balance	102.67	87.92	89.68	180.87	196.33	75.44	74.68	98.18	102.98	124.47	132.97	132.97
- Capex	8.00	8.00	8.00	8.00	8.00	8.00	8.00	8.00	8.00	8.00	8.00	8.00
- Closing Cash Balance	159.69	95.92	97.68	188.87	204.33	83.44	82.68	106.18	110.98	132.47	140.97	140.97

TRIAL BALANCE AND BALANCE SHEET – SEPTEMBER- 201X

Trial Balance	Debit Amt.	Credit Amt
Account Head	Rs. In '000	Rs. In '000
Land	1,56,170	-
Buildings	2,86,160	-
Plant & Machinery	14,36,047	-
Computers	15,675	-
Furniture & Fixtures	16,025	-
Office Equipments	1,562	-
Vehicles	6,623	-
Capital Work in Progress(CWIP)	87,765	-
Less: Up-to-date Depreciation	(8,88,251)	-
Other Current Assets	124	-
Wool Tops local purchased	622	-
Greasy Wool - Imported	60,086	-
Synthetic Fibre	3,555	-
Synthetic Yarn	7,614	-
Dyes & Chemicals	4,886	-
Man Made & Wool Tops (WIP)	98,102	-
Process Stock	1,53,795	-
Finished Goods Stock	2,67,747	-
Waste Stock	1,345	-
Store & Spare Parts	21,382	-
Fuel & Furnace	6,286	-
Cloth Purchases	12,536	-
Profit on sales return	(3,298)	-
Sundry Debtors	4,70,126	-
Export. Debtors	58,468	-
Prov. for Doubtful Debts	(33,535)	-
Prov. for Doubtful Advances	(2,025)	-
Travelling Advance	264	-
Staff Advance	217	-
Workers Advance	44	-

Advance against exp.	32,845	-
Income Tax Advance Less : Prov.	3,396	-
Cenvat Recoverable	886	-
Pre Deposit	2,102	-
Security Deposit	11,404	-
Prepaid Expenses	2,582	-
Investments	3	-
Bank 1	(164)	-
Bank 2	221	-
Bank 3	471	-
Bank 4	9,109	-
Bank 5	3,906	-
Bank 6	43	-
Bank 7	89	-
Bank 8	(7,482)	-
FD	19,840	-
Cash in Hand	216	-
Share Capital	-	11,03,000
Share Premium account	-	4,32,726
Capital Reserve	-	250
P & L A/c	-	(61,364)
Bills Admitted	-	1,63,073
Interest Accrued but not due	-	2,146
Interest Accrued & due	-	-
Wool Creditors	-	2,45,967
Pub. Creditors	-	6,809
Stores and other Creditors	-	92,365
Unclaimed Salary	-	7
Unclaimed Wages	-	67
Unclaimed Spl. Inc.	-	83
Wages/Salary payable	-	4,139
Unpaid Bonus	-	18
TDS	-	1,394
Emp. L.I.C. Premium	-	23
E.S.I. Deductions	-	186
Trustees of P.F.	-	2,006
Pension Fund A/C	-	346
Staff Security Deposit	-	5,537
Emp. Cooperative Stores	-	-
Applicable Sales Tax payable	-	306
Value Added Tax	-	-
Service Tax & Excise Payable	-	54
Trade Deposits	-	29,326

Bank Secured Loans-Term Loan	-	1,26,125
Bank Auto Loan (Secured)	-	-
Bank C/C	-	2,37,609
Sales-Domestic	-	6,10,406
Sales-Govt.	-	58,301
Sales-Instnl.	-	27,587
Sales Export	-	1,75,434
Export Benefits received	-	6,329
Misc. Yarn Sales	-	1,279
Garments, Accessories Sales	-	19
Job Work (Receipts)	-	15,662
Waste & Noils Sales	-	9,037
Turn Over Incentive	-	(35,804)
Sales Returns	-	(5,273)
Claims & Cash Discounts	-	(4,783)
Excise duty on sales	-	511
Foreign Currency Fluctuation Gain/Loss	-	6,336
Misc. Income	-	2,346
P/L on Sale of FA & FA Discard	-	8,962
Bad debts & Sundry Balances write off	-	301
Provision for doubtful debts	1,200	-
Raw Material Consumed (RM)	3,68,591	-
Job charges	12,536	
Raw Material Consumed (Dyes)	19,205	-
Stores Consumed	22,416	-
Inc/Dcs in WIP (Poly & Wool Tops)	(17,175)	-
Increase/Decrease in FG	1,66,366	-
Inc/Dcs in Waste	(953)	-
Inc/Dcs in WIP	(50,616)	-
Salary	38,665	-
Pension/Allowances	7,969	-
Wages	23,673	-
Dearness Allowance	8,381	-
Special Incentive	7,285	-
Ex-Gratia to Staff	3,058	-
Ex-Gratia to Workers	101	-
Bonus	4,056	-
Staff P.F.	4,131	-
Workers P.F.	3,995	-
E.S.I. Contribution	2,477	-
Leave Pay - Workers & Staff	668	-
Contbn.to Gratuity Fund	11,787	-
Contbn. to Superannuation	817	-

Welfare & Medical Expenses	4,476	-
LTA	1,910	-
LIC Group Insurance	519	-
Misc. Processing Charges	2,316	-
Electricity Charges	66,131	-
Fuel Consumed	30,376	-
Stationery & Printing	740	-
Postage	25	-
Courier Charges	265	-
Telephone & FAX Charges	1,004	-
Travelling Exps	6,713	-
Conveyance Exps.	2,192	-
Vehicle Running	978	-
Rates & Taxes	1,351	-
Rent	1,699	-
Books & Periodical	300	-
Fees & Subscriptions	993	-
Legal & Professional	13,173	-
Corporate Fee	10,436	-
Corporate Commission	1,902	-
One Time Management Fee	4,933	-
Gardening Expenses	19	-
Insurance Charges	507	-
Charity & Donation	385	-
Audit Fee	1,588	-
Directors' Fee	-	-
General Charges	162	-
Building Repairs	1,272	-
Machinery Repairs	1,511	-
General Repairs	114	-
Packing Charges	1,660	-
Freight & Ford Exp	3,522	-
Publicity	27,865	-
Sales Promotion. Expenses	19,173	-
Selling Commission	23,687	-
Brokerage	212	-
Royalty	10	-
Interest	21,817	2,966
Bank Charges	2,053	-
Depreciation	59,608	-
	32,71,814	32,71,814

*FA-Fixed assets, Inc-Increase ,Dcs-Decrease, WIP-Work in progress

BALANCE SHEET

Balance Sheet as at 31 Sep 201X			
		(Amounts in '000)	
EQUITY AND LIABILITIES	Notes	As at 31 Sep 201X	As at 31 Aug 201X
Shareholders' funds			
Share Capital	1	1,103,000	1,103,000
Reserves & Surplus	2	285,760	240,283
		1,388,760	1,343,283
Non-current liabilities			
Long Term Borrowings	3	126,125	127,625
		126,125	127,625
Current liabilities			
Short Term Borrowings	4	237609	143533
Trade Payables	5	345141	251810
Other Current Liabilities	6	208711	215651
		791,461	610,994
TOTAL		2,306,346	2,081,902
ASSETS			
Fixed assets	7		
Gross Block		1,918,262	1,952,877
Less: Depreciation		888,251	896,654
		1,030,011	1,056,223
Add : Capital work in progress		87,765	16,128
Net Block		1,117,776	1,072,351
INVESTMENTS	8	3	3
Current assets			
Inventories	9	615,420	669,806
Sundry Debtors	10	495,059	293,769
Cash & Bank Balances	11	20,249	24,180
Advances & Deposits	12	51,839	21,793
		1,188,567	1,009,548
		2,306,346	2,081,902
Difference		0	0

SCHEDULE FORMING PART OF BALANCE SHEET

In accounting, a schedule is defined as a supporting report or document, which constitutes detailed information, explaining the elements of the main financial report. It is a supplemental proof to all the data presented in the financial report, explaining all the numbers mentioned therein. For example, if we discuss the schedule of the balance sheet, not only are the liabilities, assets, and equities of a company presented, but a breakdown of each category is shared as a sub-category or a sub-schedule.

Rs in 000	As at 30 Sep 201X		As at August 201X	31-Mar-201X
1-Share capital	Number of shares	Amount		
Authorised				
Equity shares of Rs. 10 each	1,20,000	12,00,000	12,00,000	12,00,000
	1,20,000	12,00,000	12,00,000	12,00,000
Issued, subscribed and paid up				
Equity shares of Rs. 10 each fully paid up	11,03,00	11,03,000	11,03,000	11,03,000
2-Reserves & Surplus				
(Capital Reserve)		250	250	250
Share Premium A/c		4,32,726	4,32,726	4,32,726
Profit and loss A/c				
Balance at the beginning of the year		(61,364)	(61,364)	(48,026)
Less: Loss for the period		(85,852)	(1,31,330)	(13,338)
Balance at the end		(1,47,216)	(1,92,693)	(61,364)
Total		2,85,760	2,40,283	3,71,612
3-Long-term borrowings		Rate of interest		
Term loans from Bank (secured)	15.50%	1,26,125	1,27,625	1,48,679
Total		1,26,125	1,27,625	1,48,679
4-Short-term borrowings				
Working Capital from Bank (secured)	13.55%	2,37,609	1,43,533	1,41,681
Total		3,63,734	2,71,158	2,90,360
5-Trade Payable				
RM Creditors		2,45,967	2,51,775	
Pub. Creditors		6,809	2,742	
Stores and other Creditors		92,365	(2,707)	
		3,45,141	2,51,810	
6-Other current liabilities				
Interest accrued but not due on borrowings		2,146	1,709	2,132
Trade Deposits		29,326	29,376	29,961
Other payables:				
Employees dues		12,412	19,626	12,985
Power			8,114	3,135
Legal Expenses			2,997	51,764
Interest		2,146	3,085	24,741
Publicity & Sales Promotion		-	-	10,430
Selling Commn			13,709	35,094
TOD, Claims			23,183	32,706
Other expenses payable		1,62,681	1,13,852	(6,620)
Total		**2,08,711**	**2,15,651**	**1,96,328**

7-Fixed Assets									
Particulars	Gross Block				Depreciation				Net Block
	As on		Adjustment/	As at	Up to	For the	Adjustment/	Upto	As at
	1-Sep-201X	Additions	Deductions	30-Sep-201X	1-Sep-201X	Period	Deductions	30-Sep-201X	30-Sep-201X
Land	1,56,170	-	-	1,56,170	170	0	-	170	1,56,000
Building	2,86,160	-	-	2,86,160	71,452	767	-	72,219	2,13,941

Plant & Machinery	14,70,662	-	34,615	14,36,047	8,06,619	8,887	19,760	7,95,746	6,40,301
Computers	15,675	-	-	15,675	9,346	149	(20)	9,514	6,161
Furniture & Fittings	16,025	-	-	16,025	5,915	(181)	(1,403)	7,138	8,887
Office Equipment	1,562	-	-	1,562	694	(47)	(319)	965	597
Vehicles	6,623	-	-	6,623	2,459	40	-	2,499	4,124
Sub Total	19,52,877	-	34,615	19,18,262	8,96,654	9,615	18,018	8,88,251	10,30,011
Capital work in progress	16,128	71,637	0	87,765	-	-	-	-	87,765
Total	**19,69,005**	**71,637**	**34,615**	**20,06,027**	**8,96,654**	**9,615**	**18,018**	**8,88,251**	**11,17,776**

	As On	30-Sep-12	31-Aug-12	31-Mar
8-Investments		3	3	3
9-Inventories				
(At cost or net realizable value whichever is lower)				
Raw materials		76,763	89,457	32,066
Stores and spares		27,668	29,081	22,415
Waste (at net realisable value)		1,345	1,802	391
Work-in-progress		2,51,897	2,42,354	1,32,905
Finished goods		2,57,747	3,07,112	3,06,692
		6,15,420	6,69,806	4,94,470
10-Trade receivables				
(Unsecured Considered good)				
Receivables o/s for more than 6 months		-	-	418
Other Receivables		4,95,059	3,32,734	3,14,005
		4,95,059	3,32,734	3,14,423
11-Cash and bank balances				
Cash in hand		216	11	298
Balances with Scheduled banks				
- Current Accounts		6,193	4,329	638
- Fixed Deposits Accounts		19,840	19,840	19,815
		26,249	24,180	20,751
12-Advances & Deposits				
(Unsecured Considered good)				
Advances recoverable in cash or in kind or for value to be received		35,952	21,147	1,26,200
Advance Tax		3,396	(134)	11,162
Deposit with Government deptt. & Others		12,367	13,182	11,825
Other Current Assets		124	(12,402)	
		51,839	21,793	1,49,187
		For the Month	Upto	
		30-Sep-201X	30-Sep-201X	31-Aug-201X
13-Other Income				

Foreign Exchange gain/(Loss)		1,165	6,336	5,171
Miscellaneous Receipts		-	2,346	2,346
Profit on sale of FA (Fixed Assets)		592	8,962	8,370
Excess Liabilities./ Provision written back		8	301	293
Interest recd.		153	2,966	2,813
(Increase)/Decrease in stocks				
Opening Stock				
Finished Stock		3,07,112	3,06,692	3,06,692
Process Stock		2,42,354	1,32,905	1,32,905
Waste & Scraps		1,802	391	391
		5,51,268	4,39,989	4,39,989
Less: Closing Stock				
Finished Stock		2,57,747	2,57,747	3,07,112
Process Stock		2,51,897	2,51,897	2,42,354
Waste & Scraps		1,345	1,345	1,802
		5,10,989	5,10,989	5,15,452
		40,279	(71,000)	(75,463)
14-Consumption of Raw materials (Other than dyes & chemicals)				
Opening Stock		85,649	27,879	27,879
Add : Purchases & Expenses thereon		36,393	4,25,125	3,88,732
		1,22,042	4,53,004	4,16,611
Less : Closing Stock		71,877	71,877	85,649
Raw material Consumption		50,165	3,81,127	3,30,962
Purchases of Finished goods		12,361	12,536	175
Total Consumption		62,526	3,93,663	3,31,137
14 a-Consumption of Dyes				
Opening Stock		3,808	4,187	4,187
Add : Purchases & Expenses thereon		7,808	19,904	12,096
		11,616	24,091	16,283
Less : Closing Stock		4,886	4,886	3,808
Dye Consumption		6,730	19,205	12,475
15-Employee Cost				
Salaries, Wages, Bonus etc.		37,676	98,082	60,406
Contribution to Provident & other funds		2,852	21,249	18,397
Employees' Welfare		2,905	6,953	4,048
		43,433	1,26,284	82,851
16-Other Manufacturing, Administration & General exps.				
Power & Fuel		22,634	96,507	73,873
Other Manfg. Expenses		-	-	-
Repairs & Maintenance				
Buildings		206	1,272	1,066
Plant & machinery		172	1,511	1,339

Others		6	114	108
Stores, Spares Consumed		2,103	22,416	20,313
Packing		167	1,660	1,493
Charity & Donation		10	385	375
Insurance		(271)	507	778
Legal & Professional		1,497	13,173	11,676
Rent		21	1,699	1,678
Rates & Taxes		1,284	1,351	67
Fees & Subscription		526	993	467
Payment to Auditors		431	1,588	1,157
Directors' Fee		-	-	-
Miscellaneous Expenses		301	10,300	9,999
		29,087	1,53,476	1,24,389
17-Selling Expenses				
Advertisement & Sales Promotion		(3,172)	47,038	50,210
Brokerage, Commission & Royalty		10,249	23,909	13,660
Freight Expenses		463	3,522	3,059
		7,540	74,469	66,929
17 a-Corporate Fees & Exps.				
Fees/Commn.		3,471	12,338	8,867
Exps.		-	-	-
		3,471	12,338	8,867
17 b-One Time Management Fees				
Interim Management		-	4,933	4,933
18-Interest & Financial Charges				
On T/L		1,590	10,589	8,998
On C/C		2,095	9,407	7,312
Others		284	1,821	1,536
Bank Charges		295	2,053	1,758
		4,266	23,869	19,604

ESTIMATED PROFIT AND LOSS ACCOUNT SUMMARY

INCOME Rs 000	Upto Sep 201X	30-Sep 201X	31-Aug 201X	30-Jun 201X	30-Jul 201X	31-Aug 201X
Sales & Income from Operations	2,23,995	8,58,705	6,34,710	2,87,477	99,829	1,70,460
Other Income	1,918	20,911	18,993	11,232	1,831	5,930
	2,25,913	8,79,616	6,53,703	2,98,709	1,01,660	1,76,390
EXPENDITURE						
(Increase)/Decrease in FG & WIP Stocks	40,279	97,622	57,343	8,180	(41,037)	(35,816)
Raw material Consumed	62,526	3,93,663	3,31,137	1,67,001	73,774	98,865
Dye Consumption	6,730	19,205	12,475		3,223	3,849
Employee Cost	17,921	1,26,284	1,08,363	63,896	18,955	25,512
Manufacturing, Admn., Selling & General expenses	35,627	2,27,945	1,92,318	1,04,895	22,642	60,681

Corporate Fee & Expenses	3,471	17,271	13,800	3,613	2,920	7,267
Interest & Financial Charges	4,266	23,870	19,604	12,045	3,594	3,965
Depreciation	9,615	59,608	49,993	28,984	10,799	10,210
Total	1,80,435	9,65,468	7,85,033	3,88,614	94,870	1,74,533
Profit before Tax	45,478	(85,852)	(1,31,330)	(89,905)	6,790	1,857
Balance Brought Forward from previous years	(2,46,303)	(61,364)	(1,14,973)	(1,14,973)	(1,61,857)	(1,55,067)
Balance carried to Balance Sheet	(2,00,825)	(1,47,216)	(2,46,303)	(2,04,878)	(1,55,067)	(1,53,210)

ESTIMATED MONTHLY PROFIT AND LOSS ACCOUNT

	Upto Sep 201X	30-Sep 201X	31-Aug 201X	30-Jul 201X	30-Jul 201X	31-Aug 201X
INCOME						
Fabric Sales '000 Mtrs.						
Civil- Domestic	430	1,558	1,128		225	219
Govt/Institutional	18	155	137		18	102
Export	45	342	297		26	41
Total	493	2,055	1,562		268	362
Commission Spinning. '000 Kgs	36	168	132	103	28	-
Avg. Fabric Realisation Rs./ Mtr.	442	402	389		358	455
Fabric Sales						
Civil- Domestic	1,80,674	5,65,057	3,84,383	1,99,107	77,025	84,779
Govt/Institutional	8,400	85,888	77,488	9,361	6,780	61,347
Export	28,880	1,75,434	1,46,554	62,561	12,216	18,663
Total Fabric Sales	2,17,954	8,26,379	6,08,425	2,71,029	96,021	1,64,789
Commission. Spinning.	2,007	15,662	13,655	10,181	1,510	1,964
Other Sales	1,517	10,335	8,818	3,440	1,924	3,096
Total Sales	2,21,478	8,52,376	6,30,898	2,84,650	99,455	1,69,849
Other Income	1,918	20,911	18,993	11,232	1,831	5,930
Export Benefit	2,517	6,329	3,812	2,827	374	611
Total Income	2,25,913	8,79,616	6,53,703	2,98,709	1,01,660	1,76,390
(Increase)/Decrease in FG/ WIP Stock	40,279	97,622	57,343	8,180	(41,037)	(35,816)
Raw Material	48,448	3,68,591	3,20,143	1,42,228	71,501	95,376
Job Charges	1,717	12,536	10,819	5,222	2,163	3,434
Outside Purchases	12,361	12,536	175	10	110	55
Dyes & Chemical Consumed	6,730	19,205	12,475	8,503	3,223	3,849
Stores, Consumables & Packing	2,654	26,973	24,319	12,416	4,806	3,997
Power & Fuel	22,634	96,507	73,873	36,013	7,913	29,947
Salary, Wages & benefit	17,921	1,26,284	1,08,363	63,896	18,955	25,512
Depreciation	9,615	59,608	49,993	28,984	10,799	10,210
Total Cost of Sales	1,62,359	8,19,862	6,57,503	3,05,452	78,433	1,36,564
Gross Margin on Sales	63,554	59,754	(3,800)	(6,743)	23,227	39,826

% to Sales	28	7	(1)	(2)	23	23
Selling expenses	4,242	71,171	66,929	35,020	9,287	22,622
Corporate Fee/Commn.	3,471	12,338	8,867	1,902	-	6,965
Mgmt. Change - One Time Costs	-	4,933	4,933	1,711	2,920	302
Admn. & Other Overheads	6,097	33,294	27,197	21,446	636	4,115
Total Sales & Admin expenses*	13,810	1,21,736	1,07,926	60,079	12,843	34,004
% to Sales	6	14	17	20	13	19
Total Operating Expenses	1,76,169	9,41,598	7,65,429	3,65,531	91,276	1,70,568
Income from Operations	49,744	(61,982)	(1,11,726)	(66,822)	10,384	5,822
% to Sales	22	(7)	(17)	(22)	10	3
Interest Expenses	4,266	23,870	19,604	12,045	3,594	3,965
Profit /(Loss) before Income Tax	45,478	(85,852)	(1,31,330)	(78,867)	6,790	1,857
Income Tax	-	-	-	-	-	-
Net Income /(Loss)	45,478	(85,852)	(1,31,330)	(78,867)	6,790	1,857
EBITDA	59,359	(2,374)	(61,733)	(37,838)	21,183	16,032
% to Sales	26	(0)	(9)	(13)	21	9
EBITDA before Corp Fee, Commn. & Exp	62,830	14,897	(47,933)	(34,225)	24,103	23,299
% to Sales	28	2	(7)	(11)	24	13

COMPARATIVE PRODUCTION SUMMARY

Production	Actual	Prev Yr.	YTD
	Current year Sep	Sep	Current year Sep
-. Spinning Efficiency	80	82	79
- Weaving Efficiency	78	63	74
- Spinning (Kg)	1,70,337	2,41,628	8,28,019
- Weaving (Mtr)	3,70,220	5,06,074	17,01,410
Dispatches (Mtrs)	4,92,113	4,86,000	18,48,827
Order Received (Mtrs)	2,62,355	2,30,000	19,72,285
No of Staff	298	316	312
No of Workers Permanent	523	549	563
No of Workers Casual	562	681	476
Total No of Employees	1,383	1,546	1,351

8
Costing and Cost Analysis

Cost can be defined as expenditure incurred to produce a given good or service. Cost is the amount to be paid for a good or service or the resources given in exchange for such good or service. In commercial terms, cost is a monetary valuation of the effort, materials, risks and opportunity costs all put together. Value is measured in terms of the usefulness of the product, and cost is measured strictly in monetary terms. While cost is a very generic term, it can be classified further as prime cost, sunk cost, factory cost, direct cost, indirect cost, fixed cost, variable cost, etc. It is advisable to classify costs as it gives more information about it.

Costing is essentially a technique *via* which we assign costs to various elements of a business. It is a system of ascertaining costs. We follow certain rules and principles to guide us in ascertaining costs. Some examples of methods of costing for ascertaining costs are historical costing, standard costing, etc. Assigning variable costs according to activity levels is direct costing and assigning fixed costs irrespective of activity levels is known as absorption costing.

Usually, there are three elements involved in the concept of cost - material cost, labour cost and expenses.

MATERIAL COST

Materials used to produce the final product may be classified into direct materials and indirect materials.

Direct Materials: These are all such components or material(s) that are used to produce the final product. Direct material cost is the cost of materials entering into and becoming constituent elements of a product or saleable service. Thus, materials which can be identified with units of output or service are known as direct materials.

Examples: Cotton, polyester, wool, silk, etc. used in production of cloth, cloth in garments, leather used in the case of production of leather goods and stones in the production of cement, clay in bricks, steel in machines, timber in furniture, fruits in canning industry, glue for book binding, batteries for Transistor/radios, etc.

Indirect Materials: Materials used for the product other than direct materials are called indirect materials. In other words, material costs which cannot be identified with a specific product, job, process are known as indirect materials' costs. Indirect materials may be used at the factory, office or selling and distribution divisions.

Examples: Consumable stores, lubricants, printing and stationery material, small tools, office stationer, advertising posters, and materials used in maintenance of plant and machinery

LABOUR COST

The manpower used for conversion of materials into finished products is termed as labour. It may be further classified into direct labour and indirect labour.

Direct Labour: Labour, which plays an active and direct part in the production of a particular product, is termed as direct labour, which includes cost of wages, salaries, incentives, bonus, etc. of the employees of an undertaking. Direct labour cost can be conveniently and specifically cha ged to specific products. It is also called 'direct wages'. Direct labour cost is the cost of labour directly engaged in production operations.

Indirect labour: It is employed to carry out tasks incidental to goods produced or services rendered. Indirect labour cost cannot be practically traced to specific units of output. It may be incurred in the factory, office, or selling and distribution divisions. Indirect labour cost is the remuneration paid for labour which is engaged to help production operations.

Examples: Wages of a storekeeper, salaries of office staff and sales representatives, directors' fees, security guards, cleaning staff, etc. The remuneration paid to these persons cannot be traced back to a production job, process, or production order. Clerical and managerial staff, sales clerks, distribution employees are also included in the orbit of indirect labour.

EXPENSES

Direct Expenses: These are expenses which can be directly allocated to a particular job, product, or unit of service. These expenses can be directly identified with a unit of output, job, process, or operation. They are specifically incurred for a job, or unit or process and in no way connected with other jobs or processes. Direct expenses are also known as chargeable expenses.

Examples of such expenses are hire charges paid to some special machinery required for a particular contract, cost of designs and development or mold incurred in manufacturing, cost of blocks needed in book publishing, hire charges of special plant used for a job, Royalty on products, cost of special patterns, designs or plans for a particular job or work order, etc.

Indirect Expenses: These are expenses other than direct material and labour, which cannot be directly identified with units of output, job, process, or operation. They are incurred in common and can be apportioned to various cost centers or cost units proportionately on some basis.

Examples of such expenses are factory rent, lighting, insurance, depreciation, bank charges, advertising, office and administration expenses, selling and distribution expenses, etc.

Overheads: The aggregate of indirect material cost, indirect labour cost, and indirect expenses is termed as overheads. Thus, all indirect costs are overheads. *So,*

Overhead = Indirect materials + Indirect labour + Indirect expenses They may

be classified broadly into three types:

A. Factory or manufacturing overheads:

These are indirect costs incurred inside a factory or works. For example, factory supplies such as oil, consumable stores, lubricants, indirect labour such as factory manager's salary, timekeeper's salary, etc., and indirect expenses such as factory rent, factory lighting, factory insurance, etc. These may be variable or fixed. Variable overheads are related to the volume of production. Fixed overheads are related to time. Power, fuel etc. are examples of variable overheads whereas rent, insurance, rates, and taxes, salaries of supervisions, etc., are examples of fixed overheads. Overheads are allocated, apportioned, and absorbed by products on an equitable basis.

B. Office and administrative overheads:

These are indirect costs incurred in the general and administrative office. For example, stationery, cleaning materials etc. would be indirect materials, and salaries of office staff, directors' fees, etc., would be indirect labour, and rent, insurance, lighting, etc., of the office would be indirect expenses.

C. Selling and distribution overheads:

These are indirect costs incurred with respect to the selling and distribution of goods and services. Examples are indirect materials such as packing materials, printing and stationery materials, etc., and indirect labour such as salaries of sales staff and sales manager, Incentives, etc., and indirect expenses such as advertising expenses, insurance, godown rent, etc.

COMPONENTS OF TOTAL COST

1. **Prime Cost** – Prime cost is the sum of direct material costs, direct labour costs, and direct expenses. It is also known as Basic, First or Flat Cost.

2. **Factory Cost** – This comprises Prime Cost and factory or works overheads, which include costs of indirect material, indirect labour and indirect factory expenses. This cost is also known as Works Cost, Production Cost or Manufacturing Cost.

3. **Cost of production** – This cost includes Factory Cost and office and administration overheads.

4. **Total Cost** – This is the sum of the Cost of Production and selling and distribution overheads.

The total expenditure consisting of material, labour and expenses can further be analysed as under:

Prime Cost = Direct materials + Direct Labour + Direct expenses

Works Cost (Factory) = Prime Cost + Factory overhead

Cost of Production = Factory Cost + Administration overhead

Total Cost (Cost of sales) = Cost of production + Selling and distribution overhead

Wages and Salary

CTC: CTC i.e. cost to company is different from the employee's in-hand salary. CTC is the total expenditure by a company towards an employee which includes components such as basic salary, special allowances, HRA, employee provident fund, etc.

Gross Salary: The salary amount calculated before deduction of taxes or any other deduction is the gross salary. Gross salary includes the basic pay, dividends or bonus, or any other differentials.

Net salary: This is also the take-home salary after deducting TDS or any other deductions as per company norms.

Basic Salary: Basic salary is the base income of an employee, comprising of 35-50 % of the total salary. It is a fixed   amount that is paid prior to any deductions or increases due to bonus, overtime, or allowance.

Elements of Wages and Salary

- **Dearness Allowances (DA):** Dearness allowance is a certain percentage of the basic salary paid to employees, aimed at mitigating the impact of inflation.

- **House Rent Allowance (HRA):** A house rent allowance is that component of the salary which is paid to employees for meeting the cost of renting a home. Some companies, who provide accommodations, are not liable to pay any HRA.

- **Conveyance Allowance:** Conveyance allowance, also known as transport allowance offered by employers to compensate fort employees' travel expenses and from their residence and workplace either paying by coins or by providing the appropriate convenience facility.

- **Leave Travel Allowance:** Offered by employers to their employees to cover travel expenses when he or she is on leave from work

- **Medical Allowance:** This is a fixed allowance paid to the employees of an organization to meet their medical expenditure not covered by ESIC (Employees State Insurance Act.)

- **EPF:** This is a contribution by the employee to the provident fund governed by the Employees Provident Fund Organisation. The employer also makes a similar contribution on behalf of the employees for their retirement.

- **Annual Bonus:** A bonus is a reward, paid to an employee for their good work in the organisation. The basic objective of a bonus is to share the profits earned by the organisation amongst its employees. In India, there is a principal law relating to this procedure of payment of minimum and maximum bonus to employees.

- **Production Incentive:** This is a reward given for work performed by workers in addition to normal wages or salary. Different schemes are implemented to motivate workers for boosting production levels.

- **Over time Payment:** Payment made to workers for working beyond routine working hours as per government rules.

- **Perquisites or Fringe Benefits** Perquisites, also referred to as fringe benefits, are benefits that some employees enjoy because of their official position. These are generally non-cash benefits given in addition to the cash salary. Some examples of perquisites include provision of car for personal use, rent-free accommodation, payment of premium on personal accident policy, etc. The monetary value of perquisites is added to the salary and the employee pays tax on them.

- **ESIC:** As per the provisions of the Factories Act, if a factory falls under a certain criteria, then the employer is required to avail ESIC scheme for its employees. Both the employers and employees have to contribute as per the rules and employees are covered for availing medical facilities as prescribed therein.

- **Gratuity:** Gratuity is a lumpsum benefit paid by employers to those employees who are retiring from the organization. This is only payable to those who have completed 5 or more years with the company.

COMMON FORMATS FOR COSTING

Examples:

Cotton fiber is removed from the cotton seeds through the ginning process. Normally the recovery of lint from the cotton balls is only about 33-35%. The rest, about 65%, is cotton seeds and trash which are separated and removed during the ginning. The list is transported to the godowns or spinning mills directly after getting compressed through the press into bales of 170 Kg so as to make it commercially viable mainly from a storage and transport point of view. Cotton seeds are used for edible oil and after extraction of oil waste, trash and seeds are further used for cattle feed. Hence, while doing the costing of one kg or one bale of cotton, we need to consider the actual yield from the balls, amount recovered from the sale of extracted seeds, cost of ginning and transportation, etc.

COSTING OF COTTON (Farm to Spinner)

Prices of *Kapas* is considered only for example	
	Rs /Kg
A-Raw material (*Kapas*) cost to Ginners /Quintal	9000
B-Raw material cost to Ginners *Kapas* /Kg	90
C-Yield 33% (B*33/100) (approx.)	29.70
D-Seeds 64% (B*64/100) (approx.)	57.60
E-Wastage 3% (B*3/100) (approx.)	2.70
F-Total Cost of Waste (D+E)	60.30
G-Cost of Usable cotton (B+F)	150.30
H-Cost of Ginning	8.40
I-Cost of baling /Packing	0.25
J-Cost of ginned cotton (G+H+I)	158.95

K-Sale realisation of Cotton seeds (@Rs 20/Kg)	12.80
L-Net Cost of ginned cotton (J-K)	146.15
M-Cost of Transportation	
a-Farm to ginning plant	1.00
b-Ginning plant to end users	1.00
N--Total Cost of Transportation (a+b)	2.00
O-Selling expenses	(Not considered)
P-Total Cost of cotton upto end users (L+N)	148.15
Q-Cost of Cotton /Candy of 356 Kg	52741.40
Candy is the Unit for Cotton trading	
Note: Proportionate price is adjusted by end users if moisture content is more than the standard	

COTTON (RM) AND COTTON YARN COSTING

Per Kg				Conversion Cost
RAW MATERIALS RATE	MECH		Count	Paisa /Count
RS/CANDY (356 Kg)	54000		Up to 40s	105
RS/KGS	151.85		41s to 50	115
CST 2% (Before GST)	3.04		60s and above	120
FREIGHT	0.75		Doubling	35
INSURANCE			TFO	45
TOTAL	155.63		Reserve twist /Slub	25
ENTRY TAX (1%)				Rs. Per kg
MIXING COST	155.63		Hank 2/20s	3.0
YIELD %	86.25		Hank 2/30s	3.25
NET RATES	177.02		Hank 2/40s	4.0
RECOVERY % WASTE CREDIT	2.75		Hank 2/60s	4.75
NET RAW MATERIAL COST	179.86		Mix Yarn Per kg	45.0

COUNT	24s k	2/24s k	24s k RT	2/24s k RT	30s k	2/30s k	40s k	2/40s k
TOTAL (R M)	179.86	179.86	179.86	179.86	179.86	179.86	179.86	179.86
TOTAL R M (RS/KG)	179.86	179.86	179.86	179.86	179.86	179.86	179.86	179.86
YARN CONV.COST	25.2	33.6	31.2	39.6	31.5	42	42	56
Ex Mill Cost	205.06	213.46	211.06	219.46	211.36	221.86	221.86	235.86
BROKERAGE & COMMISSION @1.5%	3.08	3.2	3.17	3.29	3.17	3.33	3.33	3.54
CASH DISSCOUNT (1.00%)	2.05	2.13	2.11	2.19	2.11	2.22	2.22	2.36
TRANSPORT	2	2	2	2	2	2	2	2
NET SALE RATE	212.19	220.8	218.34	226.95	218.65	229.41	229.41	243.76
ROUNDED OFF	212.2	220.8	218.35	226.95	218.65	229.41	229.4	243.75

*K-Carded, RT-Reverse twist

COTTON AND YARN PRICES TREND (HISTORICAL DATA)

Count	Rs/candy	10s OE	16s OE	20s OE	20s k	16s C	20s C	30s C	40s C	40s Comp.	2/20s C	2/40s C Comp.
Apr.08 to March.09	23005	86	100	110	98	112	114	126	146	165	133	158
Apr.09 to March.10	22984	89	105	112	118	122	130	143	158	170	152	183
Apr-10	27979	100	117	123	137	145	158	170	182	202	178	230
May-10	28435	104	119	124	144	150	162	175	190	208	180	233
Jun-10	29525	102	116	121	142	153	165	177	193	210	180	235
Jul-10	29785	102	116	121	139	155	162	175	193	206	178	235
Aug-10	31080	112	122	129	150	162	175	185	215	227	195	243
Sep-10	36880	122	134	142	165	174	18	192	222	235	205	255
Oct-10	39930	127	142	152	177	185	198	205	230	242	212	260
Nov-10	43069	132	150	162	185	202	210	215	245	255	220	267
Dec-10	41717	152	164	172	190	205	215	230	238	260	210	260
Jan-11	44810	187	192	190	194	215	220	235	248	265	230	270
Fcb-11	51245	192	200	208	222	230	235	250	270	285	252	295
Mar-11	59540	195	203	210	225	235	240	255	274	290	256	305
Increase %	158.8	126.7	103	90.9	129.6	110	111	102	87.7	75.8	92.5	93

*OE-Open end, K–Carded, C-Combed, Prices in Rs/Kg

COSTING OF WOOL SCOURING

		GRADES	64'S	60'S	58'S
A	GREASY WOOL ONE LOT -KG		5000	5000	5000
B	GREASY WOOL PRICE KSH (Kenyan Shilling)		75	54	42
C	WASTE% (YIELD 70%)		30	30	30
D	COST OF WASTE (BxC/100)		22.5	16.2	12.6
E	**TOTAL COST/KG KSH(B+D)**		**97.5**	**70.2**	**54.6**
F	WATER CONSUMPTION LTR		66500	66500	66500
G	CUBIC LTRS		66.5	66.5	66.5
H	WATER RATES/CUBIC LTRS KSH		72	72	72
I	COST OF WATER KSH		4788	4788	4788
J	**COST OF WATER /KG KSH (I/A)**		**0.96**	**0.96**	**0.96**
K	CHEMICAL CONSUMPTION				
L	SOAP KG (200 kg for 18-20 mt)		60	60	60
M	RATE OF SOAP/KG		170	170	170
N	**COST OF SOAP/KG(LxM)/A**		**2.04**	**2.04**	**2.04**
O	SODA ASH KG		250	250	250
P	RATE OF SODA ASH KG		10.11	10.11	10.11
Q	**COST OF SODA ASH KG(Oxp)/A**		**0.51**	**0.51**	**0.51**
R	POWER CONSUMPTION				
S	WILLO (OPENER) (6KW) KSH		705.6	705.6	705.6

T	SCOURING M/C (156KW) KSH	18345.6	18345.6	18345.6
U	TOTAL	19051.2	19051.2	19051.2
V	COST OF POWER/KG	3.67	3.67	3.67
W	LABOURS ENGAGED(SORTING/SCOURING)	14	14	14
X	WAGES/MONTH@6500 KSH	91000	91000	91000
Y	WAGES/DAY KSH	3791.67	3791.67	3791.67
Z	LABOUR COST/KG KSH(Y/A)	0.76	0.76	0.76
AA	COST OF STEAM /KG	4	4	4
AB	TOTAL COST OF SCOURING/KG KSH(J+N+Q+V+Z+AA)	11.93	11.93	10.93
AC	COST OF SCOURED WOOL(E+AB)	109.43	82.13	65.53
AD	CONVERSION RATE 01 USD= 80 KSH	80	80	80
AE	WOOL-POLYTHENE/BALE=1KG- KSH /KG	0.76	0.76	0.76
AF	WOOL METAL STRIP/BALE=2KG/BALE- KSH /KG	1.15	1.15	1.15
AG	WEIGHT OF BALE KG	185	185	185
AH	PACKING COST/KG (AE+AF)	1.96	1.9	1.9
AI	RETE OF METAL STRIP /KG KSH	106	106	106
AJ	RATE OF POLYTHENE SHEET 96X500/KG KSH	140	140	140
AK	Rate OF POLYTHENE SHEET 96X250/KG KSH	120	120	120
AL	FINANCE COST KSH	24	24	24
AM	COST OF OH KSH	20	20	20
AN	TOTAL COST OF SCOURED WOOL/KG((AC+AH+AL+AM)	155.33	128.03	111.43
AO	TRANSPORT FACTORY TO M)MBASA 40FCL KSH	85000	85000	85000
AP	TRANSPORT COST/KG KSH (AO/AQ)	5.94	5.94	5.94
AQ	NETT WEIGHT IN 40FCL KG	14300	14300	14300
AR	WOOL+LOCAL TRANSPORT/KG (AN+AP)	161.28	138.98	117.38
AS	FOB USD (AR/AD)	2.02	1.67	1.47
AT	FREIGHT MOMBASA PORT - MUMBAI USD	1150	1150	1150
AU	FREIGHT/KG USD	0.08	0.08	0.08
AV	COST /KG C&F USD	2.1	1.76	1.55

COSTING OF SCOURED WOOL, IMPORT

Type of Wool Micron	18.5 M	17.5 M	20.00	21.00	23.50
Invoice Quantity (Kgs) [A]	13,993.0	13,577.0	7,145.3	13,282.0	7,347.0
Actual Scoured Quantity (Kgs) [B]	14,598.8	14,339.2	7,747.3	13,661.4	7,567.4
CIF Rate in Australian $	14.61	15.57	13.95	13.49	11.32
Forex Conversion Rate (At booking)	64.04	64.04	77.70	77.91	77.79
CIF Rate (INR) (D*E)	935.62	997.10	1,083.92	1,051.01	880.58
Freight	-	-	-	-	-
Custom Duty INR	49.32	52.56	61.38	58.28	48.91
Clearing Expenses INR	4.50	4.50	4.50	4.50	4.50
Finance Charges INR	1.20	1.20	1.20	1.20	1.20
Total Cost / Kg INR (Sum F to J)	990.64	1,055.36	1,150.99	1,114.99	935.19

Total Amount for Invoice Quantity(K*B)	1,38,62,087.73	1,43,28,671.52	82,24,204.50	1,48,09,244.76	68,70,826.73
Cost / Kg on actual Scoured Weight (L/C)	949.54	999.27	1,061.56	1,084.02	907.95
Round Off	950.00	999.00	1,062.00	1,084.00	908.00
Scouring Charges /Kg	13.00	13.00	14.00	14.00	14.00
Total Cost/Kg INR (N+O))	963.00	1,012.00	1,076.00	1,098.00	922.00
Forex actual conversion Rate	63.03	63.03	63.03	63.03	63.03
Actual Amount Paid (B*D*R)	1,28,86,221	1,33,24,685	62,82,886	1,12,93,797	52,42,289
Less: Amount Considered (F)	1,30,92,192	1,35,37,665	77,44,898	1,39,59,460	64,69,642
Difference (gain)	(2,05,971)	(2,12,979)	(14,62,011)	(26,65,664)	(12,27,352)

SPINNING DEPARTMENT SECTION WISE MANUFACTURING COST (worsted)

Machine-wise Cost (Expenses Per month) (in Lacs Rs)							
Particulars	Gill Box & Flyer Frame	Ring Frame	Auto Coner	Assembling winding	TFO	Steaming & Rewinding	Total Exp
1. Staff Salary & Amenities	1.48	1.48	1.48	1.48	1.48	1.48	8.88
2 Wages & Amenities (Inclu. Contract)	4.64	24.64	7.2	3.52	6.56	4.48	51.04
3. Stores & Spares	1.35	3.15	1.35	0.45	2.25	0.45	9
5. Power	4.72	22.36	8.35	0.89	17.53	1.26	55.11
6. Coal						1.5	1.5
7. Water						0.1	0.1
Total Expenses/Month	12.19	51.63	18.38	6.34	27.82	9.27	125.63
Working Day	26	26	26	26	26	26	26
Total Expenses/day Rs	46896	198563	70689	24383	107004	35644	483180
Yarn Production Per Month	96382	84084	96096	98826	76518	87464	84084
Yarn Production Per Day	3707	3234	3696	3801	2943	3364	3234
Cost Per Day Per Kg Rs	12.65	61.4	19.13	6.41	36.36	10.6	149.41
Average Count Nm	52	52	52	52	52	52	52
Cost Per Count Per Kg Rs	0.24	1.18	0.37	0.12	0.7	0.2	2.87
Distribution %	8.47	41.1	12.8	4.29	24.34	7.09	100

Amount Rs.	PRODUCTION DEPARTMENTS ANNUAL EXPENSES AND COST PER UNIT						
Particulars	Dyeing (A)	Spinning (B)	Doubling & TFO (C)	Recombing to Spinning (D) (B+C)	Weaving & G Perch (E)	Finishing, Mending & Inspection (F)	A+ D+E+F Rs.
Wages	4,969,363	28,232,374	12,464,823	40,697,197	24,710,253	25,013,762	95,390,575
Wages allocation	1,419,438	8,064,234	3,560,425	11,624,658	7,058,183	7,144,876	27,247,155
Welfare	130,237	739,915	326,679	1,066,594	647,607	655,562	2,500,000
Salary	3,819,671	3,902,408	3,697,822	7,600,230	10,222,634	7,170,698	28,813,233
Staff allocation	13,432,327	13,723,284	13,003,830	26,727,114	35,949,111	25,216,614	101,325,167
Stores	1,872,042	17,908,255	2,964,747	20,873,002	17,731,782	11,419,464	51,896,290
Distribution Stores	271,473	2,596,955	429,931	3,026,886	2,571,364	1,655,987	7,525,711
Machinery Repairs	809,795	4,126,297	1,112,003	5,238,300	1,313,863	920,420	8,282,378

Building and Other Repairs	110,448	1,056,567	174,917	1,231,483	1,046,155	673,735	3,061,822
Power	12,323,463	53,349,717	42,998,625	96,348,342	32,385,647	16,276,236	157,333,687
Coal & Furnace Oil	30,453,926	-	8,390,957	8,390,957	-	53,003,118	91,848,000
Depreciation	3,035,438	72,420,312	5,625,113	78,045,425	24,855,808	14,182,646	120,119,317
Total Expenses/Year	**72,647,622**	**206,120,317**	**94,749,869**	**300,870,187**	**158,492,408**	**163,333,117**	**695,343,334**
Total Expenses/day(348 W. days)	208,758	592,300	272,270	864,570	455,438	469,348	1,998,113
Production Kg and Mtr per Year	**1,932,916**	**2,436,380**	**2,436,380**	**2,436,380**	**Mtr 5686316**	**Mtr 5836500**	
Production Kg and Mtr per days	5,554	7,001	7,001	7,001	16,340	16,772	
Rs Per Kg	**37.58**	**84.60**	**38.89**	**123.49**	**27.87/Mtr**	**27.98/Mtr**	

DEPARTMENT WISE ANNUAL EXPENSES AND COST ELEMENTS PER UNIT OF PRODUCTION

Amount Rs	Dyeing	Cost/ Unit	Spinning	Cost/ Unit	Doubling & TFO	Cost/ Unit	Recombing to Spinning	Cost/ Unit	Weaving & G Perch	Cost/ Unit	Finishing, Mending & Inspection	Cost/ Unit
Wages	49,69,363	2.57	2,82,32,374	11.59	1,24,64,823	5.12	4,06,97,197	16.70	2,47,10,253	4.35	2,50,13,762	4.29
Wages allocation	14,19,438	0.73	80,64,234	3.31	35,60,425	1.46	1,16,24,658	4.77	70,58,183	1.24	71,44,876	1.22
Welfare	1,30,237	23.45	7,39,915	0.30	3,26,679	0.13	10,66,594	0.44	6,47,607	0.11	6,55,562	0.11
Salary	38,19,671	1.98	39,02,408	1.60	36,97,822	1.52	76,00,230	3.12	1,02,22,634	1.80	71,70,698	1.23
Staff allocation	1,34,32,327	6.95	1,37,23,284	5.63	1,30,03,830	5.34	2,67,27,114	10.97	3,59,49,111	6.32	2,52,16,614	4.32
Stores	18,72,042	0.97	1,79,08,255	7.35	29,64,747	1.22	2,08,73,002	8.57	1,77,31,782	3.12	1,14,19,464	1.96
Distribution Stores	2,71,473	0.14	25,96,955	1.07	4,29,931	0.18	30,26,886	1.24	25,71,364	0.45	16,55,987	0.28
Machinery Repairs	8,09,795	0.42	41,26,297	1.69	11,12,003	0.46	52,38,300	2.15	13,13,863	0.23	9,20,420	0.16
Building and Other Repairs	1,10,448	0.06	10,56,567	0.43	1,74,917	0.07	12,31,483	0.51	10,46,155	0.18	6,73,735	0.12
Power	1,23,23,463	6.38	5,33,49,717	21.90	4,29,98,625	17.65	9,63,48,342	39.55	3,23,85,647	5.70	1,62,76,236	2.79
Coal & Furnace Oil	3,04,53,926	15.76	-	-	83,90,957	3.44	83,90,957	3.44	-	-	5,30,03,118	9.08
Depreciation	30,35,438	1.57	7,24,20,312	29.72	56,25,113	2.31	7,80,45,425	32.03	2,48,55,808	4.37	1,41,82,646	2.43
Total Expenses/Year	7,26,47,622	37.58	20,61,20,317	84.60	9,47,49,869	38.89	30,08,70,187	123.49	15,84,92,408	27.87	16,33,33,117	27.98
Production Kg and Mtr per Year	19,32,916		24,36,380	1.00	24,36,380	1.00	24,36,380		5686316		5836500	

ALLOCATION OF WAGES AND SALARY: Salary and wages of departments, which are not directly involved in manufacturing (servicing, administration etc), are distributed by manufacturing departments proportionately on the basis of their total expenses, no. of manpower or on the basis of formulae decided by individual organisations.

DEPARTMENT WISE WAGES ALLOCATION IN ANNUAL BUDGET

Amount Rs	Dyeing	Spinning	Doubling & TFO	Recombing to Spinning	Weaving & Perching	Finishing, Mending & Inspection	Total	%age to total wages	Dist	%age to total manf. Wages
Mfg Wages	4,969,363	28,232,374	12,464,823	40,697,197	24,710,253	25,013,762	95,390,575	76.23		
Designing	270,593	1,537,319	678,739	2,216,058	1,345,531	1,362,058	5,194,241	4.15	5,194,241	5.45

R M Godown	32,178	182,815	80,714	263,529	160,008	161,973	617,688	0.49	617,688	0.65
Stores	34,508	196,051	86,558	282,609	171,593	173,700	662,411	0.53	662,411	0.69
Yarn Godown	39,604	225,002	99,340	324,343	196,932	199,351	760,230	0.61	760,230	0.80
QC Deptt	17,334	98,482	43,481	141,963	86,196	87,255	332,748	0.27	332,748	0.35
Mechanical	278,166	1,580,340	697,733	2,278,073	1,383,185	1,400,174	5,339,599	4.27	5,339,599	5.60
Electrical	187,680	1,066,265	470,765	1,537,030	933,244	944,706	3,602,660	2.88	3,602,660	3.78
R Shop	25,152	142,897	63,090	205,987	125,070	126,606	482,816	0.39	482,816	0.51
General	27,229	154,694	68,299	222,993	135,395	137,058	522,676	0.42	522,676	0.55
Warehouse	299,331	1,700,585	750,822	2,451,407	1,488,429	1,506,711	5,745,879	4.59	5,745,879	6.02
Canteen/G.H.	132,316	751,722	331,892	1,083,614	657,942	666,023	2,539,894	2.03	2,539,894	2.66
Sundries	75,346	428,060	188,992	617,052	374,657	379,259	1,446,314	1.16	1,446,314	1.52
Welfare	130,237	739,915	326,679	1,066,594	647,607	655,562	2,500,000	2.00	2,500,000	2.62
Total	6,519,039	37,036,523	16,351,926	53,388,449	32,416,043	32,814,200	125,137,730			
Wages as per budget							125,137,731			

*Mfg-Manufacturing, Dist-distribution, GH-Guest house

DEPARTMENT WISE SALARY ALLOCATION IN ANNUAL BUDGET

Salary and wages of departments who are not directly involved in manufacturing, are distributed on manufacturing departments on the basis of formulac decided by different organisations.

Amount Rs	Dyeing	Spinning	Doubling & TFO	Recombing to Spinning	Weaving & G Perch	Finishing, Mending & Inspection	Total	%age to total salary	Distribute	%age total manf. Salary
Mfg. Salary	3,819,671	3,902,408	3,697,822	7,600,230	10,222,634	7,170,698	28,813,233	23.03		
Management	3,937,790	4,023,086	3,812,173	7,835,259	10,538,758	7,392,444	29,704,250	23.74	29,704,250	103.09
Designing	593,828	606,691	574,885	1,181,576	1,589,270	1,114,798	4,479,473	3.58	4,479,473	15.55
R M Godown	119,238	121,821	115,435	237,256	319,119	223,847	899,461	0.72	899,461	3.12
Stores & Purchases	472,483	482,717	457,410	940,128	1,264,512	886,996	3,564,119	2.85	3,564,119	12.37
QC Deptt	563,331	575,534	545,361	1,120,894	1,507,651	1,057,546	4,249,423	3.40	4,249,423	14./5
Mechanical	385,596	393,949	373,296	767,244	1,031,976	723,883	2,908,699	2.32	2,908,699	10.10
Electrical	523,245	534,579	506,553	1,041,132	1,400,368	982,292	3,947,038	3.15	3,947,038	13.70
General Offic	117,146	119,683	113,409	233,092	313,518	219,918	883,674	0.71	883,674	3.07
Personnel / HR	407,085	415,903	394,099	810,001	1,089,486	764,223	3,070,795	2.45	3,070,795	10.66
Time Office	139,775	142,803	135,316	278,119	374,082	262,401	1,054,376	0.84	1,054,376	3.66
PF/ ESI	71,640	73,192	69,355	142,546	191,731	134,490	540,408	0.43	540,408	1.88
Planning	189,382	193,485	183,341	376,825	506,846	355,529	1,428,583	1.14	1,428,583	4.96
Accounts	433,851	443,248	420,011	863,259	1,161,121	814,472	3,272,702	2.62	3,272,702	11.36
Cash	46,578	47,587	45,092	92,680	124,658	87,442	351,358	0.28	351,358	1.22
Costing	105,512	107,798	102,147	209,944	282,384	198,079	795,920	0.64	795,920	2.76
Legal / Regulatory	338,019	345,341	327,236	672,576	904,644	634,565	2,549,804	2.04	2,549,804	8.85
IT	410,011	418,892	396,931	815,823	1,097,316	769,716	3,092,866	2.47	3,092,866	10.73
Warehouse & Invoice	575,140	587,598	556,792	1,144,390	1,539,253	1,079,714	4,338,497	3.47	4,338,497	15.06
Sales	2,044,348	2,088,630	1,979,132	4,067,763	5,471,315	3,837,871	15,421,296	12.32	15,421,296	53.52
R Shop	215,345	220,010	208,476	428,486	576,331	404,269	1,624,432	1.30	1,624,432	5.64
Corporate Office	698,432	713,561	676,152	1,389,713	1,869,224	1,311,173	5,268,542	4.21	5,268,542	18.29

Training Staff	259,857	265,486	251,567	517,053	695,458	487,831	1,960,200	1.57	1,960,200	6.80
Security	587,774	600,505	569,023	1,169,529	1,573,066	1,103,432	4,433,801	3.54	4,433,801	15.39
Driver	100,284	102,456	97,085	199,541	268,391	188,264	756,479	0.60	756,479	2.63
Wage & Salary	96,637	98,730	93,554	192,285	258,631	181,417	728,970	0.58	728,970	2.53
Total	17,251,998	17,625,692	16,701,652	34,327,344	46,171,746	32,387,312	130,138,400			
Salary as per Budget							130,138,400			

ALLOCATION OF DEPRECIATION

Actual Depreciation Rs	Dyeing	Spinning	Doubling & TFO	Recombing to Spinning	Weaving & Perching	Finishing, Mending & Inspection	Total	Dist	%age to Mfg Dep.
Plant & Machinery	2,239,485	60,641,116	4,932,814	65,573,930	15,063,717	12,155,521	95,032,653		
Buildings	172,397	4,263,247	173,543	4,436,790	999,511	496,144	6,104,842		
Furniture & OE	2,145			-	1,773	1,371	5,289		
Vehicles				-					
Electrical Section									
Plant & Machinery	91,690	2,482,799	201,962	2,684,760	616,746	497,677	3,890,874	3,890,874	4.09
Buildings	3,453	85,380	3,476	88,856	20,017	9,936	122,262	122,262	2.00
Furniture & OE	1,059	-	-	-	875	677	2,611	2,611	49.37
Vehicles	-	-	-	-	-	-	-		
Mechanical Section									
Plant & Machinery	62,849	1,701,848	138,436	1,840,284	422,752	341,136	2,667,021	2,667,021	2.81
Buildings	309	7,631	311	7,942	1,789	888	10,928	10,928	0.18
Furniture & OE	-	-	-	-	-	-	-	-	-
Vehicles	-	-	-	-	-	-	-	-	-
R&D &QA Section									
Plant & Machinery	6,354	172,049	13,995	186,045	42,738	34,487	269,624	269,624	0.28
Buildings	93	2,298	94	2,392	539	267	3,291	3,291	0.05
Furniture & OE	4,642	-	-	-	3,837	2,967	11,446	11,446	216.41
Vehicles	-	-	-	-	-	-	-	-	-
STORE									
Plant & Machinery	1,219	32,997	2,684	35,681	8,197	6,614	51,711	51,711	0.05
Buildings	941	23,260	947	24,206	5,453	2,707	33,307	33,307	0.55
Furniture & OE	5,593	-	-	-	4,623	3,575	13,792	13,792	260.77
Vehicles	-	-	-	-	-	-	-	-	
IT				-					
Plant & Machinery	41,654	1,127,912	91,749	1,219,661	280,182	226,090	1,767,587	1,767,587	1.86
Buildings	158	3,904	159	4,063	915	454	5,590	5,590	0.09
Furniture & OE	1,635	-	-	-	1,351	1,045	4,031	4,031	76.21
Vehicles				-				-	-
Others				-					
Plant & Machinery	12,037	325,942	26,514	352,455	80,966	65,335	510,794	510,794	0.54
Buildings	80,507	1,990,875	81,042	2,071,917	466,757	231,692	2,850,873	2,850,873	46.70
Furniture & OE	459,100	-	-	-	379,480	293,439	1,132,019	1,132,019	21,403.27
Vehicles	16,086	435,578	35,432	471,010	108,201	87,312	682,608	682,608	0.72

Warehouse									
Plant & Machinery	1,686	45,655	3,714	49,368	11,341	9,151	71,547	71,547	0.08
Buildings	4,132	102,177	4,159	106,337	23,955	11,891	146,315	146,315	2.40
Furniture & OE	15,719	-	-	-	12,993	10,047	38,759	38,759	732.82
Vehicles	-	-	-	-	-	-	-		
Grand total									
Plant & Machinery	2,456,974	66,530,317	5,411,867	71,942,185	16,526,640	13,336,012	104,261,811		
% to total P &M	2.36	63.81	5.19		15.85	12.79			
Buildings	261,988	6,478,772	263,730	6,742,502	1,518,937	753,980	9,277,408		
% to total Building	2.82	69.83	2.84		16.37	8.13			
Furniture & OE	489,893	-	-	-	404,933	313,121	1,207,947		
% to total Furniture	40.56	-	-		33.52	25.92			
Vehicles	16,086	435,578	35,432	471,010	108,201	87,312	682,608		
% to total vehicles	2.36	63.81	5.19		15.85	12.79			

Notes: Manufacturing department depreciation is allocated department-wise on actual basis. Service department depreciation is allocated on the basis on mfr dep. (manufacturing depreciation) individually.

DEPARTMENT WISE MONTHLY DIRECT AND ALLOCATED COST SUMMARY

	Budget / Month	Apr-15	May-16	Jun-15	Jul-15	Aug-15	Sep-15	Oct-15	Nov-15	Dec-15	Jan-16	Feb-16	Cumulative	FY-15 ACT
Top Dyeing														
Top Production (KG)	94198	31712	t24070	46857	80893	53652	44537	82962	55416	87256	67076	62739	637169	396856
Yarn (KG)	0	6429	5605	10683	5124	6279	2140	2246	2599	3594	2514	1892	49104	62178
Direct Cost (Rs.)	5359918	3195079	2605414	3661442	4822029	3926261	3380055	4957031	3929425	5158841	4087422	3914330	43637328	37640456
Direct Cost/ kg (Rs.)	56.90	83.77	87.80	63.63	56.06	65.51	72.41	58.18	67.73	56.78	58.74	60.56	63.59	82.00
Allocated cost/kg (Rs.)	6.09	19.14	27.21	16.73	9.58	13.11	21.19	9.08	14.72	11.16	14.84	16.29	14.32	22.36
Fabric Dyeing														
Production (MTR)	41672	16441	28107	31602	25946	43671	36161	43973	62011	49787	55822	54187	447707	478665
Direct Cost (Rs)	1452777	767317	1011437	978824	1055022	1114320	972083	1252651	1434143	1299408	1273100	1258686	12416991	13937590
Direct Cost/ MTR (Rs.)	34.86	46.67	35.99	30.97	40.66	25.52	26.88	28.49	23.13	26.10	22.81	23.23	27.73	29.12
Allocated cost/MTR (Rs.)	9.08	29.28	18.94	20.09	20.95	11.87	18.04	11.60	9.08	13.43	12.20	12.81	14.48	20.87
Spinning														
Production (KG)	96952	45420	42536	49009	84893	81469	76008	84352	81399	85779	86578	75637	793080	595837
RPM	8656	8274	8237	8175	8564	8393	8362	8517	8347	8324	8444	8432	8390	8391
Count	51	57.62	54.7	58.3	58.8	59.5	57.5	58.7	58.8	63.1	58.6	62.5	59.2	57.6
Direct Cost (Rs.)	12714449	10025612	8784134	9910483	12687189	13406637	11472874	13045676	12242486	12533521	11851854	12369855	128330321	108319288
Cost per count (Rs.)	**3.24**	**5.44**	**5.78**	**5.41**	**3.49**	**3.70**	**3.93**	**3.54**	**3.59**	**3.40**	**3.51**	**3.90**	**3.94**	**4.55**
Direct cost/ count (Rs.)	2.57	3.83	3.78	3.47	2.54	2.77	2.63	2.64	2.56	2.32	2.34	2.62	2.73	3.16
Allocated cost/count (Rs.)	0.67	1.61	2.00	1.94	0.95	0.94	1.31	0.90	1.03	1.08	1.18	1.28	1.21	1.39

Weaving														
Production (MTRs)	310943	136872	117899	150854	221347	231193	242558	282642	297253	294705	327008	331154	2633484	1937220
RPM	448	371	370	370	363	365	374	376	380	384	386	386	377	378
Pick	65	68	64	58	67	64	64	70	69	74	67	69	67	68
Cost per pick (Rs.)	**0.75**	**1.24**	**1.50**	**1.45**	**0.90**	**0.94**	**0.90**	**0.73**	**1.03**	**0.67**	**0.69**	**0.66**	**0.88**	**0.93**
Direct Cost (Rs.)	11973331	7321430	6786213	7358826	8738629	9728221	8468788	10172233	16229526	8834188	9355444	9182099	102175597	81836692
Direct cost / pick (Rs.)	0.59	0.81	0.91	0.85	0.60	0.65	0.55	0.52	0.80	0.41	0.43	0.40	0.58	0.63
Allocated cost/pick (Rs.)	0.15	0.44	0.59	0.60	0.31	0.29	0.35	0.21	0.23	0.26	0.26	0.25	0.30	0.30

Finishing														
Production (MTRs)	297083	134669	115285	118611	189189	156772	187290	266513	265688	264146	242939	251287	2192391	1820925
Direct Cost (Rs.)	9172324	5018673	4950280	5680837	6634663	8014556	6141187	7181233	7026768	8953810	6454627	6148904	72205537	72801255
Direct Cost/MTR (Rs.)	30.87	37.27	42.94	47.89	35.07	51.12	32.79	26.95	26.45	33.90	26.57	24.47	32.93	39.98
Allocated cost/ MTR (Rs.)	10.81	30.34	39.20	45.44	24.39	28.06	29.56	16.25	17.99	21.49	23.79	23.45	25.09	30.77
Power														
Units Consumed (KEB + Own)	1508000	1086750	845180	1069690	1489500	1484840	1389260	1635020	1580780	1719530	1567270	1568410	15436230	12027000
KEB Cost / HFO / HSD / Stores & Lub oil cost (Rs.)	10401111	7627229	6087964	7505376	10076909	10050318	9458676	11064065	10816973	11668288	10629957	10629368	105615123	83429019
Unit /Ltr (Rs.)	0.00	0.00	1.31	0.53	0.00	3.68	1.58	3.63	3.56	1.45	2.73	0.94	3.69	3.82
Cost/unit (Rs.)	6.90	7.02	7.20	7.02	6.77	6.77	6.81	6.77	6.84	6.79	6.78	6.78	6.84	6.94
Units/MTR	5.08	8.07	7.33	9.02	7.87	9.47	7.42	6.13	5.95	6.51	6.45	6.24	7.04	6.60
Boiler														
No. of Tonnes generated-Steam	1067.5	677	613	839	1077	991	890	1208	1211	1390	1394	1312.55	11602.55	8957
Furnace Oil (Ltrs)	76250	50690	45952	62895	80685	74250	66710	90489	90730	104135	104475	98345	869356	674615
Avg. KG per ltr of HFO	14.00	13.36	13.34	13.34	13.35	13.35	13.34	13.35	13.35	13.35	13.34	13.35	13.35	13.28
No. of Tonnes generated-WHRB	0	0	0	0	0	0	0	0	0	0	0	0	0	0
Cost of Furnace Oil (Rs.)	2668750	1379275	1543987	2065472	2418936	1930500	1573689	2127396	2133062	2247233	1800104	1656130	20875785	27531072
Avg cost per Kg of Steam (Rs.)	2.50	2.04	2.52	2.46	2.25	1.95	1.77	1.76	1.76	1.62	1.29	1.26	1.80	3.07
Litres / MTR	0.26	0.38	0.40	0.53	0.43	0.47	0.36	0.34	0.34	0.39	0.43	0.39	0.40	0.37
Thermopack														
Furnace Oil (Ltrs)	50833	32600	31035	37340	51345	52220	55610	62171	63070	70920	64345	55045	575701	389795
Rate Per Ltr	35	27.21	33.6	32.84	29.98	26	23.59	23.51	23.51	21.58	17.23	16.84	24.10	40.36
Cost of HFO (Rs.)	1779167	887046	1042776	1226246	1539323	1357720	1311840	1461640	1482776	1530454	1108664	926958	13875442	15730606
Total Production	297080	134669	115285	118611	189189	156772	187290	266513	265688	264146	242939	251287	2192391	1820925
Cost per Mtr (Rs.)	5.99	6.59	9.05	10.34	8.14	8.66	7.00	5.48	5.58	5.79	4.56	3.69	6.33	8.64
Litres / MTR	0.17	0.24	0.27	0.31	0.27	0.33	0.30	0.23	0.24	0.27	0.26	0.22	0.26	0.21

DEPARTMENT WISE DIRECT AND INDIRECT COST SUMMARY

Spinning	TARGET		Weaving		
PARTICULARS		P/KG	PARTICULARS		P/MTR
Production (Kgs)	96952		Mandays (Including Mending)	95980	
Delivery (Kgs)	95013		Production (Mtrs)	4024950	

No. of Spindle Shifts	1010880		Production sample (Mtrs)	27272	
AVG. COUNT	51		Total Production - Grey Perch (Mtrs)	4052222	
R.P.M	8656		RPM	419.17	
TPM	710		Sample Looms Run	465.78	
Efficiency %	88%		Average Picks	68.50	
No. of Mandays	6396		Working Efficiency	71%	
Fuel Consumption in Ltrs	8388	0.086	Installed Efficiency		
Steam Consumption in Kg	117425	1.21	No. of Loom shifts run	76261	
Power Units	791424	8.16	Water Consumption	0	
Water Consumption	0	0.0	Power Units	5797511	
			EXPENSES (Rs.)		
Direct Expenses (Rs.)			**Direct Expenses**		
Power Expenses	5458677	56.3	Power	33876690	8.42
Water Charges	0	0.00	Water	0	0.00
Fuel Expenses	293563	3.03	Lubricants	2944576	0.73
Stores & Spares	915000	9.63	Stores & Spares	7578481	1.88
Wages	5076210	53.43	Wages	28848438	7.17
Salaries	971000	10.22	Wages- Mending	16575160	4.12
Total Direct Expenses (A)	6962210	73.28	Salaries	13217253	3.28
			Direct Expenses- Total	103040598	25.60
Indirect Expenses			**Indirect Expenses**		
Apportioned Expenses	3311058	34.15	Apportioned Expenses	52375867	13.01
Total Indirect Expenses(B)	3311058	34.15	**Indirect Expenses-Total**	52375867	13.01
Total Expenses. (A+B)	10273268	107.4	**Total Expenses**	155416465	**38.61**
Avg. Cost Per Count Rs		2.11	**Avg. cost/ Pick Rs**		**0.56**
			Direct Cost/ Pick		**0.37**
			Allocated Cost/ Pick		**0.19**
			Stores & Spares cost / Loom Shift		**137.99**

Top Dyeing	TARGET		Fabric Dyeing		
PARTICULARS		**P/KG**	**PARTICULARS**		**P/Mtr**
Mandays	1352		Mandays	174	
			Dyed qty (Mtrs)	62011	
Production- Top Dyeing (Kgs)	94198		Overdyed quantity (Mtrs)	1703	
Production- Yarn Dyeing (Kgs)	-		Redyed quantity (Mtrs)	1788	
Steam consumption (Kgs)	213500	2.3			
Fuel consumption (Ltrs)	15250	0.2	Steam consumption in kg (15%)	181650	2.93
Power consumption (Units)	98815	1.0	Fuel consumption (Ltrs)	13610	0.22
Water consumption (KL)	2521	0.026	Power consumption (Units)	36511	0.59
			Water consumption (KL)	2330	0.04
	Rs.	**Rs.**		**Rs.**	
Direct Expenses			**Direct Expenses**		
			Dyes & Chemicals consumed	696446	11.23
Dyes & Chemicals consumption	2611667	27.73	Fuel	319959	5.16
Fuel for steam	533750	5.67	Power	249836	4.03
Power	681556	7.24	Water	69908	1.13

Water	75625	0.80	Wages	97993	1.58
Wages	1073020	11.39	**Direct expenses- Total**	794608	**23.13**
Salaries	384300	4.08			
Direct Expenses-Total	5359918	56.90	**Apportioned Expenses**	563085	9.08
			Indirect Expenses	563085	9.08
Apportioned Expenses	573637	6.09	**TOTAL**	**1357693**	32.81
Indirect expenses- Total	573637	6.09			
TOTAL EXPENSES	**5933555**	**62.99**			

Finishing	TARGET		Packing	TARGET	
PARTICULARS		P/Mtr			
Mandays	34529		PARTICULARS		P/Mtr
Production (Mtrs)	4378936		Mandays	2366	
Steam consumption (Kgs)	9888480				
Fuel for Thermopack (ltrs)	687998		Folded - Inhouse (Mtrs)	297083	
Fuel for Boiler (Ltrs)	755469		Folded - PV (Mtrs)		
Power consumption (Units)	1679845		Packed quantity (Mtr)	297083	
Water consumption (KL)	107284				
				297083	
	Rs.	Rs.		Rs.	Rs.
Direct Expenses			**Direct expenses**		
Chemicals	12616721	2.88	Packing materials - Folding	683292	2.30
Fuel for Thermopack	28046379	6.40	Packing materials- Warehouse	1188333	4.00
Fuel for Boiler	30727802	7.02	Wages	1877785	6.32
Power	9387454	2.14	Salaries	680000	2.29
Water	2682102	0.61	**Direct Expenses-Total**	4429410	14.91
Stores & Spares/others	618406	0.14	**Apportioned Expenses**		
Singeing (Gas)	1073575	0.25	**TOTAL**	4429410	
Wrapper cloth	6687759	1.53			
Wages	15532403	3.55			
Salaries	4651688	1.06			
Direct Expenses- Total	112024288	**25.58**			
Apportioned Expenses	87203278	19.91			
Total	199227565	**45.50**			

COST CALCULATION GENERATOR SET POWER VS GOVT POWER SUPPLY COST

	FY-1			FY-2		
PARTICULARS	Gen Set 1	Gen Set 2	TOTAL	Gen Set 1	Gen Set 2	TOTAL
Units Generated	17880	29090	46970	20880	44830	65710
Net units generated			**46970**			**65710**
HSD Consumed Ltrs	5514	8700	14214	8703	9189	17893
Furnace Oil (Ltrs)		1660	1660		6168	6168
Units generated per Litre	3.24	2.81	2.96	2.40	2.92	2.73
Water consumption (KL)			0			0
EXPENSES	Rs.	Rs.	Rs.	Rs.	Rs.	Rs.
Furnace Oil		190439	190439		560139	560139

HSD	206268	225205	431473	248810	29825	278635
Lub Oil	0	5148	5148	33779	0	33779
Stores, spares etc regular	44460	55668	100128	6735	0	6735
AM Charges						
Water				0		0
Maintenance charges				0		0
Maintenance Contract -				0		0
Total Expenses				727187		879288
Avg cost per Unit (Own)				**15.48**		**13.38**
Cost of HFO/Ltr				40.67		44.88
HSD						
Cost of Govt Power Supply						
			CUMULATIVE		CUMULATIVE	
Units Consumed			**21055170**			**20581650**
MD Cost @ 170/KVA for 4500 KVA			9180000			9180000
Upto 100000 Units			64719605			4660000
Above 100000 Units			38086083			100640488
Discount @ 2Paise / Unit			421103			411633
Total			**111564585**			**114068855**
Tax @ 6%			5578229			5703443
Grand Total			**117142814**			**119772297**
Unit Cost Govt Supply			**5.56**			**5.82**
Total Power Cost						
Total Units (GS + Own)			21102140			20647360
Total Cost (GS+ HFO+HSD)			117870001			120651586
Avg Cost (KEB + OWN+ IEX)			**5.6**			**5.84**

CALCULATION OF STEAM COST

PARTICULARS	FY-12 ACT	FY-13 ACT
No.of Tonnes generated-Steam	18312	14996
Furnace Oil (Ltrs)	1399016	1131558
Avg Kg per ltr of HFO	13.09	13.25
EXPENSES		
HFO Rs Ltr	40.67	44.96
Furnace Oil (Rs)	56903337	50876987
Avg cost per Kg of Steam	3.11	3.39
USAGE OF HFO IN THERMOPACK FOR FINISHING		
Furnace Oil (Ltrs)	687998	645568
Rate Per Ltr	40.77	44.88
Cost of HFO (Rs)	28046379	28973643
Total Production (Mtrs)	4383509	3993847
Cost per Mtr	6.40	7.25

ELEMENTS OF DIRECT COST OF PRODUCTION DEPARTMENTS

DYEING - TOP/ YARN	TARGET		MARCH	CUMULATIVE	
PARTICULARS	Rs	P/KG	P/KG	Rs	P/KG
Production- Top Dyeing (Kgs)	94167			553268	
Production- Yarn Dyeing (Kgs)	-			62968	
Direct Expenses (Rs.)					
Dyes & Chemicals consumption	1867292	19.83		14259501	23.14
Fuel for steam	896623	9.52		7372326	11.96
Power	582534	6.19		5900285	9.57
Water	85466	0.91		688796	1.12
Wages	765600	8.13		7629522	12.38
Salaries	470000	4.99		4561318	7.40
Direct Expenses-Total	4667515	49.57		40411748	65.58
DYEING - FABRIC	**TARGET**		**MARCH**	**CUMULATIVE**	
PARTICULARS		P/Mtr	P/Mtr		P/Mtr
Dyed qty (Mtrs)	91522			401005	
Overdyed quantity (Mtrs)	**0**			50646	
Redyed quantity (Mtrs)	0			16435	
Direct Expenses (Rs.)					
Dyes & Chemicals consumed	2882944	31.50		4777705	11.91
Fuel	672467	7.35		5529244	13.79
Power	206049	2.25		2106823	5.25
Water	81134	0.89		653879	1.63
Wages	143550	1.57		1433853	3.58
Direct Expenses- Total	3986144	43.55		14501504	**36.16**
SPINNING	**TARGET**		**MARCH**	**CUMULATIVE**	
PARTICULARS		P/KG	P/KG		
Production (Kgs)	116136			904645	
Direct Expenses (Rs.)					
Power Expenses.	5703551	49.11		55226059	61.05
Fuel Expenses	493143	4.25		4054779	4.48
Stores & Spares	833333	7.18		8411062	9.30
Wages	3907750	33.65		40102836	44.33
Salaries	928000	7.99		9513841	10.52
Total Direct Expenses	11865777	102.17		117308577	129.67
Direct Cost/ Count		1.96			**2.24**
WEAVING	**TARGET**		**MARCH**	**CUMULATIVE**	
PARTICULARS		P/MTR	P/KG		P/MTR
Production - (Mtrs)	569350			2510154	
EXPENSES (Rs.)					
Direct Expenses					
Power	2959167	4.98		21938677	8.74
Sapcosize	483948	0.81		1847571	0.74
Stores & Spares	1025000	1.73		7490332	2.98
Wages	2695550	4.54		27243135	10.85
Wages- Mending	1626900	2.74		15943551	6.35

Salaries	1054000	1.77		11135023	4.44
Direct Expenses- Total	9844565	16.57		85598289	34.10
Direct Cost/ Pick		0.25			0.49

FINISHING	TARGET		MARCH	CUMULATIVE	
PARTICULARS		P/Mtr	P/Mtr		P/Mtr
Production (Mtrs)	513667			2328430	
Direct Expenses (Rs.)					
Chemicals	1541000	3.00		7411460	3.18
Fuel for Thermopack	2420882	4.71		21938451	9.42
Fuel for Boiler	2305602	4.49		19905280	8.55
Power	861169	1.68		8342351	3.58
Water	203473	0.40		1446175	0.62
Stores & Spares/others	166667	0.32		640576	0.28
Singeing (Gas)	100000	0.19		901515	0.39
Wrapper cloth	924600	1.80		2757732	1.18
Wages	1579050	3.07		15738026	6.76
Salaries	420000	0.82		4065394	1.75
Direct Expenses- Total	10522443	20.48		83146958	35.71

ANNUAL COST ANALYSIS OF A COMPOSITE UNIT

	Value in Rs	For The Month			Year to Date		
		Nov-14		Nov-13	Nov-14		Nov-13
		Budget	Actual	Actual	Budget	Actual	Actual
1	Fabric Production (Mtrs)	5,00,000	3,15,991	4,01,731	40,82,000	25,68,242	38,35,938
2	Fabric Dispatches (Mtrs)	5,69,000	3,74,320	3,16,240	47,73,000	32,33,320	41,73,888
3	Fabric Dispatches Domestic (Mtrs)	2,85,000	2,66,497	2,58,071	26,24,000	24,67,497	24,25,974
4	Dyes & Chemicals (Value)	44,33,864	22,81,611	46,50,000	3,69,49,587	2,43,09,938	4,64,97,958
5	Dyes & Chemicals / Mtr (4/1)	8.87	7.22	11.57	9.05	9.47	12.12
a	Stores, Consumption Value	65,60,656	47,12,028	41,63,909	5,30,45,458	4,00,14,790	4,18,22,284
	Stores, Consumption/Mtr	13.12	14.91	10.36	12.99	15.58	10.90
b	Repair Maintenance Value	9,45,350	2,61,831	(70,855)	85,08,150	37,76,227	76,65,379
	Repair Maintenance/Mtr	1.09	0.83	(0.18)	2.08	1.47	2.00
c	Packing value	17,30,544	32,50,902	18,35,580	1,32,86,695	1,30,69,805	1,83,24,895
	Packing /Mtr	3.46	10.29	4.57	3.25	5.09	4.78
6	Total Stores, Consu & Pakg (Value)	92,36,550	82,24,761	59,28,634	7,48,40,303	5,68,60,822	6,78,12,558
7	Total Stores, Consu & Pakg/ Mtr	18.47	26.03	14.76	18.33	22.14	17.68
a	Power (Value)	1,44,43,140	1,03,96,298	1,28,64,065	11,79,13,798	10,18,08,935	11,54,69,857
	Power /Mtr	28.89	32.90	32.02	28.89	39.64	30.10
b	Fuel (Value)	86,20,696	62,59,172	71,70,231	7,03,79,363	5,01,39,243	5,95,07,994
	Fuel /Mtr	17.24	19.81	17.85	17.24	19.52	15.51
8	Total Power & Fuel (Value)	2,30,63,836	1,66,55,470	2,00,34,296	18,82,93,161	15,19,48,178	17,49,77,851
9	Total Power & Fuel/Mtr	46.13	52.71	49.87	46.13	59.16	45.62
10	Power No. of Units Consumed	25,81,409	16,70,320	22,08,232	2,13,96,476	1,55,77,300	2,16,62,702

11	Power no of Units/Mtr (on production mtrs)	**5.16**	**5.29**	**5.50**	**5.24**	**6.07**	**5.65**
12	No. of Workers	1,214	1,167	1,246	1,214	1,075	1,175
13	Wages (Value)	1,08,30,623	1,05,00,000	99,76,500	9,45,71,797	9,20,79,000	8,88,35,000
14	Average Wage/Worker	**8,921**	**8,997**	**8,007**	**77,901**	**85,655**	**75,604**
15	Wages Per Mtr Rs. (on production mtrs)	**22**	**33**	**25**	**23**	**36**	**23**
16	No of Staff	307	299	314	307	303	313
17	Salary (Value)	1,08,44,867	1,07,01,000	1,06,21,500	9,76,03,802	9,97,27,000	9,06,83,000
18	Average Salary/Employee	**35,325**	**35,789**	**33,826**	**3,17,928**	**3,29,132**	**2,89,722**
19	Selling Expenses						
a	Commission Value	62,46,637	45,39,055	10,16,674	5,17,75,515	3,58,73,337	3,11,96,257
	Commission/Mtr (on dispatch mtrs)	**10.98**	**12.13**	**3.21**	**10.85**	**11.09**	**7.47**
b	Sales Promotion Value	17,20,513	47,60,575	45,07,928	1,42,70,161	2,76,87,447	1,67,96,975
	Sales Promotion/Mtr (on dispatch mtrs)	**3.02**	**12.72**	**14.25**	**2.99**	**8.56**	**4.02**
c	Freight etc (Value)	3,49,000	8,57,806	8,46,585	33,27,985	60,79,351	1,01,29,966
	Freight /Mtr (on dispatch mtrs)	**0.61**	**2.29**	**2.68**	**0.70**	**1.88**	**2.43**
a	Total Com ,Sale Prom & Freight Value	83,16,150	1,01,57,436	63,71,187	6,93,73,661	6,96,40,135	5,81,23,198
b	Total Commission & Freight etc./Mtr	**14.62**	**27.14**	**20.15**	**14.53**	**21.54**	**13.93**
c	Advertisement & Publicity (Value)	1,27,27,273	1,11,244	9,23,379	10,81,81,819	6,20,01,211	5,60,03,807
d	Advertisement & Publicity /Mtr Rs.(on only domestic dispatch mtrs)	**44.66**	**0.42**	**3.58**	**41.23**	**25.13**	**23.09**
e	Royalty (8 %) (Value)	1,47,407	-	-	24,13,626	8,99,992	
20	Cash Discounts (Value)	-	9,30,566	16,87,978		77,31,574	65,03,488
21	Cash Discounts /Mtr	-	**2.49**	**5.34**	-	**2.39**	**1.56**
24	Agent Commission (Value)	13,65,632	-	1,58,435	1,09,25,056	24,04,948	96,24,130
25	Agent Commission /Mtr	**2.40**	-	**0.50**	**2.29**	**0.74**	**2.31**
26	Admin & Other Ohds (Value)	63,70,000	64,51,000	61,90,000	5,73,32,000	6,11,96,000	6,26,88,000
27	Admin & Other Ohds /Mtr (on production mtrs)	**12.74**	**20.42**	**15.41**	**14.05**	**23.83**	**16.34**

COSTING OF COTTON YARN

Cotton yarn costing						
	PARTICULAR (K-Carded)	**K**	**K**	**K**	**K**	**K**
A	**TECHNICAL PARAMETERS (Count Ne)**	**14**	**16**	**18**	**20**	**24**
1	No. of Ring Frames Allotted	4	10	4	6	2
2	No. of Spindles Per Ring Frame	476	476	476	476	380
3	No. of Allotted Spindles per Shift (8 hours)	1,904	4,760	1,904	2,856	760
4	Utilisation %	97	97	97	97	97
5	Actual Working Spindles Per Shift	1846.9	4617.2	1846.9	2770.3	737.2
6	Hence Actual Working spindles .Per Month (30days)	166219.2	415548.0	166219.2	249328.8	66348.0

7	Average Spindle Speed (RPM)	12500.0	13500.0	13500.0	14000.0	14000.0
8	T.P.I. (T.M.=4.0/4.5)	15.0	16.0	17.0	17.8	19.6
9	Gms / Spindle / Shift @ 100%	428.6	379.7	317.6	283.1	214.3
10	Efficiency Gms / Spindle Shift @ 93%	398.6	353.1	295.4	263.3	199.3
11	No. of Spindles Required / Kg of yarn / Per Shift	2.51	2.83	3.39	3.80	5.02
12	No. of spindles allotted per month	166219.2	415548.0	166219.2	249328.8	66348.0
13	Production Per Month Kgs.	66250.2	146733.9	49103.1	65654.7	13222.2
14	Packed Production Per Month Kgs.	65852.7	145853.5	48808.5	65260.8	13142.9
15	Units (Power) / Kg of Yarn (0.37/Spindle/shift)	0.928	1.048	1.252	1.405	1.857
16	Cost of Power / Kg of yarn	3.806	4.296	5.135	5.761	7.612
	Electricity Tariff Rs 4.10 Per Unit					
	DIRECT COST in Rs.					
17	Raw Material Cost in Candy(355.6 kg)	19000.0	18500.0	19000.0	18500.0	19000.0
18	Raw Material Cost / Kg	53.431	52.025	53.431	52.025	53.431
19	CST 4% (in the past)	2.137	2.081	2.137	2.081	2.137
20	Transport	0.500	0.500	0.500	0.500	0.500
21	Raw Material Cost / Kg.	56.068	54.606	56.068	54.606	56.068
22	Yarn Recovery @ 85% Cost of Clean Cotton / kgs.	65.962	64.242	65.962	64.242	65.962
23	Less Waste Sale / Kg.	2.000	2.000	2.000	2.000	2.000
24	Net Clean Cotton Cost / Kg.	63.962	62.242	63.962	62.242	63.962
25	Packing Cost / Kg.	2.000	2.000	2.000	2.000	2.000
	MANUFACTURING EXPENSES in Rs.					
26	Salaries/kg(5 lac/month)	0.691	0.780	0.933	1.047	1.383
27	Wages / kg.(370+117)	3.114	3.515	4.202	4.714	6.228
28	Stores & Spares / Kg.(Rs.0.25/Spindle /shift)	0.627	0.708	0.846	0.95	1.25
29	Interest / Kg.1%	0.640	0.622	0.640	0.622	0.640
30	Year End Expenses (5 lacs./month)	0.691	0.780	0.933	1.047	1.383
31	Administration / Kg.(5 Lacs./month)	0.691	0.780	0.933	1.047	1.383
32	Total Manufacturing Cost / Kg.	6.455	7.187	8.486	9.425	12.271
33	**Total Cost (16+25+32)**	**12.26**	**13.48**	**15.62**	**17.19**	**21.88**
34	Total yarn Cost (24+33)	76.22	75.72	79.58	79.43	85.85
35	**Add Contribution 1%**	**78.51**	**76.48**	**80.38**	**80.22**	**86.70**
	SALES in Rs.					
36	Sales Rate / Kg.	**84.0**	**82.0**	**88.0**	**86.0**	**95.0**
37	Sales Expenses / Kg.(Brokerage & CD 2.5 % of Sales rate)	2.100	2.050	2.200	2.150	2.375
38	**Tax**					
	Transport	0.2	0.5	0.2	0.5	0.2
38	**Profit & Loss / Kg of Yarn**	**5.48**	**3.73**	**6.02**	**3.92**	**6.58**

WORSTED SPINNING, SECTION WISE COSTINGS

	Particulars- Amount in Rs	Gill Box & Flyer Frem	Ring Frame	Auto Coner	Assembling winding	TFO	Steaming & Rewinding	Total Exp
1	Staff Salary & Amenities	1.48	1.48	1.48	1.48	1.48	1.48	8.88
2	Wages & Amenities (Inclu. Contract)	4.64	24.64	7.2	3.52	6.56	4.48	51.04

3	Stores & Spares	1.35	3.15	1.35	0.45	2.25	0.45	9
5	Power	4.72	22.36	8.35	0.89	17.53	1.26	55.11
6	Coal						1.5	1.5
7	Water						0.1	0.1
	Total Expenses/Month	**12.19**	**51.63**	**18.38**	**6.34**	**27.82**	**9.27**	**125.63**
	Working Days	**26**	**26**	**26**	**26**	**26**	**26**	**26**
	Total Expenses/day Rs	46896	198563	70689	24383	107004	35644	483180
	Yarn Production Per Month Kg	96382	84084	96096	98826	76518	87464	84084
	Yarn Production Per Day Kg	3707	3234	3696	3801	2943	3364	3234
	Cost Per Day Per Kg Rs	12.65	61.4	19.13	6.41	36.36	10.6	149.41
	Average Count Nm	52	52	52	52	52	52	52
	Cost Per Count Per Kg Rs	0.24	1.18	0.37	0.12	0.7	0.2	2.87
	Distribution %	**8.47**	**41.1**	**12.8**	**4.29**	**24.34**	**7.09**	**100**

WORSTED YARN COSTING–A

	PARTICULAR	Commed	Commed	Commed	Commed	Commed	Commed	Commed	Commed	Commed	Commed
A	TECHNICAL PARAMETERS (Count Nm)	24	32	40	45	48	60	72	80	90	100
1	No. of Ring Frames Alloted	2	2	2	1	2	3	2	2	2	2
2	No. of Spindles per Ring Frame	400	400	400	400	400	400	400	400	400	400
3	No. of Alloted Spindles per Shift (8 hours)	800	800	800	400	800	1200	800	800	800	800
4	Utilisation in %	95.5	95.5	95.5	95.5	95.5	95.5	95.5	95.5	95.5	95.5
5	Actual working spindles per shift	600	600	624	312	640	960	640	640	640	640
6	Actual working spdls.per month (26 days)	46800	46800	48672	24336	49920	74880	49920	49920	49920	49920
7	Average spindle speed (rpm)	7000	7000	7200	7000	7800	8000	7500	7500	7500	6500
8	T.P.I.	11	13	16	17	18	19	22	22	27	28
9	100% Gms / Spindle/ shift	323	205	137	112	110	86	58	52	38	28
10	RF Efficiency %	75	75	78	78	80	80	80	80	80	80
11	Actual Gms/Spindle/shift	242	154	107	87	88	68	46	42	30	23
12	Spindles required / kg. of yarn / per shift	4.13	6.49	9.34	11.49	11.36	14.7	21.73	23.8	33.33	43.47
13	No. of spindles alloted per month	46800	46800	48672	24336	49920	74880	49920	49920	49920	49420
14	Production per month kgs.	11347	7201	5207	2118	4396	5125	2305	2075	1503	1130
15	Production Packed per month kgs.	11177	7093	5129	2086	4330	5048	2271	2044	1480	1113
16	Units (power) reqd /kg. of yarn(0.58/slp/shift)	2.40	3.77	5.42	6.67	6.59	8.53	12.61	13.81	19.33	25.22
17	Cost of power /kg. of yarn	20.97	32.95	47.43	58.33	57.67	74.63	110.33	120.83	169.17	220.65
	electricity tariff Rs 8.75 per unit										
	DIRECT COST in Rs.										
18	Raw material cost										
	Blend	80/20 TW	70/30 TW	55/45 TW	55/45 TW	55/45 TW	55/45 TW	55/45 TW	55/45 TW	55/45 TW	55/45 TW
	Wool/Kg (Combed top)	790	790	790	790	790	790	1142	1252	1419	2243

	Poly/Kg	143	143	143	143	143	143	143	143	143	143
19	Raw material cost/kg	273	338	435	435	435	435	593	642	718	1088
20	Yarn recovery @ 93% cost of clean RM /kg.	293	363	467	467	467	467	637	691	772	1170
21	Less waste sale /kg. (recovery)	5	5	5	5	5	5	5	5	5	5
	Net Raw Material cost /kg.	**288**	**358**	**462**	**462**	**462**	**462**	**632**	**686**	**767**	**1165**
	MANUFACTURING EXPENSES IN Rs.										
22	Combing & tow to top conversion (TTC)Cost /Kg	10	11	12	12	12	12	12	12	12	12
23	Recombing Cost /Kg	30	30	30	30	30	30	30	30	30	30
24	Packing Cost /Kg	2	2	2	2	2	2	2	2	2	2
25	Salaries /Kg(8.88 lac/month)	3	5	8	9	9	12	17	19	27	36
26	Wages	19	30	44	54	53	68	101	112	155	206
27	Power	20.97	32.95	47.43	58.33	57.67	74.63	110.33	120.83	169.17	220.65
28	Stores & Spares /kg.(Rs. 0.83/ spls/shift)	3	5	8	10	9	12	18	20	28	37
29	Interest /kg	0	0	0	0	0	0	0	0	0	0
30	Administration /kg.	0	0	0	0	0	0	0	0	0	1
31	**Total manufacturing cost /kg.**	**87.97**	**115.95**	**151.43**	**175.33**	**172.67**	**210.63**	**290.33**	**315.83**	**423.17**	**544.65**
32	**Cost Per Count**	**3.67**	**3.62**	**3.79**	**3.90**	**3.60**	**3.51**	**4.03**	**3.95**	**4.70**	**5.45**
33	Total yarn cost	375.97	473.95	613.43	637.33	634.67	672.63	922.33	1001.83	1190.17	1709.65
34	**Conversion Cost (Spinning)**	**49**	**76**	**108**	**132**	**131**	**168**	**248**	**276**	**380**	**505**
35	**Cost Per Count/Kg**	**2.02**	**2.36**	**2.7**	**2.94**	**2.73**	**2.8**	**3.45**	**3.45**	**4.22**	**5.05**

WORSTED YARN COSTING–B

Particular		Rate			Rate			Rate			Rate			Rate	
Count Nm			2/48			1/60			2/60			1/60			2/60
Blend			80/20 PW			80/20 PW			80/20 PW			70/30 PW			70/30 PW
Type of Polyester			S.D			S.D			S.D			S.D			S.D
Wool Micron			22.5			22.5			22.5			22.5			22.5
Polyester Rs Per Kg	80%	105	84	80%	105	84	80%	105	84	70%	110	77	70%	110	77
Wool Rate Per Rs Kg	20%	961	192	20%	961	192	20%	961	192	30%	961	288	30%	961	288
Material Cost			**276**			**276**			**276**			**365**			**365**
Waste	8%		22	8%		22	8%		22	8%		29	8%		29
Rs/Kg															
Combing			4			4			4			6			6
Tow to Top Conversion			8			8			8			7			7
Dyes and Chemicals			23			23			23			23			23
Dyeing			33			33			33			33			33
Recombing			34			34			34			34			34
Spinning			151			171			186			186			201
Total Conversion Charge			**252**			**272**			**287**			**288**			**303**
Total Cost of yarn (Dyed Per Kg)			**550**			**570**			**585**			**682**			**697**

CONVERSION COST OF WORSTED YARN BASED ON MONTH PRODUCTION

PARTICULAR'S	MAR-15th	MAR-15th	MAR-15th	MAR-15th
PACKED PRODUCTION 60 Nm CONVERTED. [MT]	166.7			
PACKED PRODUCTION ACTUAL. [MT]	165.2			
NO OF INSTALLED SPINDLES	21580			
NO OF WORKED SPINDLES	19809			
SPINDLE UTILIZATION	**91.8%**			
SALES VALUE Rs Lacs	2089			
	Rs. Lacs	Rs. / Kgs	% on Total Cost	% on Sale
CONVERSION COST				
POWER & FUEL COST	84.8	51.3	29.2	4.1
COAL& BIOMASS	8.7	5.3	3.0	0.4
STORE & SPARE COST	24.4	14.7	8.4	1.2
REPAIR & MAINTENANCE	-	-	-	-
TOTAL MANUFACTURING EXPS	**117.8**	**71.3**	**40.5**	**5.6**
SALARIES & BENEFITS	52.1	31.5	17.9	2.5
WAGES & BENEFITS	52.8	32.0	18.2	2.5
CONTRACTOR WAGES	26.8	16.2	9.2	1.3
WELFARE EXPENSES	1.7	1.1	0.6	0.1
TOTAL MANPOWER EXPS	**133.5**	**80.8**	**45.9**	**6.4**
RENT, RATES & TAXES	0.1	0.0	0.0	0.0
INSURANCE & BANK	2.3	1.4	0.8	0.1
TRAVELLING EXPS	18.6	11.2	6.4	0.9
COMMUNICATION COST	1.5	0.9	0.5	0.1
OTHER OVERHEADS + Modern	17.2	10.4	5.9	0.8
TOTAL ADMN & OTHER EXPS	**39.6**	**24.0**	**13.6**	**1.9**
TOTAL CONVERSION COST	**290.9**	**176.1**	**100.0**	**13.9**

ANALYSIS OF COST ELEMENTS OF WORSTED SPINNING PLANT

	Wool Top	Per Kg	Preparatory	Per Kg	RF/ Spinning	Per Kg	Post spinning	Per Kg	Packing	Per Kg	Total	Total Per Kg	% age
Power	320927	2.65	513980	3.26	1868729	11.44	2599145	19.15		0.00	**5302781**	32.5	17.8
Salary	81611	0.67	267248	1.69	250828	1.54	168006	1.24	56689	0.35	**824382**	5.1	2.8
Wages	563181	4.64	1172891	7.43	1884884	11.54	2472101	18.22	280095	1.72	**6373152**	39.1	21.4
Stores	1,94,371	1.60	12,03,420	7.63	11,21,634	6.87	2,48,853	1.83	15,05,448	9.24	**4273726**	26.2	14.3
SUB TOTAL (A)	1160091	9.57	3157539	20.01	5126074	31.38	5488104	40.44	1842232	11.31	16774040	102.9	
Overheads													
Power	5,25,716	4.34	6,06,595	3.85	16,17,588	9.90	12,13,191	8.94	80,879	0.50	**4043969**	24.8	13.5
Salary	5,78,689	4.77	6,67,718	4.23	17,80,581	10.90	13,35,436	9.84	89,029	0.55	**4451453**	27.3	14.9
Wages	309602	2.55	357234	2.26	952623	5.83	714467	5.26	47631	0.29	**2381557**	14.6	8.0
Store	285489	2.35	329410	2.09	878426	5.38	658820	4.85	43921	0.27	**2196065**	13.5	7.4
SUB TOTAL (B)	1699496	14.01	1960957	12.43	5229218	32.01	3921913	28.90	261461	1.60	13073045	80.2	43.8

TOTAL (A+B)	2859587	23.58	5118496	32.44	10355292	63.38	9410018	69.34	2103693	12.91	29847085	183.2	
Production Kg	121268		157760		163378		135702		162939				
OVERALL	Wool Top	Per Kg	Preparatory	Per Kg	RF/ Spinning	Per Kg	Post spinning	Per Kg	Packing	Per Kg	Total		
Power	846643	6.98	1120575	7.10	3486316	21.34	3812335	28.09	80879	0.50	9346750	57.4	31.32
Salary	6,60,300	5.44	934966	5.93	2031409	12.43	1503441	11.08	145718	0.89	5275834	32.4	17.68
Wages	8,72,783	7.20	1530125	9.70	2837506	17.37	3186568	23.48	327726	2.01	8754709	53.7	29.33
Store	4,79,860	3.96	1532830	9.72	2000060	12.24	907673	6.69	1549369	9.51	6469791	39.7	21.68
TOTAL (A)	2859587	23.58	5118496	32.44	10355292	63.38	9410018	69.34	2103693	12.91	29847085	183.2	

*Total cost is on Packed production, RF-Ring, Figures in Rs.

Overheads: Expenses other than direct manufacturing department (Wool top to Packing) in different heads were divided to the main sections of manufacturing on the basis of percentage decided by the formula (Ratio of contribution on these sections)

Total Over heads	Rs	Wool Top	Preparatory	Spinning	Post Spinning	Packing
		13%	15%	40%	30%	2%
Wages	2381557	309602.41	357233.55	952622.8	714467.1	47631.14
Salary	4451452.9	578688.877	667717.935	1780581	1335435.87	89029.058
Power	3823638	497072.94	573545.7	1529455	1147091.4	76472.76
Stores	2196065	285488.45	329409.75	878426	658819.5	43921.3
	12852712.9	1670852.677	1927906.935	5141085	3855813.87	257054.258

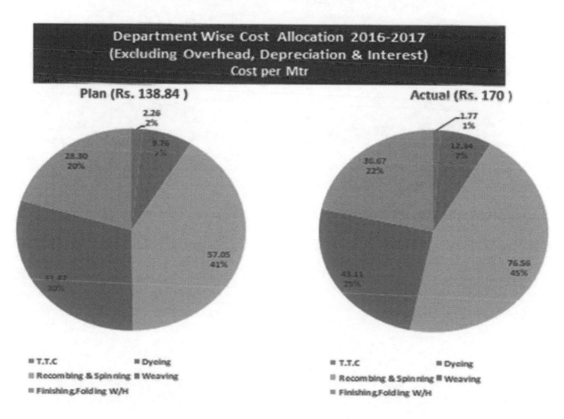

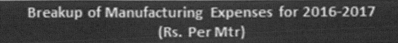

**Breakup of Manufacturing Expenses for 2016-2017
(Rs. Per Mtr)**

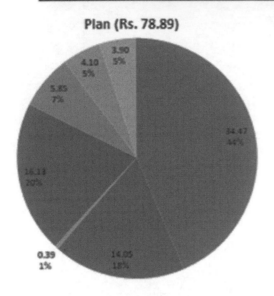

Plan (Rs. 78.89)

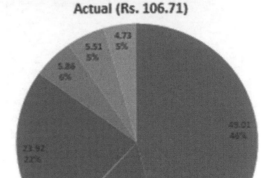

Actual (Rs. 106.71)

■ Electricity Charges ■ Coal Charges
■ Water Charges ■ Stores & Spares
■ Fabric Dyes & Chemicals ■ Packing
■ Mending Charges

**Department Wise Electricity Cost for 2016-2017
(Lac Rs.)**

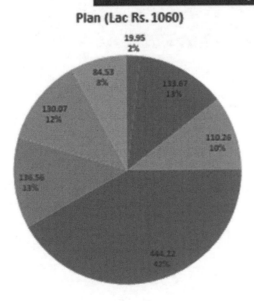

Plan (Lac Rs. 1060)

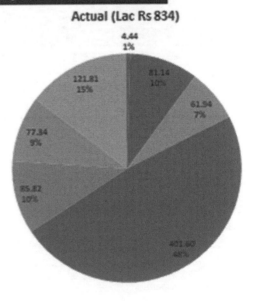

Actual (Lac Rs 834)

■ TTC ■ Dyeing ■ Recombing ■ Spinning
■ Weaving ■ Finishing ■ Other

Department Wise Coal Cost for 2016-2017
(Lac Rs.)

Plan (Lac Rs. 432)

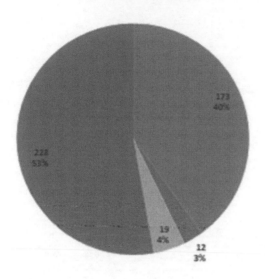

173
40%

228
53%

19
4%

12
3%

■ Dyeing ■ Recombing ■ Spinning ■ Finishing

Actual (Lac Rs. 294)

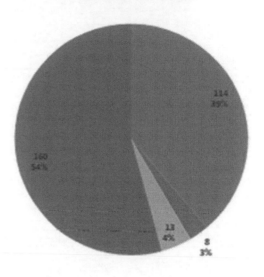

114
39%

160
54%

13
4%

8
3%

■ Dyeing ■ Recombing ■ Spinning ■ Finishing

Expenses and PBDT for 2016-2017

Plan -Lac Rs. 387

2.18
2%

7.20
6%

10.65
9%

20.02
17%

3.80
3%

14.15
12%

13.90
12%

■ Total Rm Cost   ■ Total Manufacturing Cost
■ Employment Cost   ■ Other Overheads
■ Gross Contribution   ■ Publicity & Selling Exp
■ Intrest Net   ■ PBDT

Actual -Lac (Rs. 1913)

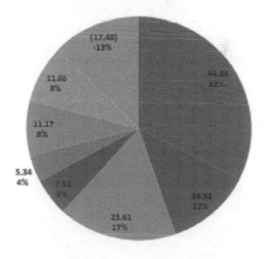

(17.48)
-13%

44.84
12%

11.06
8%

11.17
8%

5.34
4%

7.51
6%

23.61
17%

16.92
12%

■ Total Rm Cost   ■ Total Manufacturing Cost
■ Employment Cost   ■ Other Overheads
■ Gross Contribution   ■ Publicity & Selling Exp
■ Intrest Net   ■ PBDT

MONTHLY DYES AND CHEMICAL COST

Month	Total Prod/Kg	Polyester			Wool			Chemical Cost Rs	Poly Dyes Amt Rs	Wool Dyes Amt Rs	Chemical Cost Rs /Kg	Poly Dyeing Rs/Kg	Wool Dyeing Rs/Kg	Total Dyeing Rs/Kg	Average Shade %	cost of 1% depth
		Dyed kgs.	Dyes Kgs	% Shade	Dyed kgs.	Dyes Kgs	% Shade									
APRIL	116204	61381	1994	3.25	54823	1170	2.13	713556	595724.00	750920.00	6.14	9.71	13.70	17.73	2.72	
MAY	121369	61270	1617	2.64	60099	1033	1.72	864316	445189	628547	7.12	7.27	10.46	15.97	2.18	
JUNE	1,15,090	59,306	1814	3.06	55,784	1242	2.23	773491	566085	735875	6.72	9.55	13.19	18.03	2.66	
JULY	116119	60961	2113	3.47	55158	1301	2.36	689353	663109	909916	5.94	10.88	16.50	19.48	2.94	
AUGUST	109746	59361	1720	2.90	50385	1077	2.14	704837	499913	695227	6.42	8.42	13.80	17.31	2.55	
SEPTEMBER	103728	40651	1337	3.29	63077	1356	2.15	818497	382975	1061497	7.89	9.42	16.83	21.82	2.6	

ANNUAL COMPARATIVE DYEING COST

					TOP DYEING					
Year					Poly Production			Wool Production		
FY	Prod Kg	MT/ month	Total Dyes Chemical cost	Cost /kg	Quantity	Cost/ Kg	Total cost	Quantity	Cost/ Kg	Total cost
2010-11	1065170.85	88.76	32472090.90	30.49	338325.92	15.81	5350330.14	326397.46	29.37	9587011.66
2011-12	1041030.86	86.75	33788830.65	32.46	370554.78	18.84	6981488.98	398061.17	36.11	14372184.24
2012-13	991789.02	82.65	32720399.61	32.99	354054.2	21.11	7475701.05	358067.3	37.51	13429871.49
2013-14	1172561.78	97.71	36909123.88	31.48	404084.6	20.92	8455203.95	473818.9	33.48	15865110.11
2014-15	1147171.74	95.60	37914696.97	33.05	341768.54	22.47	7679205.17	511444.84	32.67	16711125.4
2015-16	1074427.13	89.54	30678290.54	28.55	321091.15	19.54	6274325.86	430201.64	26.14	11245667.96
2016-17	1122750.39	93.56	26026666.74	23.18	357866.15	16.15	5781303.63	359147.49	24.66	8856457.58
2017-18	764760.07	69.52	19038091.89	24.89	259298.65	17.84	4625646.14	293719.1	26.93	7911002.52

	YARN DYEING			FABRIC DYEING									
Year	Yarn Production			Fresh			Redyeing			Over dyeing			
FY	Kg	Cost/ Kg	Total cost	Quantity	Cost/ Kg	Total cost	Kg	cost/ Kg	Total cost	Kg	Cost/ Kg	Total cost	
2010-11	73054.57	34.48	2518573.75	302073.40	43.04	13001660.26	2645.40	18.23	48235.12	25319.50	77.66	1966279.93	
2011-12	38157.61	29.02	1107349.1	218562.7	46.3	10119767.62	2617.4	18.84	49306.52	15694.6	37.83	1158734.24	
2012-13	59189.42	29.94	1772066.2	214063.6	44.21	9463807.14	2631.5	21.59	56810.27	6414.5	81.4	522143.45	
2013-14	58316.88	23.92	1394704.9	231032	46.85	10823899.49	1585.7	21.82	34600.31	5309.4	63.21	335605.15	
2014-15	65996.6	27.80	1834768.5	220257.16	50.38	11097164.45	5940.8	26.05	154762.03	7704.6	56.81	437671.43	
2015-16	81789.07	28.62	2340511.4	221246.87	42.28	9354788.62	16542.2	25.8	426841.09	20098.4	51.55	1036155.57	
2016-17	112510.3	28.37	3191829.1	265212.7	27.72	7350794.39	6861.2	11.66	79975.71	28013.8	27.35	766306.39	
2017-18	63472.92	27.16	1723737.2	133404.1	32.72	4364587.86	6159.4	16.31	100465.17	14865.3	21.03	312653.01	

FABRIC COSTING FOR EXPORT

Example:

Quality No /Cost in Rs	Sample 1 (All Wool) (2/64X 2/64) Nm	Sample 2 (P/W - 55/45) (2/60X 1/40)Nm	
Overall Finish Weight Grams	235	275	
Overall Grey Weight Grams	235	275	
Grey Weight Grams	235	145	130
Picks Per Inch	58	72	
Weave	Plain	Plain	
Yarn Count Nm	Warp & weft 2/64	Warp 2/60	Weft 1/40
Blend	100% Wool - 19.5Mic	55%P,45%W - NP / 22.5Mic	55% P,45%W - NP / 22.5Mic
Cost of Polyester Rs/Kg		138	138
Cost of Wool Top Rs/Kg	1056	947.8	947.8
Cost of Polyester In Blend	0.00	75.90	75.90
Cost of Wool In Blend	1056.00	426.51	426.51
Blended Cost Rs/Kg	1056.00	502.41	502.41
Dyeing Cost /Kg	78	65.9	65.9
Wastage in spinning dept @ (1056+78)/0.85*0.15	200.12		
Wastage in spinning dept @8.00%		49.42	49.42
Spinning Cost per count Rs	4.23	3.23	2.67
Spinning Cost Rs/Kg	271	194	106.8
Recovery of Spinning waste	-3	-3	-3
Yarn Cost Rs/Kg	1601.8	808.5	721.5
Yarn Cost Rs/Mtr	376.43	117.24	93.80
Remnant waste @ 2.00 Rs/Mtr	7.53	4.22	
Cost of Selvedge Yarn	2.25	2.25	
Wastage in weaving @ (376.43+7.53+2.25)/95.5*1.5	18.20	10.25	
Weaving Charges @ 45Paisa/Pick		32.40	
Weaving Charges @ 60 Paisa/ pick	34.80		
Processing Charges/Mtr	30	30	
Shrinkage Cost (376.43+7.53+2.25+18.20+34.8+30)/95*5	24.70	15.27	
Packing cost /Mtr	4.5	4.5	
Direct Costs / (sum of yarn cost to packing cost)	**498.40**	**309.93**	
Working Capital Interest @ (498.4*14%/365*60)	11.47	7.13	
Total Cost per Meter	509.87	317.06	
Selling Price USD / Mtr	**10.04**	**6.41**	
Selling Price INR / Mtr @ USD 62.00	622.30	397.73	
Less Value Loss (622.30*2%)	12.45	7.95	
Less Freight	5.00	5.00	
Less Agent Commission @5% + S.T	35.47	22.67	
Less Bank Charges	3.00	3.00	
Less Royalty @3.5% + S.T	24.89	15.91	

Marketing Over Heads (622.30*2%)	12.45	7.95
Net Selling Price / Mtr	529.05	335.24
Net Contribution / (529.05-509.87)	19.17	18.18
DEPB/Duty Draw Back / Mtr	5.83	6.82
Contribution /Mtr	25.00	25.00

FABRIC COSTING FOR EXPORT (Summary)

(Cost/Mtr Rs) Blend Description	Quality No 2/40 Nm Poly - Wool 55:45 (24.5 Mic)	Quality No 2/54 Nm Poly wool 55:45 (22.5 Mic)	Quality No 2/60 Nm Poly Wool 55:45 (22.5 Mic)
Weight of the fabric Gram/Mtr	405	395	260
Picks per inch	56	60	54
Raw material cost (INR)	177.89	201.98	131.69
Spinning Cost (INR)	13.53	20.70	15.52
Weaving Cost (INR)	10.01	10.84	9.65
Finishing Cost (INR)	17.96	17.96	17.96
Dyeing Cost (INR)	10.67	10.41	6.85
Overheads (INR)	21.79	23.59	21.02
Depreciation (INR)	16.62	17.99	16.03
Packaging Cost (INR)	2.50	2.50	2.50
Interest on WC (INR)	9.15	10.33	7.47
Flagging and Defective Cost (INR)	19.82	20.73	14.32
Sales Expense (INR)	30.54	34.48	24.93
Profit (INR)	47.65	53.80	38.90
Export Benefit (INR)	-18.27	-20.62	-14.91
Net Ex-Mill Price per meter (INR)	347.04	391.87	283.31
Conversion rate INR to USD	45.90	45.90	45.90
Net Ex-Mill Price per meter (USD)	7.56	8.54	6.17
Agent Commission (USD)	0.50	0.50	0.50
Total Ex-Mill price per metre	8.06	9.03	6.67
Net Ex-Mill Price per yard (USD)	7.37	8.26	6.10

FABRIC COSTING OF LUXURY SUITINGS FOR DOMESTIC MARKET

	BRAND NAME		
	Quality No: --------	Warp	Weft
	Luxury Suitings for Higher Segment	Ends/CM	Picks/CM
		34.64	31.12
A	Reed Space CM	167.64	
B	Finish Weight of Fabric Grams/Mtr	239	
C	Grey Weight of fabric Grams/Mtr	244	
D	Wevg Length Contraction %	5	
E	Grey Weight Grams	135.47	108.53
F	Picks Per Inch	79.04	
G	Weave	2X1 FANCY	
H	Yarn Count Nm	'2/90	1/50

I	Count Nm	90	50
J	Blend	P40%,W60% - LP / 17.5 M	P40%,W60% - LP / 17.5 M
K	Polyester %	0.4	0.4
L	Wool %	0.6	0.6
		Rs/kg	Rs/kg
M	Cost of Polyester Per Kg	148.00	148.00
N	Cost of Wool Per Kg	1030.06	1030.06
O	Per Kg Cost of Polyester In Blend (O*40/100)	59.20	59.20
P	Per Kg Cost of Wool In Blend (N*40/100)	618.03	618.03
Q	Per Kg Cost of Blend (O+P) Rs	677.23	677.23
R	Dyeing Charges/kg Rs	71.2	71.2
S	Cost of Dyed Blend/Kg (Q+R) Rs	748.43	748.43
T	Wastage in spinning dept @15.00%		
U	Wastage in spinning dept @8.00%(S*8/100) Rs	59.87	59.87
V	Spinning Cost Per count Rs/Kg /Count	4.35	2.66
W	Spinning Cost (V*I)	391.5	133
X	Recovery of Spinning waste	-3	-3
Y	**Yarn Cost per Kg (Sum Q toX)**	**1945.24**	**1686.74**
		Rs /Mtr	Rs /Mtr
Z	Yarn Cost/Mtr (Y*E/1000)	263.52	183.06
AA	Remnant Yarn Waste @2.00% (on Warp+weft Weight)	9.11	
AB	Selvedge Yarn	2.25	
AC	Wastage in weaving @4.50%(on Warp+weft Weight)	21.58	
AD	Weaving Charges Per Pick 45 Paisa (F*45/100)	35.57	
AE	Processing Charges /Mtr	30	
AF	Piece dyeing Charges		
AG	Shrinkage Cost (@5.50%)	31.73	
AH	Packing cost/Mtr	66.8	
AI	Production Per Loom Per Shift(Mtrs)	50.90	
AJ	Total Production in 96 Looms Mtrs	4886.68	
AK	Direct Costs Per Mtr (Sum Z to AH)	643.62	
AL	Fixed Admin/Sales Overhead Per Mtr of Prodn	14.28	
AM	Working Capital Interest @14.00%(AK*14%/365*180)	44.44	
AN	**Total Cost per Meter (Sum AK to AM)**	**702.34**	
AO	**Selling Price INR per Meter**	**1865**	
AP	Less Value Loss @4.00%(AO*4/100)	74.60	
AQ	Less Stock lot Loss @ 3.5%(AO*3.5/100)	65.28	
AR	Less Bonus & Incentive @ 16.25% (AO*16.25/100)	303.06	
AS	Less Agent Commission @3.5% + ST(AO*4/100)	74.60	
AT	Less Royalty @3.5% + S.T (AO*4/100)	74.60	
AU	Marketing Over Heads @5% (AO*5/100)	93.25	
AV	Discounts / Claims @ 1 % (AO*1/100)	18.65	
AW	Extended Credit Period @ 1.5% (AO*1.5/100)	27.98	

AX	Publicity Charges 6.25% (AO*6.25/100)	116.56	
AZ	**Net Selling Price per Meter(AO- (Sum of AP to AX))**	**1016.43**	
AAB	Contribution (AZ -AN)	314.09	
AAC	Contribution /Loom/Shift(AAB*AI)	15987.87	
AAD	Contribution (EBDAI) (AAB)	314.09	
AAE	EBDAI % (AAD/AO)	17%	
AAF	EBDAI Per Loom Per Shift (AAD* AI)	15987.87	
AAG	Interest Per Mtr of Production 100%	48.04	
AAH	Depreciation per Mtr of Production 100%	48.04	
AAI	Contribution per Mtr after Depn. and Interest	218.01	

COTTON FABRIC COSTING

Weave	Twill 2/1	Plain	Plain	Twill 2/1	Twill 2/1	Twill 3/1	Twill 2/1	Twill 3/1	Twill 2/1
Quality Number	**1**	**2**	**3**	**4**	**5**	**6**	**7**	**8**	**9**
CONSTRUCTION WARP Count Ne	**2/60c**	**2/40c**	**2/40**	**2/40c**	**2/40c**	**16c**	**20c**	**30c**	**30c**
WEFT Count Ne	**2/60c**	**2/40c**	**20c**	**20c**	**2/40c**	**12oe**	**16oe**	**10oe**	**10oe**
YARN RATES /Kg WARP	340	285	270	285	285	235	240	220	220
WEFT	340	285	240	240	285	205	210	190	190
FOR CALCULATION WARP COUNT	**30**	**20**	**20**	**20**	**20**	**16**	**20**	**30**	**30**
FOR CALCULATIONWEFT COUNT	**30**	**20**	**20**	**20**	**20**	**12**	**16**	**10**	**10**
ENDS PER INCH	144	108	108	114	128	104	108	132	132
PICKS PER INCH	86	60	60	64	68	56	52	68	64
SORT WIDTH - INCHES **GREY**	63	63	63	63	63	63	63	63	63
REED SPACE INCHES	64.5	64.5	64.5	64.5	64.5	64.5	64.5	64.5	64.5
WEIGHT (KGS/MTR) WARP	**0.201**	**0.226**	**0.226**	**0.239**	**0.268**	**0.272**	**0.226**	**0.184**	**0.184**
WEFT	**0.109**	**0.114**	**0.114**	**0.122**	**0.130**	**0.178**	**0.124**	**0.259**	**0.244**
TOTAL WEIGHT Kg	0.310	0.341	0.341	0.361	0.398	0.450	0.350	0.443	0.428
MTRS/KG	3.223	2.937	2.937	2.772	2.515	2.222	2.857	2.255	2.336
ESTIMATED COST									
YARN COST / Mtr - WARP	68.38	64.48	61.09	68.07	76.43	64.00	54.30	40.56	40.56
-WEFT	37.13	32.57	27.43	29.25	36.91	36.44	26.00	49.21	46.32
YARN COST -Total (RS/Mtr)	**105.51**	**97.05**	**88.52**	**97.32**	**113.34**	**100.44**	**80.30**	**89.77**	**86.88**
HARD WASTE S-2% , A-3.5% (Rs / Mtr)	3.69	3.40	3.10	3.41	3.97	3.52	2.81	3.14	3.04
WVG.COST (22 Paisa /Pick & 24 Paisa for Dobby)	18.92	13.20	13.20	14.08	14.96	12.32	11.44	14.96	14.08
TOTAL GREY COST (RS/MTR)	**128.12**	**113.65**	**104.81**	**114.81**	**132.26**	**116.28**	**94.55**	**107.87**	**104.00**
DYEING COST @ Rs 90/-/Kg	27.93	30.65	30.65	32.46	35.79	40.51	31.50	39.90	38.53
PIGMENT BLOTCH / Mtr									
SHRINKAGE 2.5%	3.90	3.61	3.39	3.68	4.20	3.92	3.15	3.69	3.56
VALUE LOSS 2.5%	3.90	3.61	3.39	3.68	4.20	3.92	3.15	3.69	3.56
TRANSPORT	1.00	1.00	1.00	1.00	1.00	1.00	1.00	1.00	1.00
TOTAL COST OF PROCESSED FABRIC/ Mtr - Rs	**164.85**	**152.51**	**143.23**	**155.63**	**177.45**	**165.63**	**133.36**	**156.17**	**150.66**

S- Sulzer, A-Air Jet

COSTING OF BLANKETS

BRAND NAME/ SIZE OF BLANKETS IN INCHES	MOTO	MOTO	MOTO	SEAGUL	SEAGUL	ELDO	ELDO	369	MANYOY	MANYOY	KKAY	KKAY
	50 X 75	54 X 80	60 X 90	54 X 80	60 X 90	54 X 80	60 X 90	54 X 80	54 X 80	60 X 90	54 X 80	60 X 90
FABRIC LENGTH/PC MTRS	2	2.13	2.38	2.13	2.38	2.13	2.38	2.13	2.13	2.38	2.13	2.38
WARP 300 D Poly RW	30	30	30	30	30	30	30	30	30	30	30	30
WEFT-SODDY Nm	1.5	1.5	1.5	2	2	2	2	1.5	1.5	1.5	1.5	1.5
MILL GREY (Natural)%	73	73	73	0	0	0	0	90	40	40	0	0
COLOURED SOLID %	27	27	27	100	100	100	100	10	60	60	100	100
EPI	16	16	16	28	28	28	28	16	16	16	28	28
PPI	14	14	14	22	22	22	22	14	15	15	18	18
REED SPACE INCHES	56	61	67	61	67	61	67	61	61	67	61	67
TOTAL ENDS	930	990	1100	1680	1900	1680	1900	990	990	1900	1680	1900
WEIGHT OF WARP/ MTR- Grams	32.55	34.65	38.5	58.8	66.5	58.8	66.5	34.65	34.65	66.5	58.8	66.5
WEIGHT OF WEFT/ MTR- Grams	554.03	603.49	662.85	711.26	847.55	711.26	847.55	603.49	646.60	710.20	775.92	852.24
WEIGHT/MTR -Grams	586.58	638.14	701.35	770.06	914.05	770.06	914.05	638.14	681.25	776.70	834.72	918.74
WARP WEIGHT/PIECE- Grams	65.10	73.80	91.63	125.24	158.27	125.24	158.27	73.80	73.80	158.27	125.24	158.27
WEFT WEIGHT/PIECT- Grams	1173.15	1359.25	1669.22	1640.23	2175.44	1640.23	2175.44	1359.25	1451.06	1848.55	1777.95	2186.60
WEIGHT/PC OF BLANKET- Kg	**1.24**	**1.43**	**1.76**	**1.77**	**2.33**	**1.77**	**2.33**	**1.43**	**1.52**	**2.01**	**1.90**	**2.34**
STANDARD WEIGHT Kg/PIECE	1.1	1.25				1.7	2.2	1.7	1.7	2.2		
COST OF WARP KSH	11.07	12.55	15.58	21.29	26.91	21.29	26.91	12.55	12.55	26.91	21.29	26.91
COST OF WEFT KSH	81.69	94.65	116.23	175.50	232.77	175.50	232.77	91.37	95.48	121.63	97.79	120.26
YARN COST KSH	**92.76**	**107.20**	**131.81**	**196.80**	**259.68**	**196.80**	**259.68**	**103.92**	**108.03**	**148.54**	**119.08**	**147.17**
WEAVING WASTE 5% (TAKEN2%) KSH	1.86	2.14	2.64	3.94	5.19	3.94	5.19	2.08	2.16	2.97	2.38	2.94
COST OF YARN KSH	94.61	109.34	134.45	200.73	264.87	200.73	264.87	105.99	110.19	151.51	121.46	150.11
RASING WASTE Gms (4%)	46.93	54.37	66.77	65.61	87.02	65.61	87.02	54.37	58.04	73.94	71.12	87.46
COST OF RAISING WASTE	3.28	3.81	4.67	7.02	9.31	7.02	9.31	3.53	4.56	5.80	3.91	4.81
TOTAL COST OF YARN KSH	**97.90**	**113.15**	**139.12**	**207.75**	**274.18**	**207.75**	**274.18**	**109.53**	**114.74**	**157.32**	**125.37**	**154.92**
FINAL WT OF BLANKET Kg	1.14	1.32	1.62	1.62	2.22	1.62	2.22	1.32	1.40	1.85	1.75	2.23
COST OF LABLE KSH	0.45	0.45	0.45	0.45	0.45	0.45	0.45	0.45	0.45	0.45	0.45	0.45
POLY BAG SIZE INNER	16X29	17X34	20X30	20X30	20X30	17X34	20X30	17X34	17X34	20X30	20X30	20X30
POLY BAG SIZE OUTER	26X38X65	26X40X60		26X36X70	26X36X70	26X36X70	26X36X70	26X40X60	26X40X60	26X36X70	26X36X70	26X36X70
COST OF POLY BAG INNER/PC KSH	3.38	3.94	4.4	4.4	4.4	3.95	4.4	3.95	3.95	4.4	4.4	4.4
BLANKET IN OUTER BAG	30	25	15	20	20	20	20	25	25	20	20	20
PRICE OF POLY BAG OUTER KSH	60	60	65	62	62	62	62	60	60	62	62	62
COST OF P/BAG OUTER/PC	2.00	2.4	4.33	3.1	3.1	3.1	3.1	2.4	2.4	3.1	3.1	3.1
THREAD WHPG +OVER LOCK KSH	2.8	3	3.5	3	3.5	3	3.5	3	3	3.5	3	3.5
COST OF STITCHING/ PACKING KSH	**8.63**	**9.79**	**12.68**	**10.95**	**11.45**	**10.50**	**11.45**	**9.80**	**9.80**	**11.45**	**10.95**	**11.45**
YARN RATES /Kg KSH												
POLY 150/48/2 D	170	170	170	170	170	170	170	170	170	170	170	170

FANCY MILL GREY 1.5 Nm	65.8	65.8	65.8	65.8	65.8	65.8	65.8	65.8	65.8	65.8	65.8	65.8
COLOURED 1.5Nm(54 & 88.2) AVE-	80	80	80	80	80	80	80	80	80	80	80	80
COLOURED 2 Nm	107	107	107	107	107	107	107	107	107	107	107	107
MULTI FANCY 1.5	55	55	55	55	55	55	55	55	55	55	55	55
WHIPPING YARN/KG @310												
OVER LOCK YARN/10000 @110 (USD to KSH)												
TOTAL COST OF BLANKET/PC KSH	104.53	120.54	147.47	215.60	282.53	215.15	282.53	116.93	122.14	165.67	133.22	163.27
COST OF A BALE KSH	3195.85	3073.41	2277.08	4374.04	5712.65	4365.04	5712.65	2983.19	3113.59	3375.31	2726.44	3327.46
COST OF BLANKET/PC IN BALE KSH	106.53	122.94	151.81	218.70	285.63	218.25	285.63	119.33	124.54	168.77	136.32	166.37
COST OF BLANKET/PC WITH OH KSH	163.53	179.94	208.81	275.70	342.63	275.25	342.63	176.33	181.54	225.77	193.32	223.37
SALES PRICE KSH	130.00	150.00			340.00	250.00	310.00	130.00	140.00	190.00	240.00	310.00
CONTRIBUTION KSH	23.47	27.06	-151.81	-218.70	54.37	31.75	24.37	10.67	15.46	21.23	103.68	143.63
OH KSH	57	57	57	57	57	57	57	57	57	57	57	57
P/L KSH	-33.53	-29.94	-208.81	-275.70	-2.63	-25.25	-32.63	-46.33	-41.54	-35.77	46.68	86.63
SALES PRICE B GRADE. KSH	115	125						125	125			

*PC- Piece, KSH- Kenyan Shiling, SODDY YARN RATES FOR WEFT:

MOTO MOTO MILL GREY/HEATHER	COLOUR	1.5 NM	SELF MADE	40+23=63/KG
MOTO MOTO MILL GREY/HEATHER	SOLID	1.5 NM	SELF MADE	57+23=80/KG
KAY KAY MULTI FANCY		1.5 NM	IMPTD	55/KG
ELDO SOLID COLOUR		2.0 NM	IMPTD	107/KG

COSTING OF GARMENT

Rates in KSH (Kenya Shilling 01 USD= 110 KSH)	JACKET			TROUSER			SAFARI TOP			SKIRT			SHORT		
	QTY	RATES	COST/PC	QTY	RATES	COST/PC	QTY	RATES	COST/PC	QTY	RATES	COST/PC	QTY	RATES	COST/PC
FABRIC MTR	2.2	120	264	1.25	120	150	1.25	120	150	1	120	120	0.65	120	78
TEFLON/MTR										1.5			1.5		
LINNING MTR	2	50	100		50		0	50	0		50			50	
CHEST PC MTR	0.25	100	25		150		0	150	0		150			150	
SHOULDER PC NOS	3	45	135		45			45			45			45	
COLLAR LINING MTR	0.15	250	37.5		250		0.15	250	37.5		250			250	
SHOULDER ROLLS NOS	1	40	40		10			10			10			10	
BODY FUSING MTR	0.75	220	165		150			150			150			150	
PAPER FUSING/CANVAS	0.04	100	4	0.05	9.5	0.475	0.06	9.5	0.57	0.05	9.5	0.475	0.05		0
BELTING		3.2		0.85	3.2	2.72		3.2			3.2		0.75	3.2	2.4
ZIPPER	1	2	2	1	2	2		2		1	2	2	1	2	2
THREAD MTR	200	0.011	2.2	300	0.011	3.3	180	0.011	1.98	180	0.011	1.98	200	0.011	2.2
BUTTONS BIG	3	1.15	3.45	2	1.15	2.3	5	1.15	5.75	0	1.15	0		1.15	

BUTTONS SMALL	6	0.57	3.42	0	0.57	0	0	0.57		1	0.57	0.57	2	0.57	1.14
HOOKS				1	2	2	0			1	2	2	1	2	2
PACKETING MTR		55		0.24	55	13.2	0						0.24	55	13.2
POLY BAG NOS	1	30	30	1	1	1	1	1	1	1	1	1	1	1	1
LABLES	1	2	2	1	2	2	1	2	2		2	1	1	2	2
CELLO TAP		1			1			1			1				1
HANGER NOS	1	80	80	0	80	0	0	80	0		80			80	
CARTON@150/15PC	1	10	10			1			1			1			1
EMBROIDERY COST		200	0												
TOTAL COST/PIECE KSH			904.57			180.995			200.8			131.03			105.94
SELLING PRICE/PIECE KSH			1250			320			800			250			185
CONTRIBUTION/ PIECE KSH			345.43			154.005			347.6			118.98			79.06

COST COMPARISON OF DIFFERENT OPTIONS OF WEAVING MACHINES

There is continuous advancement in technology related to machineries. Before finalising a machine, it is not only the basic cost of the machines which are taken into account but a detailed techno-commercial comparison is also conducted based on the quality to be produced, rate of production, waste generation, versality, materiality, maintenance cost, after-sales service, technical support, expected usable life of machine and per unit cost of production. And above all, the final appearance and finish of fabric after dyeing and finishing is considered. However, at times, in addition to these parameters, decisions are taken based on global feedback, record of accomplishment, technician's preference, and its brand value.

EXAMPLE:

In the following table, a comparison is done between 40 nos. of Air-jet weaving machine and two different Flexible Rapier Machines of same brand manufactured in China and Europe. Product is Fancy Suitings with name in selvedge of fabric. The wages and salary is on assumption basis. Warping, drawing in, pin dropping, folding, mending etc. are considered on a fixed piece rate. The working is only upto weaving stage. Weaving charge is considered 24 paisa per pick as per market rate i.e. Rs 16.80 Per Mtr (24 Paisa per pick *70 Picks per inch /100=16.80). Billing means weaving charges earned for fabric produced meters and earning is the difference of billing and expenses.

A–COST OF PROJECT

FEATURES						
Model of Looms						
Type of looms	Airjet		Flexible Rapier		Flexible Rapier	
Working width CM		190		190		190
Speed PPM (Picks/minute)		750		480		575
Electronics Dobby	Yes		Yes		Yes	
Electronics Let off	Yes		Yes		Yes	
Electronics Take up	Yes		Yes		Yes	
Electronics Name Writer	Staubli		China		Staubli	
Selvedge Tucked in	YES		YES		YES	
Heald frames levers Installed		14		16		16
Levers Effective		11		16		16

PROJECT COST (In Rs. lacs)	EUROPE	CHINA	EUROPE
	Air Jet	Flexible Rapier	Flexible Rapier
No of Looms	40	40	40
Looms	1683	942	1141
Other Accessories	620	175	175
Name Writer (Selvedge Jacquard)	239	90	182
Compressor	54		0
Humidity Plant & Electricals	96	84	84
Knotting Machine	12	12	12
Warping Machine	50	50	50
TOTAL PROJECT COST	2754	1353	1644

B–PRODUCTION, EXPENSES, EARNING AND PER UNIT COST OF WEAVING

Types of Looms	Airjet	%	Rapier	%	Rapier	%
	New		China		European	
No of Looms	40		40		40	
PPM (RPM)	750		480		575	
Average Picks /Inch	70		70		70	
Efficiency (%)	85		85		85	
Production/Loom/Day Mtr	333		213		255	
Production Per Day Mtr	13324		8527		10215	
Production /Month Mtr	399724		255824		306455	
Billing/ Month Rs	6715367		4297835		5148448	
Billing Per Pick Rate Rs	**0.24**		**0.24**		**0.24**	
Selvedge Waste (%)	2		2		2	
Expenses Per Pick(without Intr & Dep)	0.10	42.45	0.12	51.42	0.11	44.83
Expenses Per Pick(with Int & Dep)	**0.23**	**95.43**	**0.24**	**98.63**	**0.22**	**92.75**
Earning Per Pick without Interest & Dep)	0.14	57.55	0.12	47.23	0.13	53.64
Earning Per Pick with Int & Dep	**0.01**	**4.57**	**0.00**	**1.37**	**0.02**	**7.25**
PROFIT & LOSS ACCOUNT		Rs. Per		Rs. Per		Rs. Per
		Mtr		Mtr		Mtr
Production Billing (Earning)	6715367	16.8	4297835	16.8	5148448	16.8
Expenses						
Wages	204290	0.51	188721	0.74	188721	0.62
Warping Charges	149857	0.37	94655	0.37	113388	0.37
Common Charges	181200	0.45	181200	0.71	181200	0.59
Contract Labourers	140908	0.35	93061	0.36	109896	0.36
Mending	399724	1	255824	1.00	306455	1.00
Folding	15989	0.04	10233	0.04	12258	0.04
Power	1139281	2.85	782204	3.06	782204	2.55
Store	162000	0.41	162000	0.63	162000	0.53
Salary	305924	0.77	305924	1.20	305924	1.00
Insurance	32814	0.08	18853	0.07	28953	0.09
Year End Expenses	87145	0.22	85648	0.33	85648	0.28
Yarn waste						
Water	31500	0.08	31500	0.12	31500	0.10
TOTAL	2850632	7.13	**2209822**	**8.64**	**2308148**	**7.53**
Expenses per pick without Int & Dep	0.10		0.12		0.11	

Capital expense (Cost of Project) Rs lacs	2372		1363		1645	
Interest on Capital @ 12% Rs	2372081	5.93	1352850	5.29	1644705	5.37
Depreciation @ 6% Rs	1186040	2.97	676425	2.64	822352	2.68
Total of Int. & Dep Rs	3558121	8.90	2029275	7.93	2467057	8.05
Grand Total Rs	**6408753**	**16.03**	**4239097**	**16.57**	**4775205**	**15.58**
Cash Generation Rs	**306614**	**0.77**	**58738**	**0.23**	**373243**	**1.22**
Percentage of cash generation %		4.57		1.37		7.25

*Assumptions:

Common Charges includes design development, sampling, engineering and utility expenses, stores, purchase, personal, security gardening, cleaning, peon, workshop, Admin Overheads)

Contract Labourers Drawing in and Pin Droppings @ Rs 38/1000 Ends

Mending Charges @ Rs 1 Per Mtr for Fancy Fabric

Folding + Grey Checking Rs 0.04

Spare and Stores - Rs 45 Per Loom Per Shift

Insurance – Rs. 166 Per Lac /month - (2000 Per Lac/Year)

Year End - Bonus, Gratuity, LTA, EL of Staff and Workers

Water Consumption 30KL @Rs.35/per KL for Humiliation Plant

C–DETAILS OF WAGES

	Air Jet						Rapier China						Rapier Europe					
Particulars	Ist	2nd	3rd	Total	Rate	Amount	Ist	2nd	3rd	Total	Rate	Amount	Ist	2nd	3rd	Total	Rate	Amount
	Shift	Shift	Shift	per day	(26 days)	Rs.	Shift	Shift	Shift	per day	(26 days)	Rs.	Shift	Shift	Shift	per day	(26 days)	Rs.
Operator	4	4	4	12	3573	42876	4	4	4	12	3573	42876	4	4	4	12	3573	42876
Operator Helper	2	2	2	6	3573	21438	2	2	2	6	3573	21438	2	2	2	6	3573	21438
Fitter	2	1	1	4	3633	14532	2	1	1	4	3633	14532	2	1	1	4	3633	14532
Beam Gaiter	2	2	2	6	3464	20784	2	2	2	6	3464	20784	2	2	2	6	3464	20784
Oil Man	1	0	0	1	2929	2929	1	0	0	1	2929	2929	1	0	0	1	2929	2929
Roll Doffe Coolie	2	2	2	6	2929	17574	2	2	2	6	2929	17574	2	2	2	6	2929	17574
Winder	1	1	1	3	3009	9027	1	1	1	3	3009	9027	1	1	1	3	3009	9027
Weft Carrier	1	1	1	3	3009	9027	1	1	1	3	3009	9027	1	1	1	3	3009	9027
Jobber	0	0	0	0	0	0	0	0	0	0	0	0	0	0	0	0	0	0
Compressor Operator/ Humidity boy	1	1	1	3	3800	11400				0		0				0		0
TOTAL	16	14	14	44		149587	15	13	13	41		138187	15	13	13	41		138187
PF				13.61%		20359					13.61%		18807					13.61%	18807
ESIC				4.75%		7105					4.75%		6564					4.75%	6564
Off days wages (4 days)						27239							25163						25163
Wages / month						204290							188721						188721
Avg. wages / worker						155							143						143

D–DETAILS OF SALARY

PARTICULARS	Toyota NOS	RATE	AMOUNT	China NOS	RATE	AMOUNT	Europe NOS	RATE	AMOUNT
Weaving Manager	1	50000	50000	1	50000	50000	1	50000	50000
Shift In-charge	3	30000	90000	3	30000	90000	3	30000	90000
Shift Supervisor	0	0	0	0	0	0	0	0	0
Maintenance In-charge	1	20000	20000	1	20000	20000	1	20000	20000
Maintenance Supervisor	3	15000	45000	3	15000	45000	3	15000	45000
Grey Checker	2	3200	6400	2	3200	6400	2	3200	6400
Data Clerk	1	3200	3200	1	3200	3200	1	3200	3200
Quality Round Checker	3	4000	12000	3	4000	12000	3	4000	12000
Total	14		226600	14		226600	14		226600
HRA	20%		45320	20%		45320	20%		45320
P.F. %	13.61%		30840	13.61%		30840	13.61%		30840
E.S.I.C.	4.75%		3163.5	4.75%		3163.5	4.75%		3163.5
G.Total			305924			305924			305924

E–CALCULATIONS FOR POWER COST

Machine Features	Air Jet	Rapier China	Rapier Europe
Main Motor	3.5 KW	7.5 KW	7.5 KW
Let off Motor	0.5 KW	0.75 KW	0.75 KW
Take Up Motor	0.5 KW	0.75 KW	0.75 KW
Jacquard - Name Writer	0.5 KW	0.5 KW	0.5 KW
		0.5 KW	0.5 KW
TOTAL KW	**5**	**10**	**10**
Description			
No of Looms	**40**	**40**	**40**
Production / Day mtr.	13324	8527	10215
Production / Month,mtr.	399720	255810	306450
Power Units / Day (Looms) @65% Load	2652	5304	5304
Lighting Load unit/day(10kw)	216	216	216
Humidification plant unit/day (75kw as per Botlibai)	1188	1188	1188
Compressor Unit / Day	5832	0	0
Total Power - Unit / Day	**9888**	**6708**	**6708**
Total Power - Unit / Month	296640	201240	201240
Power Cost per unit - Rs.	5.30	5.35	5.35
Power per month cost Rs.	1572257	1075934	1075934
Power Unit / Mtr	0.74	0.79	0.66
Power Cost / Mtr	3.93	4.21	3.51
Compressor Details			
Compressor (Kaeser make) - CFM	1707	0	0
Compressor required for 40 looms@35 CFM /M/c.	1400	0	0

Unit Consumption / Month	296640	201240	201240
PER UNIT CALCULATION *			
Fix cost 90% of KVA @ Rs 220 / KVA	89100	69300	69300
Energy Charges @ 5.5/Unit	1631520	1106820	1106820
Meter Rent	1440	1440	1440
Duty @ 5% on Energy charges	81576	55341	55341
Cess @ 10 Paise per unit	29664	20124	20124
Power factor incentive 4% of energy charges	65261	44273	44273
Load factor incentive 12% of energy charges	195782	132818	132818
TOTAL COST	1572257	1075934	1075934
Per Unit Cost	5.30	5.35	5.35

CAPITAL COST AND COST OF COTTON YARN

MANUFACTURING COST COMPARISON AT DIFFERENT NOS. OF RING FRAME SPINDLES AND SPEED

The following formats, option 1 and option 2 indicate the tentative yarn cost per count and capital expenses. Amount of working capital, machines, and equipment are approximate (lump sum) and conversion cost is determined at different speeds keeping ring frames no. of spindles constant and yarn count almost same. Finance cost on machines and equipment is taken at 13% as well as at 8% per annum along with 5% interest subsidy. Capital cost is also taken with and without 10% capital subsidy under the special scheme. **The objective of this exercise is to calculate profitability and pay-back period deciding on installing the Spinning factory.** Raw material price and yarn sale rates influenced by market. The challenge before the business head is to minimise manufacturing and overhead expenses. Raw material yield, employment cost, utility, and overheads expenses are elements of yarn cost. Hence, the ring frame capacity, speed and utilisation of other resources plays a key role in optimising the total cost. Break even analysis must be done before deciding the size of the plant to ensure its viability.

OPTION 1

Projections												
No. of Ring Frames	10											
No of RF Spindles	12000											
Ring Frame Speed -RPM	16000			18000			20000			22000		
Avg. Count Ne	26.86			26.72			26.87			26.93		
Prod /Day Kg	5800			6550			7205			8000		
Prod./Month MT	174			196.5			216.15			240		
	Rs. Lacs	Rs/Kg	Per count	Rs. Lacs	Rs/Kg	Per count	Rs. Lacs	Rs/Kg	Per count	Rs. Lacs	Rs/Kg	Cost per count
Wages General	12.76	7.33	0.27	12.76	6.49	0.24	12.76	5.9	0.22	12.76	5.32	0.2
Salaries	4	2.3	0.09	4	2.04	0.08	4	1.85	0.07	4	1.67	0.06
Stores and Spares	3.12	1.79	0.07	3.1	1.58	0.06	3.09	1.43	0.05	3.14	1.31	0.05
Power Expenses	49.47	28.43	1.06	49.47	25.18	0.94	49.47	22.89	0.85	49.47	20.6	0.77
Packing Materials	2.09	1.2	0.04	2.36	1.2	0.04	2.59	1.2	0.04	2.88	1.2	0.04
Year End Expenses	3.5	2.01	0.07	3.5	1.78	0.07	3.5	1.62	0.06	3.5	1.46	0.05

	Rs. In Lacs	RS/KG	ps/count	Rs. In Lacs	RS/KG	ps/count	Rs. In Lacs	RS/KG	ps/count	Rs. In Lacs	RS/KG	ps/count
Admin. & Factory OH	4.2	2.41	0.09	4.2	2.14	0.08	4.2	1.94	0.07	4.2	1.75	0.06
A-Mfg Expenses	**79.14**	**45.48**	**1.69**	**79.39**	**40.4**	**1.51**	**79.61**	**36.83**	**1.37**	**79.35**	**33.3**	**1.24**
B-Monthly Interest on Capex @ 13.00% PA	39.23	22.54	0.84	39.23	19.96	0.75	39.23	18.15	0.68	39.23	16.3	0.61
C-Deprecation @ 10.34% Per month	31.2	17.93	0.67	31.2	15.88	0.59	31.2	14.43	0.54	31.2	13	0.48
Total Expenses (A+B+C)	**149.57**	**85.96**	**3.2**	**149.82**	**76.24**	**2.85**	**150.04**	**69.41**	**2.58**	**150.4**	**62.7**	**2.33**
Capital Rs Lacs												
D-- Working Capital	478.57			539.55			594.04			659.8		
E- CAPEX -Cost of New Machines and Equipment	3621			3621			3621			3621		
F-Total Capital (D+E)	**4099.57**			**4160.55**			**4215.04**			**4281**		
As per Special Scheme												
G-Manufacturing Expenses	79.14	45.48	1.69	79.39	40.4	1.51	79.61	36.83	1.37	79.95	33.3	1.24
H -Monthly Interest on Capex @8.00% PA(13%-5% subsidy)	21.73	12.49	0.46	21.73	11.06	0.41	21.73	10.05	0.37	21.73	9.05	0.34
I-Deprecation @ 10.34% Per month	28.08	16.14	0.6	28.08	14.29	0.53	28.08	12.9	0.48	28.08	11.7	0.43
Total Expenses(G+H+I)	**128.95**	**74.11**	**2.76**	**129.2**	**65.75**	**2.46**	**129.42**	**59.87**	**2.23**	**129.8**	**54.1**	**2.01**
Capital												
J-Working capital	478.57			539.55			594.04			659.8		
K- Capex -Cost of New Machines and Equipment	3621			3621			3621			3621		
L -Subsidy @ 10%	362.1			362.1			362.1			362.1		
M-Actual Capex Amount (K-L)	**3258.9**			**3258.9**			**3258.9**			**3259**		
Total Capital (J+M)	3737.47			3798.45			3852.94			3919		

OPTION 2

Projections												
No. of Ring Frames	20											
No of RF Spindles	25000											
Ring Frame Speed RPM	**16000**			**18000**			**20000**			**22000**		
Avg. Count Ne	26.7			26.86			26.85			26.87		
Prod /Day Kg	11600			12950			14250			15900		
Prod./Month MT	**348**			**388.5**			**427.5**			**477**		
	Rs. In Lacs	RS/KG	ps/count	Rs. In Lacs	RS/KG	ps/count	Rs. In Lacs	RS/KG	ps/count	Rs. In Lacs	RS/KG	ps/count
Wages General	18.93	5.44	0.2	18.93	4.87	0.18	18.93	4.43	0.16	18.93	3.97	0.15
Salary	6	1.72	0.06	6	1.54	0.06	6	1.4	0.05	6	1.26	0.05

Spares and Stores	5.2	1.49	0.06	5.2	1.34	0.05	5.16	1.21	0.04	5.22	1.09	0.04
Power Expenses	72.75	20.91	0.78	72.75	18.73	0.7	72.75	17.02	0.63	72.75	15.25	0.57
Packing materials	4.18	1.2	0.04	4.66	1.2	0.04	5.13	1.2	0.04	5.72	1.2	0.04
Year End expenses	3.5	1.01	0.04	3.5	0.9	0.03	3.5	0.82	0.03	3.5	0.73	0.03
Admin & Factory Over heads	4.2	1.21	0.05	4.2	1.08	0.04	4.2	0.98	0.04	4.2	0.88	0.03
A-Mfg. Expenses	**114.76**	**32.98**	**1.23**	**115.2**	**29.66**	**1.11**	**115.7**	**27.06**	**1.01**	**116.3**	**24.39**	**0.91**
B-Monthly Interest on Capex @ 13.00% PA	63.68	18.3	0.69	63.68	16.39	0.61	63.68	14.9	0.55	63.68	13.35	0.5
C-Deprecation @ 10.34% (per month)	50.65	14.55	0.55	50.65	13.04	0.49	50.65	11.85	0.44	50.65	10.62	0.4
Total Expenses(A+B+C)	**229.09**	**65.83**	**2.47**	**229.6**	**59.09**	**2.2**	**230**	**53.8**	**2**	**230.7**	**48.35**	**1.8**
Capital Rs. Lacs												
D- Working Capital	676.44			756.1			831.4			928.3		
E-Capex (Cost of New Machines and Equipment	5878			5878			5878			5878		
F-Total Capital (D+E)	**6554.4**			**6634**			**6709**			**6806**		
As per Special Scheme												
G-Mfg . Expenses	114.76	32.98	1.23	115.2	29.66	1.11	115.7	27.06	1.01	116.3	24.39	0.91
H-Monthly Interest @8.00% PA (13%-5%)	35.27	10.13	0.38	35.27	9.08	0.34	35.27	8.25	0.31	35.27	7.39	0.28
I-Deprecation @ 10.34% per month	45.58	13.1	0.49	45.58	11.73	0.44	45.58	10.66	0.4	45.58	9.56	0.36
Total Expenses(G+H+I)	**195.61**	**56.21**	**2.11**	**196.1**	**50.47**	**1.88**	**196.5**	**45.97**	**1.71**	**197.2**	**41.34**	**1.54**
Capital Rs Lacs												
J-Working Capital	676.44			756.1			831.4			928.3		
K- Capex (Cost of New M/C)	5878			5878			5878			5878		
L-Subsidy @ 10%	587.8			587.8			587.8			587.8		
M-Actual CAPEX Amount(K-L)	**5290.2**			**5290**			**5290**			**5290**		
Total Capital (J+M)	**5966.6**			**6046**			**6122**			**6218**		

From the above two exercises (Option 1 & Option 2) it is evident that keeping the number of ring frame spindles constant, cost per count is reduced with the increase in spindle speed. The cost further reduces by increasing the total no. of spindles and spindle speed together. Theoretically, it looks quite promising. But in practice overall results may not be favourable.

With the higher spindle speed, we need to observe the quality of the yarn, machine break downs spare parts consumption, maintenance cost, rate of ends breakages, waste generation, power consumption, efficiency of machine and actual production achieved, etc.

It is a wise decision in the interest of the management to keep in mind all the above factors and use the optimum spindle speed. Payback period must be calculated with the optimum speed and not with the theoretical speed of machine.

Cost Reduction

The primary objective of any business is to make profits. Improving profits is also the top priority of business leaders. But it appears, most professionals fail to truly understand how to move beyond basics, especially when it comes to cost reduction for profit improvement. Usually, business leaders rely and focus on reducing costs primarily by reducing workforce. However, this ends up being a mistake in most situations or merely a short-term solution.

In the author's opinion, there must be consistent and sincere efforts to build an inherent cost reductive culture in the organisation. Cost reduction drives should be practiced as a routine aspect in any organisation to sustain its competitiveness in the market. Every individual of the organisation should be involved in this drive, given appropriate tasks and encouraged to participate and share their own creative ideas towards this. **A business leader should try and have a positive approach towards involving employees, giving them reasonable chances, and encouraging participation for developing cost reduction strategies so that profit improvement is perpetual and consistent.**

Many companies don't have long-term strategies for profit building and whenever they lose profits, they cut costs in the only way they know and find instant solutions i.e. the most significant resource of the management - the skilled and experienced workforce - is targeted first.

Organisations need to view profit building as a process similar to the processes that drive marketing, finance, and HRD, etc.

Before discussing cost reduction drives, it is the responsibility of a good leader to organise awareness programs at different levels for each employee in the organisation and inform them about the importance of cost reduction and valuable contributions they may add at individual levels for the benefit of the company.

Daily performance reports of individual departments is a basic tool to understand their contribution in the overall performance of the company. Every employee should have access to performance reports of their respective departments or sections. The first and foremost responsibility is to follow targets. If results are below the set targets, an analysis should be done for the reasons behind such losses and solutions that can improve the same to arrest further losses. All concerned employees should be explained the nuances of these monetary losses that may be caused due to negative variations in results in their respective departments or sections. For example, they may be explained the impact and monetary loss caused due to one percent loss in efficiency or one percent higher defective production etc. However, in case results are better than the set targets, the next targets can be uplifted to optimum levels.

It is also important to remember however that, there is a limit to saving costs within the periphery and the same should never compromise the quality of products. Before deciding cost cutting one should understand the cost elements.

COST REDUCTION PROCESS

The cost reduction process is a systematic method comprising of developing workforce, analysing situations and coming up with creative solutions towards reducing overall costs in an organisation. It involves making teams of different groups of frontline workers, lower and middle staff and senior management representing different sections of the company and motivating them to share their innovative ideas to save costs in their respective areas. Training in innovation management, a unique brainstorming method and developing action plans are the backbone of a cost reduction drive.

In this age of globalisation, the whole world has come together and is more or less dependent on each other. Due to open markets, customers now have multiple choices. They demand value for money. In most cases, it's a buyer's market and every sales and marketing representative ends up focusing on satisfying them and trying to meet their demands. In such a competitive market, every manufacturer has to make best efforts for their survival.

However, one needs to identify the areas that can be addressed for cost reduction drives so as to not only survive but also achieve sustainable growth in a business.

A market demands quality products at a minimum price, with complete assortment in choice, within a minimum lead-time. Usually, a customer looks for and analyses the following features in a product before buying it, in different priority depending on individual needs.

- **Quality of Product**: Presentation, design, features, material, durability, suitability for specific use and its brand value.

- **Price**: Needs to match with the product quality, value for money.

- **Assortment**: Various options in the same quality in terms of colours, design, size, etc.

- **Delivery**: Commitment of product availability within scheduled time.

- **Brand value:** Consumer product or belongs to lower, middle, upper, premium, or luxury segment, etc.

To achieve the above features, a manufacturer needs to be a system driven company and may adopt following methods:

- Company's policy must focus on customer satisfaction

- Transparent and defined ales policy prepared in consultation with all concerned departments and stake holders

- Regular efforts to support Image/Brand building

- Transparent and defined procurement policy

- Timely upgradation of machines and equipment

- To achieve optimum utilization of machines and equipment (Optimum speed and efficiency)

- To achieve optimum yield of material

- Implementation of customised preventive maintenance check list and schedule

- Implementation of SOP and process control to achieve assured quality

- Focus on 'NO' rework and 'ZERO' defects culture

- Regular innovation for development and process engineering

- Focus on design and product development to meet market trends

- Efficient production planning and control

- Production and timely supply of products in full assortment

- To achieve minimum lead-time

- To focus on efficient and cost-effective logistic system

- Adopting a policy of continuous cost optimisation drive

- Allocation of efficient and optimum work force

- Defined recruitment policy and focus to retain them

- Regular training, skill development, and motivational programs for entire work force

- To work continuously in the direction of keeping the work force happy. Keep in mind *"The first brand ambassadors of any company are its own employees"*.

With time, the cost of raw materials and other input cost increases continuously while the market remains reluctant to absorb increments. A portion of cost increase is passed on to the customer by increasing prices based on market size, competition, demand, buying capacity of customer, etc. Price increase is never welcomed by the market. However, it is the responsibility of an efficient sales and marketing team to convince the consumer and market to absorb and adjust to such increases.

MAJOR COST ELEMENTS AND WAYS TO CONTROL

A. **Raw material cost**: Cost increase due to low production of crops in case of natural fibers, seasonal cost increase depending on high demand, increase in input cost to produce raw material, increase in custom duty structure in case of import or export, increase in freight etc. In textile manufacturing process, out of total cost of finished product, contribution of raw material is the highest.

Ways to control raw material cost:

- Design and implement appropriate procurement policy and follow strictly

- Identify appropriate source and follow defined process of selection of raw material

- Follow smart buying plan i.e. coverage of raw material at lowest price, regular buying and averaging the raw material price

- Maintain optimum inventory level of raw material to minimise interest cost

- Check the raw material specifications and compare with norms approved by quality assurance

- Use appropriate quality of raw material for desired quality of finished product

- Use mixing or blending of different grades of raw material to achieve desired mixing or blending of raw material suitable for specified finished product E.g., Mixing of different staple length and micronaire of cotton fibers, mixing of different microns of wool fibers within a limited range for specific yarn count and quality

- To achieve the standard yield of raw material by optimising manufacturing process and implementing effective process control

- Work in the direction of establishing 'No Rework' and 'Zero Defects' culture within the company

- Establish a system of checking the moisture content, oil content of raw material and reduce the same, in case found on higher side against standard

- Verify the strength of dyes and chemicals and impurities in lubricants, etc. Work out appropriate recipes to achieve required shades of fibre, yarn or fabrics, check and compare the actual depth of the dyes with invoiced depth and adjust the price of dyes in proportion with the actual depth. Same procedure applied to chemicals and lubricants.

- Establish a system of counter checking the net weight of raw material and compare with the invoiced weight. Appropriate action to be taken in case of deviation.

B. **Employment cost:** Employment cost or cost to the company (CTC) includes salary and fringe benefits of senior management, middle and lower staff, and wages for workers including other allowances. The main reasons for increase in employment or labour cost are disproportionate numbers of workforce, inefficient, and low productive workforce, high investment cost of workforce due to frequent turn out of trained workforce, low efficiency of machineries and low productivity due to various reasons, outdated technology and low utilization of machines, etc.

Ways to control employment cost or labour cost:

- Implement effective HR Policy for selection of right candidates for the right job.

- Regular skill development training and motivational programmes to create belongingness, increasing productivity and focus on retaining them in the organisation.

- Educate the workforce, share with them their contribution in total cost of the product, and invite their participation to increase productivity levels and optimise the cost.

- Implement performance incentives, profit-sharing chemes, etc. to impart a sense of ownership amongst employees.

- Engage optimum workforce as per industry norms. Conduct regular time and motion study through industrial engineer and set internal norms.

- Implement continuous technology upgradations to increase production and productivity.

- Keep in mind, despite adopting all these measures, the employment cost may be controlled only to a limited extent.

C. Packaging Material cost: Packaging consists of enclosing or protecting products for distribution, storage, sale, and use. Materials used for packaging are known as Packing/Packaging Materials. Packing material is required for containment, protection, handling, presentation, delivery, transporting, warehousing, logistics, sale, and end-use of the product. Types of packing material required depends on the type of product to be packed. Primary Packing Materials hold the product, keeping the contents clean, fresh, sterile and safe for the intended shelf life and sale. Secondary packing material is the term used to describe larger cases or boxes that are used to group quantities of primary packaged goods for distribution. Special packaging materials are also used to make the saleable product more attractive according to its positioning and brand value.

Ways to control packaging material cost:

- Selection of proper size of packaging material is important. It should neither be excess in size to avoid waste of space nor be smaller to avoid tight fitting and damage of materials packed.

- Quality specifications of packaging materials should be decided based on types of materials to be packed and expected duration of its storage in the plant as well as at the customer's godown and shops.

- The appearance of the packaging materials helps in attracting customers and reflects the goodwill and signature trade dress of manufacturers. Hence, it is good to ensure that the chosen packing material is presentable, matches the quality of product and its brand image, does not get easily damaged in transit or affect the material itself.

- While sustainability and environment friendly methods should always be given priority and incorporated as much as possible, it is also important to balance sustainable ways of packaging against high costs involved in adopting this approach.

D. Energy Cost: Energy cost is the cost of power and fuel used directly and indirectly for manufacturing a product as well as in other areas of a company including administration, sales and marketing, stores and purchase, guesthouse, accommodation, etc.

Ways to control energy cost:

- Initial design and layout of a manufacturing plant plays a crucial role in energy saving. For instance, the selection of the types and capacity of boiler, air compressor, generator sets, humidifier or air conditioning, electric cables and other appliances can play an important role in energy consumption.

- Distance of boilers and air compressors from the plant, the layout and sizes of distribution pipes, their insulation, etc. are other key factors that influence energy consumption.

- The architecture and design of a building should be such that natural light is utilised to the maximum extent and heat loss is minimised.

- Installation of measuring devices for electric, air, water and steam consumption in each section and ideally for each machine to record and monitor, helps in controlling daily energy consumption.

- Machine-wise ratio of energy consumption and production should be assessed.

- Regular analysis of machine and section wise energy consumption including humidication plant, air conditioning, lighting, etc. should be done.

- Inverter drives, energy saving motors and lighting in the plant should be installed. Necessary precautions should be taken to control transmission losses in air, steam, and power upto the highest extent.

- A fool-proof process should be established for checking quality and quantity of coal and other fuels for each lot received in the plant as per the norms.

- In textile manufacturing, energy cost contributes significantly after raw material costs. Therefore, it is important to conduct energy conservation audits at regular intervals so as to identify areas of leakages and adopt means to control.

E. **Maintenance cost**: There is regular wear and tear in plant and machineries with time. Maintenance cost is the expense towards keeping these in an efficient usable condition.

Ways to control maintenance cost:

- Holistic implementation of customised preventative maintenance schemes is very important to control maintenance costs.

- Maintenance team members must be highly skilled, trained, and sensitive about machine cleaning, appropriate lubrication and its precise settings.

- Maintenance senior team members must be provided proper training of machines by the machine manufacturers and the same should be passed on to every individual in maintenance team as well as in production team so as to avoid unwanted breakdowns and downtime.

- It is good to conduct internal exercises for upto five years, keep records of individual breakdowns and spare part consumption, and determine the life of individual items. This will give a fair idea of monthly and annual maintenance expenses. This exercise should be continued in order to review the consumption pattern with life of plant and machineries.

- Keep record of individual spare price and machine wise consumption and relate the same with production unit. To avoid the cost difference because of foreign currency and custom duty fluctuation, it is better to monitor the number of items consumed instead of its value.

F. **Manufacturing process cost:** Cost of manufacturing process includes direct costs involved in employment, energy, and maintenance.

The cost of manufacturing process is considered satisfactory if these elements of cost are under control. Major reasons for cost increase in manufacturing are:

- Underutilization of machines' capacity and lower efficiency

- Using machines at lower speeds compared to what's recommended or considered in budget

- Losses due to higher and unscheduled break down of machines

- Losses due to higher end breakages in Winding, Warping, Weaving, Ring frames, and other machines

- Losses due to higher waiting time to attend to machine stoppages due to various reasons

- Losses due to higher energy cost of air conditioning, humidification, steam, etc. because of machine stoppages

- Higher employment cost due to lower allocation of machines per hand

- Higher maintenance cost because of improper implementation of preventive maintenance schedule

Ways to control cost of manufacturing process:

- Optimum capacity utilization of machines: Capacity utilization, efficiency, and speed of machineries depend on lot sizes and quality of product being processed. Only a few Spinning mills manage to produce limited count runs continuously with big lot sizes, use appropriate quality of raw material, and achieve the highest utilization, speed, and efficiency. However, the mills producing multiple counts and fancy yarns run with smaller lot sizes, slower speed, lower efficiency and higher change-over and lower utilization.

- It also applies to dyeing, weaving, knitting, finishing, and garmenting machines. In fancy and fashion fields, lot sizes are small, hence it is not practical to utilize the capacity/speed of the machineries as per their size, design or as recommended by manufacturers. Therefore the achievable utilization, efficiency and speed considered in budget and losses due to these factors are inbuilt in the product's costing.

- It is to up to the concerned department heads to tackle these challenges and achieve the budgeted utilization, efficiency, and speed of the machines. Lower efficiency and utilisation of machines increases the process cost and must be addressed critically by concerned department heads.

- If the rate of stoppages of any machine is higher than acceptable norms, the losses due to excess stoppages should be calculated and shared with the machine operators and shop floor workers.

A few examples of ways to explain and convey the message in a convincing manner to all concerned are shared below. The losses may be calculated and shared with them as follows:

Example 1:

Installed efficiency of a machine achieved	= 80 %
Total losses due to various reasons	= 10%
Unaccountable Losses 100 - (80+10)	= 10%

The unaccountable loss is known as interferences because of waiting hours to attend to the machine's breakages by operators.

There are standard interference losses fixed by conducting time and motion study, based on lot sizes, quality of products produced, allocation of machines per worker, climatic condition and over all work environment. The difference between actual and standard loss of interferences must be shared with all concerned so as to find ways of reaching standard levels.

Example 2:

Total no. of Looms - 64, Weaving Loom speed - 600 PPM, Average Picks / Mtr - 2362 (60 /Inch), Average Installed Effici ncy – 85%, Weaving charges - 24 Paisa/Pick (24*60/100) – Rs 14.4 / Mtr, Fabric Sale Rate – Rs 100/Mtr

Production /day @ 100% Efficiency = (600*24*60* 64)/ 2362 = 23410 Mtrs

Production /day @ 85% Efficiency = 19900 Mtrs

Earning for weaving charges 19900 mtrs * Rs 14.4 = Rs 2,86,560 /Day

1% =Efficiency = 234 Mtrs fabric

Weaving charges /day = 234 * Rs 14.4/mtr = Rs 3,369.6 /Day

Weaving charges/day = Rs 3369.6 * 30 days = Rs 1,01,088

Weaving charges/year = 101088 * 12 months = Rs 12 .13 Lacs

If a weaving mill is depending on job work, one percent loss of efficiency will cause loss of **Rs 12.13 / Year**

If company sells the fabric, the loss of sales due to loss of 1% Efficiency = 234 Mtrs * 30 days *12 Months * Rs 100/Mtr = **Rs 84.24 Lacs /Year**

Example 3:

In example no. 2 above, weight of the fabric per linear meter – 250 Grams (4 mtrs/kg), Cost of yarn /kg = Rs 250/Kg

Production /year: 19900 mtrs /day * 30 days * 12 months = 7,16,4000 mtrs

Yarn consumption /year = 7, 16, 4000 / 4 = 17, 91,000 kg

Cost of 1% waste of Yarn = 17910 kg * Rs 250 = **Rs 44 .78 Lacs**

If waste generation in above weaving department is 1 % excess than norms, there will be loss of approx. **Rs 45 lacs /year**

G. Cost of Logistics:

Ways to control Logistics cost:

- This is one of the important cost elements. Proper compactness or density of the materials, efficiently stuffing of materials within cartons, bags, etc quantity loaded in a truck or container can greatly affect transportation costs.

- Proper monitoring and recording of lead time for each consignment from stuffing at the plant to loading in the ship or air should be done and should be critically analysed to minimise the same.

- Selection of local transport and shipping agency, clearing agent for import and export activities is important for hassle free shifting of goods to the plant and customer end.

- It is recommended that an auditing exercise is conducted from time to time in order to identify areas of leakages and ways of cost reduction.

H. Cost of Overheads: Management expenses, sales and marketing expenses, finance cost, logistics expenses, conveyance expenses, guesthouse, gardening, housekeeping, IT, telephone, etc. expenses are all elements of overhead cost.

Ways to control Overhead cost:

- There must be specific accountability and contribution from senior management in revenue growth. The base should always be associated with a large number of employees and the lowest number of leaders.

- Expenses on top management should not be out of proportion. It is essential to assess and keep a balance between the net earning and expenses.

- Promoting virtual conferences and meetings can also contribute to cost reductions on account of travel and other related expenses.

- Efforts should be made towards receiving competitive prices and credit terms for the materials procured. Purchase credit terms should be more or equal than the sales credit terms but it should not be less than the sales credit terms to avoid the burden of interest.

- Sales team needs to monitor and control credit facilities to save interest cost and support cash flo , focus on speeding up business cycles.

- Focus to minimise raw materials, WIP and finished goods' inventory to control interest cost.

- Explore potential disposal of non-core assets to optimise carrying cost.

- Smart banking and negotiation of finance costs and critical monitoring and controlling other overhead expenses is helpful.

- Capacity expansion plan must be critically assessed for its viability and payback period. It should only be approved and implement until it is mandatory to promote business.

To conclude, cost reduction drives should be practiced as a routine process in any organisation, so as to sustain its competitiveness in the market. Every individual in this drive should be given appropriate tasks towards this end and they should be encouraged to participate and share their creative ideas. However, there is always going to be a limit to saving costs and it should never compromise the quality of products. Before deciding on cost cutting one should understand all cost elements comprehensively.

COST REDUCTION REPORT FOR A WORSTED COMPOSITE PLANT

Under a company's proposed cost reduction drive, each department should be assigned specific tasks to contribute towards this exercise. Individual department heads should prepare groups of staff and workers in each section and conduct internal training and motivational programs to educate them about the objective and benefits of the intended cost reduction drive. The members of these groups should in turn convey this message and interact with individual workers, create awareness and an appropriate atmosphere so that everyone comes forward to participate and give their positive contribution. Inviting participation of individual employees significantly helps in making a strong team and creates a feeling of belongingness within the organisation. Individual departments should be asked to submit reports of their respective areas; commercial heads may compile it, verify, and submit to the management. Once monetary savings are confirmed and established, the part of saved amounts may be distributed amongst concerned team members which will further motivate the entire workforce and create competitiveness among themselves. Slowly, the organisation can head towards a reform through positive cultural changes.

EXAMPLES OF COST REDUCTION REPORT

Dept.	ITEM	AREA OF COST SAVING	Starting Month	Actual Saving/ Month from Apr 15 Rs.	Saving up to March – 16 Rs.	Expected Annual Saving Rs	Frequency
Dye House	Dyeing M/c Carriers	Extra Carriers procured for three dyeing machines (10Kg -1 M/c & 20 Kg -2 M/c) to reduce down time (saving of One hour/batch = 3 extra batch/day), increased production 25 Kg /Day= 625 Kg/Month) Saving of Wages @Rs 2.55/Kg =Rs 1594/month	Sep	1,594	11,156	19,125	Recurring
	Dyes & Chemicals	Buffer agent, Ammonium Sulphate (5%) replaced with Sodium Acetate (2.5%) in Wool Top Dyeing. Dyeing 70 MT / month. Avg consumption of sodium acetate (Rs. 15/Kg x 1400 kg = Rs. 21,000) and ammonium sulphate (Rs. 11.63/Kg x 2800 kg = Rs. 32,564) Saving Rs. 11,504 / month	Jul	11,564	1,04,076	1,38,768	Recurring
Spinning	Ring Frame	Warp yarn count & fabric construction changed keeping fabric weight same. (52 Nm against 56 Nm for a specific quality, resulted 300 RPM higher spindle speed of RF & better performance. Weaving efficiency also increased approx. 10 %. (50000 Mtrs Fabric /Year, Weight 285 gram/mtr, warp weight =177 grams/mtr. Total Warp yarn used = 8.8 MT Warp Count 4 Nm Courser x Spinning cost @ Rs 2.25/count/Kg = Rs. 9/Kg x 8800 kg = Rs. 79200 PM . Saving on account of weaving is not considered.	Sep	6,600	46,200	79,200	Recurring
	Autoconer	Travelling of Overhead Blower of Autoconer changed after study from continuous to intermittent thereby saving of power. (9 m/c x 1.5kw x 2.5 shifts x 8 hrs x Rs. 6.8 / unit x 26 days x 50% saving)	Apr	23,868	2,86,416	2,86,416	Recurring
	Autoconer	By stopping Polythene Bag for packing of single yarn to deliver to post spinning for assembly winding(within the factory) Considering 15% single yarn & remaining 85% double yarn production. Approx. 30 kg saving of polythene bags @120/kg	Apr	3,600	43,200	43,200	Recurring
	Lighting	80 Tube lights from empty place which were not essential , were stopped in spinning department (80 nos. x 0.04kw x 20 hrs x 26 days x 6.8 unit Rate)	Sep	11,315	79,206	1,35,782	Recurring
	Waste	Conversion of usable waste to Yarn for all wool, bigger lots of more than 2000 kgs. So far 567 kg finish top is ready from the waste of 22.5 Mic Wool (Wool cost @800/kg less waste cost @300/kg x 567 Kg)	Mar	2,83,500	2,83,500	2,83,500	One time
	Ring Frame	Rationalization of TPM of different blend within same count to increase spindle utilisation in Ring Frame	Mar	Yet to be quantified			
	Ring Frame	Count range has been increased to incorporate different shades of same count in one machine to increase the spindle utilization.	Mar	Yet to be quantified			

	Wax	Use of Indian Wax instead of Imported. Cost of Indian Wax Rs.81/kg .Cost of Imported wax Rs 155/kg, Average consumption 200 kg / month (after trial, found same results)	Aug	14,800	**1,18,400**	1,77,600	Recurring
	Chemicals	Change of Ecodiser Tex (Acid) @51.86/kg with Ladre Sorbe T (Green Acid) @57.46/kg but strength 40% stronger, hence saving of Rs. 17.39/kg (Consumption reduced from 1000 kg to 600 kg/month)	Oct	17,385	**1,04,310**	2,08,620	Recurring
	Chemicals	Machine washable Chemicals replaced from supplier A to Supplier B (12000 Mtr Fabric) Cost of A was Rs. 7.02 Lac & cost of B, is Rs. 5.14 Lac	Sep	1,88,000	**1,88,000**	1,88,000	One Time
	Chemicals	By replacing cris... ECO (Rs 147.41/kg) with Fibro... ECO (Rs. 135.11/Kg) saving of Rs 12.3/kg (consumption 6060 kg for 5 Lac mtrs TV fabric / annum)	Nov	6,212	**31,058**	74,538	Recurring
	LPG	By putting steam heated water trays under gas, saving of 16 cylinders / month (Will be workable for 6 months in winter) @Rs. 1700/cylinder	Sep	27,200	**1,63,200**	1,63,200	Recurring
Engineering	Open width Washing M/c	72 nos. of Bearings, which were fitted inside the chambers of continuous scouring machine were fitted outside after modification, hence life increased from four to twelve months. Cost per Bearings Rs 152/= Total Cost of bearings 72x152x3= Rs. 32832/= per year. After modification cost of bearing 72X152=Rs 10944/= per year. Saving Rs 21888/= per year	Jul	1,824	**16,416**	21,888	Recurring
	Water Supply	Reduction of working hours of Submersible Pump for supplies of water from 24 Hours to 20.5 Hours. Saving of 33 Units/day (33 Unit x Rs. 6.80 x 30 days)	Aug	6,732	**53,856**	80,784	Recurring
	Water Supply	Stopping of Booster Pump by System modification. Saving of 264 electricity units/day (264 units x Rs. 6.80 x 30 days)	Aug	53,856	**4,30,848**	6,46,272	Recurring
	ETP	System modification from mechanical to pneumatic air at FMT Tank. Stopped one Motor of 7.5HP Total saving 120 unit/ day (120 units x Rs. 6.80 x 26 days)	Aug	21,216	**1,69,728**	2,54,592	Recurring
	Lighting	Energy saving by replacing conventional lights in to CFL Lights. Total 93 fittings replaced. Saving of 125.5 KWH/day (125.5 units x Rs. 6.80 x 30 days)	Mar	25,602	**25,602**	3,07,224	Recurring
	PF Rebate	Monitoring and maintaining most economic power factor (PF)	Apr	(upto Feb16)	**15,90,181**	15,90,181	Recurring
	Additional PF Rebate	Rebate in monthly electricity bill by maintaining & improving power factor from 0.98 to 0.99 (2.25% of the energy charges)	Oct	(upto Feb16)	**78,198**	78,198	Recurring
	Transformer	Saving "No load" losses of transformer by switching them "OFF" and rationalizing load on under loaded transformers. Two transformers of 1500 kva kept OFF	Oct	19,584	**1,17,504**	2,35,008	Recurring
	PLEC	Reduction in peak load exemption charges by reducing MD from existing 4100 kw to 3561 kw		87,000			**Proposed**
HR	Canteen/G-House	Reduction in Canteen & Guest House Expenses by reducing subsidy and close monitoring	Jun	23,000	2,30,000	2,76,000	Recurring
	Credit Period	Increased Credit period of various suppliers from 30 to 60 days and 60 to 90 days. Average credit period reached 45 days from 30 days, hence saving of interest on working capital.	Aug	To be quantifie			
	Coal Inventory	Reduced inventory of Coal from 3 Months to 30 days. Saving of working capital and interest cost.	April	To be quantifie			
Purchase	Cardboard	New vendor developed for packing cardboard. Reduction of landed price from 25/kg to 20/kg. Monthly consumption 10000 kg	Jan	50,000	**1,50,000**	6,00,000	Recurring
	Spiral Paper Tube	New vendor developed for Spiral Paper Tube and reduction of price from Rs 34.66 to Rs 31.64. Net saving Rs 3/pc Monthly consumption 1000 Pcs	Feb'13	3,000	**6,000**	36,000	Recurring
	Packing	Provide Plastic cord in place of Jute cord for ware house bale packing. Consumption reduced from 85 kg Jute cord to 10 Kg Plastic cord. Rate of Jute cord@50/kg and Plastic cord @150/kg. Saving/month Rs. 2750	Jan	2,750	**8,250**	33,000	Recurring
	Etc...			Total	43,35,305	59,57,096	

COST OF REPROCESS OR REWORK

If the output of any process is not as per expectation, it is rejected or necessary corrections are made to achieve desired results. Rework in any process is an additional cost. During the manufacturing process, there may be possibility of human error, imperfections related to material, or machineries involved or in the process.

Common reasons and impact of rework in a textile manufacturing process are:

1. Mixing at fibers, slivers, or roving stage by manual error in small quantity or in a big volume due to improper identification. This is not repairable in a majority of cases and if not detected at early stages, it may damage huge volumes of fabric, cause heavy monetary loss, and failure of delivery on time.

2. Mixing at yarn stage is the result of improper identification system. If it is detected at single yarn stage, it could be sorted out. If two wrong yarns are twisted, it is a waste for the desired product it is produced for.

3. If yarn fails in its specifications, it may be useless for the specified product. In case of dyed yarn, it may be rejected if shade is off. And these shall result in dead inventory.

4. At weaving stage, defective fabric is produced at times due to human or machine error or use of wrong material. Few defects are repairable but some are not. In such cases, the whole process from fiber to fabric may need to be repeated to reproduce the material.

5. In dyeing of fibre, yarn or fabric, the main reasons of rework is usually wrong assessment in sample matching, recipes, weighing of material and dyes/chemicals, improper time, temperature and pressure controls or quality of water and steam etc.

6. Finish of the fabric is a subjective matter. Main reasons of rework in finishing process are wrong weighing of material, chemicals, or variation in process, temperature, pressure, timing, improper quality of water, steam and interrupted power supply etc.

7. Any rework in manufacturing process not only causes waste of material but also increases the cost and extends the delivery period and at times leading to cancellation of orders. The quantum of loss usually depends on the stage of mistake and volume of material involved.

8. An organisation must aim at minimising rework optimally and finally achieve perfection by adopting a policy of 'FIRST TIME CORRECT'.

9. Selection of proper raw materials, perfect testing of materials, right selection of technology, implementation of strict process control, preventive maintenance of machineries and supervision are key measures that can help in arresting and minimising rework.

10. Proper record keeping of rework in every process or department is essential for assessing the root cause and its monetary implications should be quantified and shared with all concerned in the respective departments. Reasons should be identified according to category of failures due to material problems, machines faults or human error. Accordingly, there should be continuous efforts to minimise losses due to rework by individual departments as well as the management.

EXAMPLE OF REPROCESSING COST IN FINISHING DEPARTMENT

Month	Production	Reprocessed	%	K/D Mark	Blowing	Singeing	Crease	Stains	Short Width	Others	Additional Cost
	Mtrs	Mtrs			Mark	Mark	Mark				Per Month
Cost/Mtr Rs				8.77	8.77	8.87	10.09	8.87	1.76	8.77	Rs
Month											
Apr	145204	24800	17.08	Mtr 5700	1400	100	100	4800	1500	11200	
				Rs 49989	12278	887	1009	42576	2640	98224	207603
May	175421	34600	19.72	Mtr 6300	1300	0	500	4200	1300	21000	
				Rs 55251	11401	0	5045	37254	2288	184170	295409
Jun	115637	22700	19.63	Mtr 3100	1400	100	1300	1100	1900	13800	
				Rs27187	12278	887	13117	9757	3344	121026	187596

REPROCESSING COST IN DYEING DEPARTMENT

TOP/FIBER DYEING				
Month	**Total**	**Reprocessed**	**%age of Reprocessing**	***Reprocessing @ Rs 30 / Kg)**
	Production Kg	**Quantity Kg**		
April	56612	3042	5.4	91260
May	58754	3530	6	105900
June	77204	3029	3.9	90870
July	76924	3084	4	92520
Aug	70168	2432	3.5	72960
Sept	41666	1218	2.9	36540
Oct	34027	1085	3.2	32550
Nov	34864	963	2.8	28890
Dec	45436	1167	2.6	35010
Jan	40304	1323	3.3	39690
Fab	70579	1717	2.4	51510
March	70711	1475	2.1	44250
Total	**677249**	**24065**	**3,55**	**721950**
*Dyes/Chemical Cost = 8 Rs/Kgs, OH/Cost = 22 Rs/Kgs				

FABRIC DYEING

Month	Total	Reprocessed	%age of Reprocessing	Reprocessing @ Rs 17 / Kg Rs,
	Production	Quantity Kg		
April	27897	219	0.8	3723
May	37078	600	1.6	10200
June	33930	517	1.5	8789
July	15613	0	0	0
Aug	15835	400	2.5	6800
Sept	33351	0	0	0
Oct	23679	0	0	0
Nov	33508	621	1.9	10557
Dec	53382	354	0.7	6018
Jan	47812	550	1.2	9350
Fab	17404	0	0	0
March	54283	660	1.2	11220
Total	393772	3921	0.99	66657

EXCESS COST OF DYEING DUE TO UNDER UTILIZATION OF DYEING M/CS CAPACITY

	Two M/c - capacity -112kg/M/c				Three M/c - Capacity- 220kg/M/c				Total
	Should be Dyeing if in full capacity Kg	Actual quantity Dyed Kg	Excess cost per kg (due to under capacity) Rs	Excess cost (due to under capacity) Rs	Should be dyeing if in full capacity kg	Actual quantity dyed Kg	Excess cost per kg (due to under capacity) Rs	Excess cost (due to under capacity) Rs	Excess cost (due to under capacity) Rs
April									

May								
June								
July								
August								
September								
October								
November								
December								
January								
February								
March								
Total Loss								
Total								

DYEING RECIPES

REDUCTION OF DYEING COST THROUGH ALTERNATE RECIPES

Setting recipes for dyeing is a highly skilled process. Mixing of different colours is undertaken to achieve different shades and tonal differences in solid shades. It is advisable to mix minimum nos. of colour to achieve desired shades. These days there is a computerised system which provides different options of recipes for a specific shade. Same shades can be achieved through different recipes without compromising the quality. In the following chart, different recipes were developed for a 65:35 percentage polyester /viscose blended fabric. Jet dyeing for polyester dyeing and Open Jigger for viscose dyeing were used for 330 mtrs of fabric lot size equivalent to 100 kg of fabric weight.

Saving for 100,000 mtrs of fabric

Difference between recipes no 1 and 1A: KSH 13 - 6.15 = 6.85/mtr, Total Saving KSH 685,000/- Difference between recipes no 3 and 3A: KSH 10 - 5.7= 4.3/ mtr Total Saving KSH 430,000/= Recipe no 8A is lowest compared to 8 and 8B and difference is KSH 7.4 - 6.2= 1.2 /mtr Total Saving KSH 120,000/-

SHADE NO.	POLYESTER	%	GMS	RATE	AMT	COST/MTR	VISCOSE	%	GMS	RATE	AMT	COST/ MTR	TOTAL
1	Terasil Black MAW	2	2000	190	380	1.151515	Cibacron Black W-NN	3	3000	550.15	1650.45	5	6.15
1A	Terasil Black MAW	2.1	2100	190	399	1.209091	Cibacron Black WNN	3	3000	589	1767		
	Terasil Rubine 2GFL	0.2	200	380	76	0.2303	Cibacron Red FN-R	0.46	460	1280	588.8		
	Terasil Yellow 4G	0.4	400	380	152	0.46061	Cibacron Yellow FN-2R	0.88	880	1360	1196.8		
							Cibacron Yellow 4G	0.16	160	380	60.8		
	TOTAL				627	1.9	TOTAL				3613.4	11	13
2	Terasil Yellow 4G	0.42	420	380	159.6	0.48	Cibacron Yellow F4G	0.76	760	2010	1527.6		
	Terasil Rubine 2GFL	0.14	140	380	53.2	0.16	Cibacron Yellow WR-200%	0.39	390	915	356.85		
	Terasil Blue 3R1-02 150%	0.88	880	420	369.6	1.12	Cibacron Navy WB	1.3	1300	358	465.4		
	TOTAL				582.4	1.76	TOTAL				2349.9	7.12	8.9
3	Terasil Yellow 4G	0.17	170	380	64.6	0.2	Cibacron Yellow HR-200%	0.17	170	762	129.54		

	Terasil Red 3BL-01 150%	1.7	1700	1040	1768	5.36	Cibacron Red HB-150%	1.1	1100	701	771.1		
	Terasil Blue 3R1-02 150%	0.29	290	420	121.8	0.37	Cibacron Brill. Blue HGR	0.15	150	3249	487.35		
	TOTAL				**1954.4**	**5.92**	**TOTAL**				**1388**	**4.21**	**10**
3A	Terasil Yellow 4G	0.144	144	380	54.7	0.17	Cibacron Yellow FN-2R	0.26	260	1267	329.47		
	Terasil Rubine 2GFL	1	1000	380	380	1.15	Cibacron Deep RedS-B	0.66	660	985.8	650.6		
	Terasil Blue 3RL-02 150%	0.158	158	420	66.4	0.2	Cibacron Blue FN-R	0.2	195	1991	388.24		
	TOTAL				**501.1**	**1.52**	**TOTAL**				**1368.3**	**4.15**	**5.7**
8	Terasil Yellow 4G	0.07	70	380	26.6	0.08	Cibacron Red H-D 200%	0.5	500	832	416		
	Terasil Red 3R1-0L 150%	0.043	43	1040	44.7	0.14	Cibacron Navy H-2G	0.85	850	1101	935.85		
	Terasil Navy BRW	2.8	2800	320	896	2.72							
	TOTAL				**967.3**	**2.93**	**TOTAL**				**1351.9**	**4.1**	**7**
8A	Terasil Yellow 4G	0.2	200	380	76	0.23	Cibacron Yellow H-R 200%	0.12	123	711	87.45		
	Terasil Rubine 2GFL	0.21	210	380	79.8	0.24	Cibacron Red H-D 200%	0.6	600	771.3	462.78		
	Terasil Navy BRW	0.84	840	320	268.8	0.81	Cibacron Navy H-2G	1.07	1070	1006	1076.3		
	TOTAL		0		**424.6**	**1.29**	**TOTAL**				**1626.5**	**4.93**	**6.2**
8B	Terasil Yellow 4G	0.18	180	380	68.4		Cibacron Yellow WR 200%	0.2	200	915	183		
	Terasil Rubine 2GFL	0.37	370	380	140.6		Cibacron Red WB 150%	0.68	680	755	513.4		
	Terasil Navy BRW	0.72	720	320	230.4		Cibacron Dark Blue WR	0.95	950	1360	1292		
	TOTAL				**439.4**	**1.33**	**TOTAL**				**1988.4**	**6.03**	**7.4**

*KSH- Kenya Shilling. Exercise was conducted for various shades in the year 2004 in presence of Dyes Supplier's representative

COST SAVING BY REDUCING DEFECTIVE PRODUCTION

Yarn is the final   product for sale in a spinning mill and fabric for a composite textile mill.

There is usually a quality control and assurance department in a mill which conducts testing and process control from raw material to finished product to ensure the desired quality of final products. Natural textile fibers, such as cotton and wool, etc., get exposed and influenced by nature during their production. Hence, unlike man-made fibers, natural fibers may vary individually in their various properties like length, fineness, strength, elongation, maturity etc. Their properties are derived by an average bunch of fibers and used for commercial and production purpose. Despite automation, 100% uniformity is difficult to claim in the final product. There are always variations during various processes, within and from machine-to-machine due to speed, temperature, pressure, climatic conditions and human involvement. If these variations are beyond norms, products are rejected and it is called defective production. Fibers like cotton, wool, etc. and yarn are sold on the basis of properties and specifications defined by internationally accepted quality agencies and these are commercially accepted. The buyer can counter check these specifications at his end for verification.

But in case of fabrics, certain physical and chemical attributes such as weight per square meter, width, breaking strength, shrinkage, dyeing fastness, etc. are defined but its physical appearance, finish etc. are subjective. It depends on customers choices. Quality standard of fabric is based on its segment of buyers, price level, positioning and brand image etc. Accordingly, the norms for inspection of fabric at different stages is decided and final grading is done for the market. Defective fabrics are sold at different discounted prices depending on the nature of defects and length. A value loss report is prepared which provides the difference between fresh sales value and defective or second sales value. This is an important report for individual department heads and the management to study and decide on corrective measures which can control generation of defective production and minimise value loss.

Example:

Fabric For	Sound Rate / mtr (Rs)	Cuts and defective Ave Rate / mtr (Rs)	Cuts & Defective % age Actual (2011-12)	Actual (2012-13)	Prodn Mfg (Lacs, Mtrs)	Cuts & Def Mtrs C&D (Lacs, Mtrs) 2011-12	C&D (Lacs, Mtrs) 2012-13	Value loss (Lacs, Rs) 2011-12	(Lacs, Rs) 2013-13	Saving (Lacs, Rs)
Domestic	372	186	6.77	5.37	26.14	1.77	1.40	329.16	261.09	68.1
Export	374	187	4.84	2.67	4.12	0.20	0.11	37.29	20.57	16.7
Total					30.26	1.97	1.51	366.45	281.66	84.8
Govt	468	234	0.51	0.18	2.32	0.01	0.00	2.77	0.98	1.8
G. Total			5.7	4.66	32.58	1.98	1.52	369.22	282.64	86.6

Sales and Marketing

SALES AND MARKETING MANAGEMENT

Every organization needs liquidity to function. Liquidity is generated by selling products in the right market at the right price and collecting generated amounts within a specified period to support the organisation's cash flow and ensure its smooth operation. Sales and marketing are two business functions within an organization that generate revenue. The term sales per se refers to all activities that lead to the selling of goods and services. Whereas, marketing is the process of getting people interested in the goods and services being sold.

It is essential to have a plan to enable the achievement of goals & objectives. Sales and marketing strategies are prepared to achieve a predetermined long-term aim. A sales strategy is usually the 'blueprint' for the success of any organization.

KEY AREAS OF SALES STRATEGY

A sales strategy broadly comprises of:
- Setting targets
- Identifying measures
- Sales reviews
- Coaching & motivation

The first stage of any sales strategy is to understand the organization's goals for the short term (0 to 6 months), mid-term (6 to 12 months) as well as long term (12 to 24 months or more). A good tool that can be used to define objectives is a smart methodology, which should be specific, measurable, achievable, result oriented, and time bound. The success of a strategy depends on the executional effectiveness of the sales force. An important step is the positioning of the product and identification of the appropriate market. Once you know your potential customer, you need to identify competitors, their strengths, weaknesses, and means to combat these.

Sales potential is the scope for future results. The gap between performance and potential is where a scope for improvement lies. While it is marketing and management's job to expand potential, it finally depends on the sales frontline team to convert that potential into actual performance. Therefore, the role of the frontline sales team is most critical.

ROLE OF SALES EXECUTIVE

- Setting targets
- Fixing measures to achieve targets
- Analysis and managing network of dealers
- Identifying weak markets
- Visiting dealers and checking/verifying their stocks
- Building relations with dealers
- Appointment and cancellation of dealers
- Understanding counter sales

- Collecting competitors' information

- Market size/Market share estimation

- Coordinating with the manufacturing and back office

- Representation and implementation of the company's policies and procedures

- Revenue collections

- Addressing and settlement of complaints and claims, preferably on the spot

- Pricing communication & scheme execution

- Ensure Profitability: ROI

In short, a Sales Executive's role has four major dimensions:

1. To do Primary Sales

2. To drive Secondary Sales

3. To ensure In-Shop Brand Visibility

4. Revenue Collections & Credit Control monitoring

PRIMARY SALES

A. Market Mapping: This is a systematic and scientific survey of the market to understand market characteristics. It involves:

- Physical location of dealers and markets

- Presence and sale of other brands

- Share of brand-wise sales - market estimation

- Presence of exclusive brand outlets

- Shopping Malls & High Street presence

- Data collection, maintenance and analysis to assess products, dealers' and territory's current and historical performance and facilitate negotiations

- Network analysis identifying weak markets

- Approaching targets

- Scientific target setting, target communication

- Target buy-in

- Pricing communication & scheme execution

- To ascertain profitability: RO

B. Sales Planning:

Target setting should be done keeping in mind the following factors:

- Set dealer targets judiciously

- Network capacity

- Dealers' realistic stretch without going overboard

- Launching Incentive Schemes - Target setting for dealers

- Seasonal sales

- Festivals

- Marriage dates for the year

- Special local events in the year

- Take ownership of target

- Planning for achieving the sales targets - network plan & activity plan

- Planning for network expansion

- Communication of target/plan and ensuring dealers buy-in

- Prefer direct addressability route in urban markets

- Keep profile match in mind, while selecting dealers

- Service the network efficiently

Most importantly, the dealer target should be set with mutual understanding on logical grounds. Dealers should also be educated and explained the benefits   and ways to achieve desired targets.

C. Understanding Network

This involves identifying the percentage of different dealer categories present in the total network.

A bottom-heavy network is defined as a network having higher percentage of small dealers than normal. In such a network, the number of A-category dealers are very less, and B-category dealers are also not sufficient in numbers. Such a network does not have any gloss because of the absence of image creating counters (A class dealers). Handling such a network will be comparatively much easier, as small dealers are easy to handle. Presence of too many small dealers in a territory may prove to be a negative point for approaching big dealers, as big dealers require operating space. Some people adopt this approach as an easy option since appointment of big dealers requires a lot of persistence, tact and hard work. Continuing with such a network for long term will not help in volume growth, image and brand building.

A top-heavy network is defined as a network having a higher percentage of big dealers i.e. more of super A+ and A class dealers which makes the sales over-dependent on a few big dealers. Major sales coming from very few dealers is a major characteristic of top-heavy network. Such a network is not advisable for long-term as over-dependence on big dealers may prove costly due to the following reasons:

- Retaining big dealers is a costly affair in terms of money outflow

- Big dealers are expensive. They demand higher monies for advertising and other support.

- It is difficult to have consistency in sales because of over-dependence on very few dealers.

- It may lead to widespread wholeselling and price disturbances.

- It may restrict the survival of small dealers against big dealers, as a result of which dependence on big dealers is likely to increase and delay the network expansion, which may have a negative impact on business.

A balanced Network is one having a good representation of both categories of dealers. Such a network is not over-dependent on any particular category of dealers.

D. Sales Option: Mainly, companies conduct sales through wholesaler or retailers.

Advantage of Wholesales

- Bulk sales

- Convenience of dealing with just one party

- Easy logistic

- Bulk revenue collection

Drawbacks of Wholesale Channel

- Over dependence on one party or a cluster of parties

- Wholesalers are major outstanding holders

- Company has lesser bargaining power

- Wholesalers generally work in a limited area with limited parties

- They prefer to buy lower priced goods which can be sold in bulk

- Less preference to promote value added product

- Some wholesalers work like investors and disturb the pricing mechanism in the market

- Prefer to promote established products

- Company's direct control of the market is relatively lesser

- Company has high risk in case the wholesaler decides to close the business due to any reason

- Wholesaler may provide high credit to the market and Company may end up losing money and market both, in case of default

Advantage of Retail Sales

Retail sale is through retailers, company's retail shop, franchise shops i.e. exclusive brand store (EBO) or Multi brand outlet (MBO)

- MBO's buy value-added goods

- Promote exclusive and expensive materials too

- Promote new products

- Selling through counters enhances brand image

- Effective credit control and monitoring is easy

- Company can monitor its share in a given market

- Reliable primary market feedback is received about sales

E. Sales Policy and prices

Companies prepare a price list for individual products together with a sales policy every year, which includes detailed terms and conditions related to sales, payment terms, goods return, claims, interest on late payment, etc. Sales policy, price-list, and promotional schemes, if any, should be shared with the trade on time. Area sales executives should ensure that their dealers have a clear understanding of the sales policy. Dealers should also be explained the profit against their investment (ROI).

Dealer profitability can be increased by allocating exclusive rights for specific products, area for sales and providing special performance incentives and efficient supply chain by the company so that the dealer may perform with minimum stock levels and investment, and the wholeseller maintains his market outstanding within control.

Return on investment (ROI) can be defined as profit expressed in terms of % of total investment in the business.

$$R.O.I. = Profit/Total\ Investment$$
$$Profit = (operating\ profit - expenses)$$

$$Operating\ profit = (sales - net\ purchase\ value)$$

$$Net\ purchase\ value = (total\ invoice\ value - total\ credit\ notes)$$

$$Total\ investment = \{(stock\ value + outstanding) - credit\ given\ by\ the\ company\}$$

SECONDARY SALES

Drive Secondary Sales

To achieve desired sales results, it is imperative to have an efficient secondary sales tracking mechanism in place. An efficient sales tracking mechanism has following benefits

- It gives ground information about sales movement in the market
- It helps keep a track of dealer investments out of the business so as to alert the company and ensure recovery
- Gives an idea about realistic achievements to help the company plan for bridging gaps

Hence, the starting point of driving secondary sales is tracking of status quo.

Tracking Secondaries Sales:

Efforts should be made for tracking secondary sales movements from the shop on a regular basis through daily reporting, visits to counters, dealers and shop salesmen feedback. Secondary sales tracking gives an opportunity of filling up depleted stocks even if total stock is higher; and an opportunity for range selling might also be present. Based on gathered data, one can then convince dealers for potential growth opportunities for his shop and justify a higher shelf space for products. Secondary mapping can be an eye-opener in most cases and will throw up new focus areas for sales efforts.

Secondary mapping helps in identifying gaps for volume growth by monitoring and identifying key secondary sales days, peak traffic hours etc. In short, qualitatively it significantly helps in realising untapped brand/shop strengths and opportunities.

IN-SHOP BRAND VISIBILITY

It is experienced that approximately 30% customers have strong pre-fixed choices and 70% take their purchase decisions on the shop floor. They can be greatly influenced by in-shop brand visibility at retail shops.

Three key components of in-shop visibility include:

- Creative display stands, vinyls and well-designed interfaces
- POPs (Points of purchase)
- Product Display (Range & quality)

REVENUE COLLECTION AND CREDIT CONTROL MONITORING

A sales cycle gets completed with money coming back to the company on time. Any delay in this collection stage, upsets the whole cycle.

A disciplined and rational approach is needed to handle collections.

Credit-Industry Scenario: The manufacturing industry faces tough competition from other industries and reduces its profit margins to increase sales volume. Overall, there is an increasing trend of credit sales in the fabric market, which affects the company's performance negatively. Further, there is a high risk of default payments. Credit breeds inefficiencies; Inefficiency breeds disputes and disputes result in claims. Eventually, a company loses money and business both. So, it is important to stick to policy terms of credit and enforce them strictly.

Credit Limit Fixation

- Credibility & financial strength of the dealer

- Past payment record not only with our company but also with other companies.

- Stakes with company - lower the stakes, the more risky it is

- Security of credit - availability of blank cheques

Example: A dealer who is not a very good paymaster and has a limit of 5 lacs with a company and does a business of 4-5 lacs a month has started growing rapidly and commits a business of approx. 10 lacs a month and asks for a limit of 10 lacs. What should a company do in such a situation?

The answer should be to go step-wise instead of getting tempted to increase volume.

Rotation Concept

When business grows, sales executives should work on a faster rotation of credit limits by billing in small quantities but more frequently and collecting more frequently thereby rotating money faster. This rotation will help in doing a greater volume of business with even a lower limit and keeping risk factors low. Daily monitoring points:

- Credit execution

- Tracking of invoices

- Individual invoices should be tracked for due dates of payment

- Dealers should be informed 4-5 days in advance before the due date

- Executives should know the derived collection for each dealer in the territory

- Derived collection can be worked out from the collection tracker for a territory and its dealers

Approach towards Rotation

Rotational approach towards dealer stock and investment has a multiplier effect on profitability and subsequently on sales.

Collection Planning: In the beginning of the month, collection tracker should be completed for the month by putting the sales value of targeted sales volume.

Derived Collection: This will give the derived collection for the month i.e. the collection required from dealer to achieve his target with the existing credit limit.

Rotation Concept: Since credit limits are fixed, if volume plan is higher than the limit, a sales and collection phasing has to be formulated jointly with the dealer.

Collection Tracking

Credit limit Communication: This is a very critical prerequisite of being successful in collection from a dealer. It should be ensured that,

- Dealer knows his credit limit

- Understands the logic of credit limit fixation

- He knows that it is fixed and is not going to change unless his annual sales plan is changed

- Once dealer understands this, the rotation concept will work and a system of collection will be established

Operating Cycle and Significance of Working Capital Management

- Consistency in revenue collection is one of the most important steps for maintaining the business cycle and returns on investment (ROI)
- It requires continuous, day-to-day managerial supervision. Delays in fund collection affects working capital management and the company's smooth functioning and profitability
- Daily monitoring of individual sales representatives of their area's performance is important
- Strict follow up of sales policy to avoid negative issues
- Long sustained efforts should be maintained
- Appropriate mental make-up of front-end sales person is critical
- Ownership for results should be ensured by everyone involved in every sense, independent from the rest of the organisation
- Consistent approach is needed to achieve ambitious sales targets as well as timely collections to avoid pressure at month's end
- Low sales and collection on the first half of the month creates a panic at month end and can cause unnecessary stress and anxiety to everyone.

THE FIVE "P"s OF SALES & MARKETING

Product, Price, Promotion, Place, and People are key marketing elements that decide the business strategy of a company. The 5 'P's are variable tools of business, used by management to satisfy customers and their market, add value to their business, and help differentiate their business from competitors.

1. PRODUCT

Product refers to the products and services offered by a business, which include quality, function, packaging, appearance, warranty, etc. Customers need to understand the features, advantages, and benefits that can be enjoyed by buying the company's goods or services. When thinking about a product, consider the key features, benefits, and the needs and wants of customers.

Product profile example:

Yarn - 100% Cotton Yarn – 10 Ne to 60 Ne Karded, Combed, and compact yarn for the use of medium and higher segments. Blended yarn with special fibers for fashion fabrics.

Fabric - 100% Cotton Bottom weight in Plain, Twill, Sateen, & Dobby Designs in coarse, medium, and fine count range with special finishes

2. PRICE

Price refers to the pricing strategy for products and services and the way it will affect customers. Pricing does not mean just the selling price, but also performance incentives, discounts, credit terms, and value for money.

When deciding a pricing policy, it is important to consider the business's position in the current marketplace.

3. PLACEMENT

Placement means where the product is visible, made available for sale or distribution. In essence, placement decisions are associated with distribution channels and ways to reach key customers.

The product or service must be available to customers at the right time, at the right place, and in the right quantity. According to the quality and price, positioning of the product is decided for lower, middle, higher or luxury segments and sold through dealers, retailers, company's shop, franchise, e-commerce, etc.

4. PROMOTION

Promotion refers to activities that make the business known to consumers, i.e. branding of the product. Branding is done by advertising in a number of ways through electronic and print media, sponsorship, billboards, posters, carry bags, social media, sales conferences, performance incentives, etc. The cost for such promotion is substantial, hence, it is important to work out the break-even analysis i.e. the expenses towards promotion and its impact in revenue growth. Promotions should be in proportion to the positioning of the product.

5. PEOPLE

People refers to the team, workforce and management involved in the business process. The quality assessment of people is done by the satisfaction level of their customers.

TRADEMARKS, BRANDING & MARKETING
TRADEMARKS

A trademark is of major significance in marketing efforts. It helps distinguish between competing products. Trademarks include logos, unique packaging designs, slogans or tag lines and overall trade dress. The producers have to create a distinctive identity and reputation. A trademark is the tool that creates a link between a producer and the product. Trademarks also provide an essential focus that allows the company to channel advertising and other forms of market promotion. From the perspective of a newly established company, the use of trademarks signals the company's intent to become a permanent fixture in the market, and one that aims to meet the needs of its consumers.

Every business wants to create an emotional relationship with the consumer by creating and maintaining a good reputation and image of its products through its trademarks.

MARKETING

Marketing involves creating, delivering, managing, and evaluating brand messages, which information and experiences collectively impact how a brand is perceived i.e. the messaging around the sum total of activities involved in the transfer of goods or services from the producer or seller to the consumer or buyer. Like Peter Drucker, the management visionary states *"Business enterprises have two basic functions: marketing and innovation. Marketing and innovation are what produce results, all the rest are costs."*

Functional Areas of Marketing: Advertising, Customer Service, Direct Marketing, Sales Promotion, Events Sponsorships & Packaging

BRANDING

Brand is a perception resulting from experience with, and information about, a company or line of products, an asset, a guarantee of quality, a relationship between the source of the brand and the consumers. Brand is a marketing concept while trademark is a legal concept. The power of a brand is inversely proportional to its scope.

A brand becomes stronger when you narrow its focus. The creation of a brand is generally through publicity and rarely through advertising. Advertising plays an important role in the growth of a brand. Quality is important but is not the sole element that creates a brand. It is also to be remembered that 'No brand lives forever' on its own and has to be reinforced consistently through various marketing and advertising efforts.

Branding is the process of creating a brand image that engages the hearts and minds of customers. It is the process of brand recognition, brand acceptance, brand preference and brand insistence.

Three steps involved in the creation of a brand:

- Determining the desired Brand Position

- Developing Brand Identification

- Creating Brand Image: Through personal experience, word of mouth, packaging, point of sale display and advertising

Regional Branding is the creation of a concentration in one area of specialization in which the city, region, country or group of countries has world-renowned expertise. It defines a geographical location which build upon its history a set of distinctive features, knowledge and practices.

Examples: France, Italy have established textile brand images as producers as well as designers of high-quality fashion textiles. Swiss watches, German/Japan's Cars, etc.

Advantages of Branding:

For the customer: Offers assurances regarding the quality of the product. It also reassures the consumer about a function or an aesthetic such as comfort, performance in use, handling etc. Also, one can give consumers benchmarks for their purchase decisions. For example, Gore-tex® has become a reference brand for waterproof breathable clothing against which other branded or unbranded products are compared, Raymond for worsted fabric & apparels for men in India, Nike's shoes for specific sports uses, Maruti / Toyota Cars for low maintenance, etc.

For the manufacturer: It gives manufacturers control and advantage over their distribution channels, increases profit margins, and creates customer loyalty throughout the supply chain. For example, Raymond sells its worsted & PV suiting at premium prices compared to others in India. Apple products, Sony appliances, etc. are sold at higher prices in comparison to their competitors.

Strong brands may also allow a company to expand its product range on the back of the initial brand and product, however this should be done carefully. For instance, Nike, Adidas with their portfolio sportswear developed sleepwear, casual wear clothing, bags, etc.; Raymond sells accessories along with textile and apparels, Unilever is similarly known for most diversified products; Patanjali sells a range of products from Ayurvedic medicines to various other products, etc.

Trade brands generally promote their products to users at the next level of production at trade fairs. For example, yarn suppliers market their products to weavers and knitters via yarn fairs and trade advertising. There is very less direct consumer promotion.

Consumer brands are promoted through consumer advertising and are more expensive to maintain. For instance, Woolmark had focused its efforts on pulling demand through the supply chain by creating consumer interest and demand for wool products; Reliance have advertised so as to popularize Polyester as a durable fiber for masses, etc.

Why Branding is necessary: In the global market, building brands helps in increasing the value of products. For sustaining in the consumer market, it is essential to create a strong brand. A reputed brand besides being a style statement for a consumer also provides something more such as assurance for quality. The status, goodwill, and reputation associated with a brand helps in promoting goods among brand conscious customers.

A Lifestyle Brand: A lifestyle brand often caters to a particular section of the society defining the identity of a group or culture. Such a brand is a complement to the identity of an individual, which allows one to connect with the brand. Examples: Nike, Wrangler, Armani, H&M, Raymond, Van Huesen, Zodiac, Calvin Klein, Montblanc, Rado, Apple, BMW, etc. Brands are more than goodwill. It also includes elements such as image and reputation.

Challenges for Indian companies: Will Indian companies be able to defend their market against the growing invasion of foreign global brands? Can Indian companies develop strong global brands? Can Indian companies defend their domestic market? Foreign competitors are not only after the high-end market in India, they are also targeting the middle and eventually the low-end segments. The main defense for India is to develop stronger skills in innovation, differentiation, branding, and service.

Is Marketing more important than Production?

Example: The Indian manufacturer of a reputed international brand's shirt gets only $12, or 10% of the final price of $120 that is paid by a customer of high-end market. While the retailer gets approximately 60% ($72) and the brand company gets 30%, $36 respectively.

The Indian manufacturer has no defense if the International brand owner wants to switch to another manufacturer to whom he will pay $8 and keep $2 or pass it to the retailer to get more retail support. India has been known to pay more attention to product engineering than marketing "engineering". But India's future success will require smart investing in marketing and branding.

STRATEGIC TRAJECTORY FOR INDIA

- Low cost, average quality domestic products
- Low cost, good quality domestic products
- Indian high-end products made for other companies.
- Indian branded products (regional)
- Indian branded products (global)
- Indian dominant brands (global)

Case Study: Haier developed its market through three stages to compete with established competitors, which included
1. Fixing quality- Huge investment towards setting quality standards before launching in the market
2. Diversifying their products - Microwaves, toasters, air, conditioners, dishwasher, vacuum cleaners, etc.
3. Entering the Global market - Asia, U.S., Europe
4. Also, Haier entered with a U.S. partner and became a strong competitor for Whirlpool and GE. It is promoted as a global brand, not a Chinese one (Many people think it is German). Segregated lower price models in low price stores and higher price models in top stores.

FIVE C's FAVOURING INDIA

Capital: India has and can attract capital

Cost: Another few decades of low cost production

Capability: Large number of trained workers, engineers, scientists and business people

Consumers: Immense domestic market

Calm and stability: In a world of turmoil and uncertainty

MARKETING STRATEGIES

It is important to replace old marketing styles with a new marketing style that is holistic, lateral & Hi-Tech Holistic Marketing. It is much more than selling and promotion. We need to take a more holistic view of the consumer's activities, social ambience and then design products/services to satisfy their needs keeping in view risk factors . For example, products for defense sectors, agriculture equipment, automobiles sector, aviation sector, etc.

While the traditional marketing strategy involves product differentiation, cost cutting, & price-cutting, the new approach should be to innovate products and improve customer experience. Holistic marketing also requires strong software and technological support.

Lateral Marketing

Thinking beyond the conventional usages of a product in order to develop a new satisfying experience. 'Cool Plus', 'Teflon coating', Nano-tex, fire retardant, wrinkle free, easy care, anti-bacterial, stain free, cell phones with different features, etc. are good examples of lateral marketing.

Hi-Tech marketing

Marketing which combines information technology, analytical capacities, marketing data, and marketing knowledge to face the following challenges:

- Responding to low margins and economic slowdown, commoditization and rapid imitation leading to shorter product life cycles
- Competition with cheaper brands
- Raising selling and promotion costs and decreasing sales effectiveness
- Shrinking margins
- Proliferation of sales and media channels
- Power shifting to giant retailers who are demanding lower prices
- Recession, lower incomes and purchasing power
- Mergers, larger company bankruptcies, etc.

Improving Marketing Efficiency and Effectiveness: Buying inputs more efficiently, hunting down excessive communications and sales travels expenses, closing unproductive sales offices, cutting back on unproven promotion programs and tactics, putting advertising agencies on pay-for-performance basis, etc.

Replacing higher cost channels with lower cost channels, selling through websites, social media, shifting advertising money to better uses, reducing the number of brands, improving supply chain responsiveness, etc.

Responding to the Economic Slowdown: In a situation of economic slowdown there are uncertainties all around and it becomes difficult to survive. Business leaders still try different means to face the challenges such as revaluate their current resource allocations, geographical mix, market segment mix, customer mix, product mix, channel mix & promotion mix, etc. Decide whether to attack to regain market share rather than retrench. Be sure to maintain the value proposition by their brand. Try to add value instead of cutting the price when marketing strategies are showing diminishing returns because product differentiation is harder to achieve, etc.

At the time of economic slowdown, acquisitions and mergers have as many failures as successes. Globalization offers fewer opportunities because either the good markets are overcrowded or the poor markets have no money. New products unfortunately fail more times than they succeed. Price cutting does not work because competitors may match. Price rising does not work since there is not enough differentiation to support it & cost cutting has eliminated much of the fat but may now risk cutting the muscles.

FIVE WINNING STRATEGIES & EXAMPLES

- **Cost reduction:** Domestic airlines, tourism, Walmart, Rent a Car- OLA /Uber, OYO
- **Improved customer experience:** Improved product quality, better options and after sales service
- **Innovative business model:** Many apparel brands, Apple, Samsung, Sony, Barnes & Noble, Charles Schwab, Maruti, etc.
- **Improved product quality:** P&G, Samsung, Toyota, IT services, banking, etc.
- **Niching:** Progressive insurance, etc.

Planning for today:

- Defining the business
- Shaping the business to meet needs of today's customers
- Improving alignment between functional activities and business definitio
- Organization mirrors current business activities
- Optimizing current operations to achieve excellence

Planning for tomorrow:

- Redefining the business
- Reshaping the business to complete for future customers and markets
- Making bold moves away from the existing ways of doing business
- Reorganizing for future business challenges

MANAGING CHANGE TO CREATE FUTURE OPERATIONS AND PROCESSES
SOME VERTICAL METHODS

Modulation: Juice manufacturers vary the sugar content, fabric manufacturers vary fiber blending / yarn density, low/high TPI in yarn, special finishes in the fabric, etc.

Sizing: Multi-size packaging of products such as tooth paste in smallest size for a day to bigger sizes, suit length in boxes, trouser length, trouser and shirt pack, best price stores, etc.

Packaging: Nestlé's chocolates in different sizes such as low cost paper box for the grocery trade and premium metal box for the gift trade. Premium suit length in special packaging such as brief case / leather bags, perfumes in designer bottles, etc.

Design: Designing of fabric with different combination of weaves and threads, style suits / ladies dress / children wear with embroidery, styling or features of cell phone, motor cycles, shoes, etc.

Complementary: Suit length with special bags, additional gift for buying items beyond a specific limit, insurance with car, etc.

Building Brand Equity: *Marketing is the art of brand building. If you are not a brand, you are a commodity. Then, price is everything and a low-cost producer is the only winner!*

For example, despite their higher pricing, Apple products are in popular demand globally, Raymond suit lengths are first preference in India, etc.

A Brand is more than a name: A great brand representation is a promise of value such as Apple, Sony, Tata, Samsung, Maruti, Toyota, Nike, Google, Armani, Raymond, etc.

The ultimate brand builders are your employees and operators i.e. your performance, not your marketing communications only.

HOW DO YOU DEVELOP A BRAND CONCEPT?

"The brand must be an essence, an ideal, an emotion." It must be supported by beautiful logos, clever tag lines, creative turns, edgy names, The brand should have a target group in mind and be positioned to solve one of their problems better than competitive offerings. Furthermore, the brand's reputation is ultimately based on product quality, customer satisfaction, employee communications, social responsibility, etc.

Branding components: Name of the brand (should be short, suggestive, memorable, unique, and easily pronounceable), slogan, logo and typeface, colors, music, theme lines, stationery and business cards, offices, dress code and happiness of its employees are the branding components. Brand slogan is also an important component of branding. For example, American Express- 'Don't Leave Home Without It', Budweiser- 'King of Beers', Ford- 'Go Further", VISA-'It's everywhere, you want to be', Adidas-'Impossible is nothing', Raymond- 'A Complete Man', Reid & Taylor- 'Bond with the Best', etc.

CHARACTERISTICS OF STRONG BRANDS

- Provides superior delivery of desired benefits (Starbucks, FedEx, Amazon)
- Maintain innovation and relevance of the brand (Gillette, Charles Schwab, Bose)
- Establish credibility and create appropriate brand personality and imagery (Apple, Mercedes)
- Communicate with a consistent voice (Coca-Cola, Accenture)
- Strategically design and implement a brand hierarchy and portfolio (BMW, Toyota , Gap, Sony)

WHAT ARE THE MOST FREQUENT CAUSES OF BRAND FAILURE

It is easy to build a brand than to maintain its position and good will. Major reasons for brand failures are :
- Failure to live up to the brand's promise in terms of quality, value for money, timely and consistent delivery
- Failure to adequately support the brand
- Failure to adequately control the brand.
- Improper timing and Inconsistency in advertisements
- Failure to do brand equity measurement and management etc

SALES POLICY

A sales policy is a document representing the company's sales rules and procedures. This is a guide for developing, coordinating and monitoring the various decisions that have a direct bearing on the company's sales. It outlines the uniform operating procedures for the sales team to follow in routine selling situations.

Usually, a sales policy is prepared by the sales team in consultation with concerned members of senior management and approved by the CEO for a financial year and necessary changes are made according to market trends and business policy. The sales team is responsible for its effective implementation on ground.

Basic elements of sales policies are selling-activity, terms-of-sales, and services. Selling policies need to be flexible yet interpreted and enforced in a firm manner. A company relies heavily on its sales policy to help in achieving its goals and increase its profitability.

A typical sales policy consists of the following major elements:

1. **Product Policy:** Product sales policy focuses on design and product development to meet with market trends and customer satisfaction. Depending on the type of product, the sales team needs continuous support of the design and development team for creative and innovative ideas to convert them into commercially viable products for sale.

2. **Distribution Policies:** Based on the positioning of products and category of customers.

3. **Pricing Policies:** The prices of products must have value for money. Price optimization is one of the key factors in every business operation. Pricing policy has a direct impact on total business revenue and as such, must be carefully thought out.

   Following common practices are followed to fix the pricing of products:

   - **Cost-based pricing policy:** It is based on a fixed profit percentage of the overall cost of a product. The selling price covers total costs of products and attains a certain level of profit. These prices don't reflect the true value of the market, and are opposite to strategic prices as there is no consideration of market conditions.

- **Value-based pricing policy:** In this case, the pricing is decided based on a combination of market research and cost of the product. Market research depends on customer demands, expectations, preferences, financial resources, and competition. Value-based pricing increases profitability by creating customer satisfaction through product's value attributes.

- **Demand-based pricing policy:** A demand-based pricing is based on the demand and level of customer satisfaction with the quality and service. The prices are decided based on cost and projected revenues from sales.

- **Competition-based pricing policy:** This is based on competitors' prices. The company fixes prices lower, higher, or equal in comparison to their competitors' prices. As is the case with cost-based pricing, this policy can be set up quickly as it does not rely on thorough market data, which also means it is not as accurate as demand-based pricing. A company can quickly attract and influence customer perceptions of their products because they already have a pre-established customer base.

4. Promotional Policies: Various promotional methods are used for the sale of products. Sales promotion includes advertisement, campaigning through print and electronic media, design of posters/billboards /displays on social media, video clips/brand books, design & color of carry bags, etc. Promotional policies should be restricted under the approved budget. The Branding Head should decide the content, platform, and timing in consultation with the sales & marketing department as well as Business head.

EXAMPLE OF A SALES POLICY

CONDITION OF SALES: All sales contracts are subject to the company's laid down conditions (unless otherwise agreed in writing).

A. Order Acceptance

(a) The seller reserves the right to accept or refuse any order.

(b) Verbal order or order by cable, telegraph, telephone, telex, fax, email, etc must be followed by confirmation in writing from the buyer.

(c) Acceptance of order will not be deemed valid unless confirmed in writing by company.

B. Selling Price

(a) All prices are strictly as is, unless otherwise stated.

(b) Prices are subject to change without any notice.

(c) The company's prices are based on current, ex-warehouse costs, ocean and inland freight rates, insurance rates, import duties, government taxes and present rate of foreign exchange, but in the event of any changes in any of the foregoing the company reserves the right to adjust prices, on the whole, undelivered/partly delivered goods.

(d) Prices will be as per the company's declared price list.

C. Insurance and Freight

(a) Insurance charges will be added to the bill as per the formula agreed mutually.

(b) Freight charges & other taxes/levies - To be borne by buyer as per the prevailing rates at the time of delivery.

D. Delivery

(a) The company shall employ all reasonable means to effect delivery within the time stated. But it will not be liable for any loss or damage arising due to late delivery or non-fulfilment of contract by any reasons such as strikes, political strikes, government regulations, delay in transport, whether or not owned by the Company, other force majeure reasons or any reasons beyond the control of the Company.

(b) Where damage/breakage of goods occurs whilst in transit in Company's own transport or the transport hired by the Company, delays caused by such damage/breakage, and the delays caused by any other reasons, will be regarded as a matter beyond the control of Company and the Company does not accept any liability whatsoever.

(c) If delivery of goods, for which deposit has been paid, is not made within three months or the transaction is not concluded within that time limit, customers are entitled to request or ask for the refund of such deposit made.

E. Payment Terms

(a) The Company shall be entitled to charge interest at the prevailing bank lending rates on all overdue accounts. Company at all times reserves the right to refuse to deliver to a customer in case the accounts of that customer extends beyond the approved credit limit whether in connection with that particular transaction or any other transaction.

(b) For credit sales, documents will be provided to the buyer through the bank and direct to the buyer in case of advance payment or as initially agreed.

DP Terms (Documents against payment)

(i) Interest free period will be allowed until the 30th day from the date of L/R (loading receipt)

(ii) In case the payment is not made within 30 days from the date of L/R, Interest @ 18% p.a. shall be charged from 31st day.

(iii) Interest @ 24% p.a. from the date of L/R shall be charged if the documents are returned.

(iv) If documents are required to be presented through buyers' bankers, additional charges, of the buyers' bank will also be payable by the buyer.

DA Terms (Documents against acceptance)

(i) This facility shall be extended to the wholesale dealers for annual net purchases more than say, Rs.10 lacs.

(ii) Free of interest period will be allowed until 60 days from the date of L/R.

(iii) After expiry of interest-free period, interest @ 18% p.a. shall be charged from 61st day until the date of payment.

(iv) Interest @ 24% shall be charged on returned /documents from 61st day from the date of L/R.

Documents through Agent or Direct

At the request of a dealer, the company may agree to send documents through agent or directly to the dealer. On such documents, the interest-free period of 60 days for wholesalers and 30 days for retailers from the date of L/R will be allowed.

Early Payment Scheme

(i) If payments are made before the due date, prompt payment Incentive (PPI) @ 2% per month will be allowed on the agreed terms, i.e., 30 days for retailers and 60 days for wholesellers.

(ii) However, in case of early delivery of material or extra FOI (free of interest) period allowed by company, and the party makes early payment, the EPI (early payment incentive) will be allowed at 1 ½% per month. In other words, they will get PPI (prompt payment incentive) @2% per month on 30/60 days as per agreed terms but for balance FOI (free of interest) period the EPI will be allowed @ 1 ½% per month.

(iii) Credit notes will be issued for interest amount earned, but it will be adjusted against accounts. Deducting this amount from the payment against invoice will not be allowed.

(iv) All payments must be made through net banking (NEFT/RTGS), bank draft or cheques.

(v) Date of payment to the bank will be considered as the date of credit in company's account.

(vi) Payments must be made, as per exact invoice amount and no part or on account payment will be accepted. Dealers must furnish invoice wise adjustment details along with the payment details. Interest credit note for early payment and interest debit note for delayed payment will be settled on monthly basis in case of wholesalers, retailers and franchisees.

(vi) If payments are made after the allowed credit periods, interest will be charged @ 18% to both wholesellers and retailers.

(vii) In case of delayed payment beyond FOI period, the payment is to be made along with overdue charges.

(viii) The facility of direct documents (DD) will not be allowed to those dealers who are defaulters in payment.

F. Turn over Incentives

(a) The turn over incentive (TOI) scheme will be prepared in consultation with dealers.

(b) Net value of sales will be considered after deducting excise duty, rebates/claims & value of the returned goods.

(c) Performance Incentive or Bonus - TOI & Bonus will be released on quarterly or half-yearly basis for wholesalers and for retailers it will be released on yearly basis, subject to terms & conditions.

G. Goods Return/Claims

(a) No returned goods are to be accepted until & unless there is a genuine reason. If goods need to be returned for any reason, the regional sales in-charge needs to be informed, who will try to resolve the issue on the spot. In case of any disputes, returned goods will be accepted only after approval of the competent authority in the company within a specified period.

(b) Any returned goods received in the plant should be kept in a separate warehouse and investigated to identify and confirm the reason claimed by the customer.

(c) A monthly report to be generated for all returned goods as per prescribed format. Any issue raised by any agent, dealer or retailer regarding quality, delivery, etc., must be addressed by the concerned area manager instantly on the spot and he must try to resolve the issue as soon as possible in consultation with the sales head.

(d) Complaints related to quality issues, e.g. yarn fault, weaving fault, improper finish etc., should be informed to the Quality Control In-charge through a prescribed format (with quality number and piece number, etc. specified) so as to investigate it properly for corrective action.

(e) If the Quality Control In-charge finds the above reasons genuine, appropriate claim may be decided by the Sales Head & to be approved by Business Head before forwarding to Accounts for settlement.

(f) Any Debit Note for the claim has to be recommended by Sales Head and finally approved by the Business Head and should be forwarded to CFO.

H. Sales Promotion

(a) A joint publicity policy will be framed in consultation with dealers and expenses may be shared as per mutual agreement.

(b) Sales promotions such as advertisement, campaigning through print and electronic media etc. should be restricted under the approved budget. Branding Head should decide the content, platform, and timing in consultation with Sales & Marketing Head and Business Head. This includes design of posters/billboards / display on social media, video clips/brand books, design & color of carry bags etc.

(c) Sales conferences have to be planned well in advance & written information is to be circulated to dealers and Business Head.

(d) The agenda of the conference place/no. of agents/dealers expected to attend, tentative expenses etc. to be discussed & approved by Business Head.

(e) Production Head /Design and Development In-Charge /Dyeing and Finishing Heads may also be part of the conference to understand the traders' views and comments about the products.

I. Reconciliation of quantity and value of sales must be done every month between store, marketing and accounts. Any discrepancies must be sorted out immediately and duly corrected.

J. Arbitration Clause

Any dispute, difference, or question, which may arise at any time in connection with any issue, related with sales between the company and the customer related to the true construction of the conditions of Sales or the rights and liabilities of the parties hereto shall be referred to an independent arbitrator under the Arbitrations & Conciliations Act and under such jurisdiction as may be decided by the company.

Note

(i) Any alteration to the Sales Policy must be made in consultation with the dealer and to be approved by the Business Head.

(ii) Price list will be discussed among Sales & Marketing Head, Costing In-charge and CFO & approved by the Business Head.

GUIDELINES FOR OUTSOURCED FABRIC PROCUREMENT

(a) **Criteria for vendor selection**

    (i) Historical performance of existing vendor in terms of quality of the material (construction, grading, design, shades, etc.), delivery time, etc.

    (ii) New Vendor: Market reputation, manufacturing facilities available with him, credibility, location, infrastructure available, etc.

(b) **Selection of designs**

  (i) The vendor will show various designs and selection will be made by the company's sales representative.

  (ii) In case of monopoly designs, it will not be sold or leaked to any other market before the season is completed.

  (iii) Specifications of fabric e.g., weight of the fabric, blend percentage, width, selvedge details will be decided by company's Design and Development Head.

(c) After selection, the requirement will be forwarded to the Procurement Department, which will prepare the purchase order and will submit the costing sheet through Costing Department for final approval from concerned authority.

(d) Before finalizing the order, an internal costing sheet will be prepared considering the current yarn price, current weaving rate, current processing rate, etc., to compare with the vendor's offered price.

(e) Pre-shipment sample of each quality will be sent to the Company for approval of shade and finish before bulk supply.

(f) Vendor will be responsible for quality checking and grading of the fabric as per 4-Point System or as approved by company.

(g) Fabric will be inspected again in Folding Department as per Company's quality norms & will convert into pieces as per sales instructions. Any additional defect found by the Company's inspector, will be informed to the vendor and appropriate deduction will be made through debit notes.

(h) In case the finish of the fabric is not as required, it will be refinished at the plant and charges for re-finis will be informed to the supplier and will be debited.

(i) Fabric will be supplied in the tube roll form covered with polythene and HDPE bags.

(j) Delivery period must be strictly adhered to. In case of any delay, management will have the right to raise claim as per the agreement.

(k) Vendor will not produce any excess quantity than the ordered quantity for direct sale by him. If any such instance is brought to the notice of the Company, appropriate legal action will be taken, and that vendor will be blacklisted.

Note: Outsourcing of materials is a highly skilled job & needs in-depth knowledge of the product & current market rates of yarn /weaving charges & processing charges to be fully updated & may take appropriate decision before the procurement. Such data will also justify the decision taken from the audit point of view. Hence this job should be allotted to a qualified and skilled person.

EXAMPLES OF FORMATS FOR COMMON SALES REPORTS

SALES VARIANCE REPORTS

Objectives of Variance Report: Sales Variance in terms of volume and value should be analysed to take corrective measures. In fact, the sales team should review the variance daily and weekly and take timely actions to stop slippages. Negative variance may be due to low demand in the market as a whole or for a specific product due to changes in trend, in competitive price and sales and marketing policy, quality issues, insufficient, untimely and unassorted availability of products, etc.

VARIANCE FOR SALES QUANTITY (Mtrs) FOR THE FINANCIAL YEAR (APRIL, 2002 - MARCH 2003)						
	March-03			YEAR TO DATE		
DESCRIPTION	BUDGET	ACTUAL	VARIANCE	BUDGET	ACTUAL	VARIANCE
(Quality Name)	A	B	(B-A)	C	D	(D-C)
ANKARA - 45"	1000000	804282	-195718	14500000	7646703	-6853297
DYED SHIRTINGS	500000	307645	-192355	5500000	3675661	-1824339
SCHOOL CHECK	5000	4005	-995	150000	101217	-48783
TOP DYED SUITING	5000	1262	-3738	145000	44534	-100466
POLYESTER	30000	13249	-16751	555000	471704	-83296
COTTON DRILL 60"	10000	6309	-3691	135000	179930	44930
COTTON DRILL 45"	10000	2866	-7134	125000	59301	-65699
FURNISHING	10000	30847	20847	120000	248349	128349
FOAM MATERIAL OTHERS	100000	208549	108549	1200000	1598974	398974
OTHERS		2996	2996	0	37834	37834
TOTAL	1670000	138210	-287990	22430000	14064207	-8365793

RAGS & FENTS		1294	1294		8180	8180
SALES OF YARN		0	0		46970	46970
SEWING THREAD		0	0		1710	1710
JOB WORK		20962	20962		134783	134783
ACHIEVEMENT		83%			63%	

VARIANCE FOR SALES VALUE FOR THE FINANCIAL YEAR

VARIANCE FOR SALES VALUE FOR THE FINANCIAL YEAR APRIL, 2002 - MARCH 2003 (Rs '000)						
	March-03			YEAR TO DATE		
DESCRIPTION	BUDGET	ACTUAL	VARIANCE	BUDGET	ACTUAL	VARIANCE
(Quality name)	A	B	(B-A)	A	B	(B-A)
ANKARA - 45"	95500	75830	-19670	1382745	725975	-656770
DYED SHIRTINGS	30700	18654	-12046	337625	225765	-111860
SCHOOL CHECK	420	332	-88	12600	8434	-4166
TOP DYED SUITING	525	135	-390	15225	4703	-10522
POLYESTER	3300	1404	-1896	61050	50850	-10200
COTTON DRILL 60"	1480	873	-607	19980	26447	6467
COTTON DRILL 45"	1170	332	-838	14625	6999	-7626
FURNISHING	770	2318	1548	9240	19233	9993
FOAM MATERIAL OTHERS	7800	19080	11280	93600	145291	51691
OTHERS	0	317	317	0	4811	4811
TOTAL	141665	119275	-22390	1946690	1218508	-728182
RAGS & FENTS		6890	6890		41933	41933
SALES OF YARN		0	0		10710	10710
SEWING THREAD		0	0		126	126
JOB WORK		363	363		2839	2839
ACHIEVEMENT		84%			63%	

DEALER WISE COMPARATIVE ANNUAL SALES PERFORMANCE (Value)

Objectives: The objective of dealer and area wise sales analysis is to find out the reasons for both positive and negative variance so as to take corrective measures for the future. Reasons for variations should be discussed with dealers. There may be low demand for a specific product in his area, short supply, or untimely availability of booked products, local issues like unrest, etc.

DEALERS NAME	APRIL - MAR 2003		APRIL - MAR 2002		OUTSTANDING	OUTSTANDING	SALES	COLLECTION	OUTSTANDING
(Amt Rs 000)	Sales	Collection	Sales	Collection	31st Mar 2003	31st Mar 2002	Variance	Variance	Variance
A	0	2	0	1	14	16	0	1	-2
B	79074	72895	118661	114648	2733	3137	-39587	-41753	-404
C	14866	13686	0	50	786	445	14866	13636	341
D	162	145	544	487	13	-73	-382	-342	86
E	23562	18969	23589	23755	5262	3385	-27	-4786	1877
F	0	0	0	100	11741	11740	0	-100	1
G	0	300	0	0	114	415	0	300	-301
H	5324	5101	10188	10050	256	257	-4864	-4949	-1

I	15866	13755	30160	27794	128	-632	-14294	-14039	760
J	515	210	4245	4665	543	248	-3730	-4455	295
K	275	260	666	570	740	759	-391	-310	-19
L	0	0	0	40	385	385	0	-40	0
M	110	115	7969	8086	76	82	-7859	-7971	-6
N	0	0	119	120	-5	-4	-119	-120	-1
O	0	100	1031	1287	705	704	-1031	-1187	1
P	1344	1310	37503	35546	2069	2025	-36159	-34236	44
Q	34199	31814	43884	42599	-16	27	-9685	-10785	
R	82540	72444	67451	64062	1605	952	15089	8382	
S	386	356	307	358	1844	1812	79	-2	
T	19189	16991	3999	3857	159	226	15190	13134	
U	8703	7456	11748	11374	205	45	-3045	-3918	
V	22533	19104	6754	5523	6168	5045	15779	13581	
W	0	200	0	20	932	1131	0	180	-199
X	0	0	0	0	-1	-1	0	0	0
Y	103008	94187	140750	136427	3655	1654	-37742	-42240	2001
Z	0	0	0	0	71	-71	0	0	142
AA	21106	17567	24792	24394	8100	5873	-3686	-6827	2227
AB	5237	5000	10836	10500	146	-111	-5599	-5500	257
AC	0	1300	7929	10700	461	1838	-7929	-9400	-1377
AD	25413	24931	635674	63779	1708	3069	-38151	-38848	-1361
AE	11175	11131	8825	9420	859	1526	2350	1711	-667
AF	765	756	5680	5734	160	231	-4915	-4978	-71
AG	2993	2622	2341	2317	148	95	652	305	53
AH	0	0	0	0	0	0	0	0	0
AI	0	0	0	0	286	288	0	0	-2
AJ	50390	49859	109554	100718	6262	10477	-59164	-50859	-4215
AK	0	0	0	0	12	12	0	0	0
AL	27052	24500	0	0	518	-42	27052	24500	560
AM	0	1250	0	3500	6655	7302	0	-2250	-747
AN	19119	18045	46495	46058	3346	3678	-27376	-28013	-332
AO	881	805	2309	2328	2	-9	-1428	-1523	11
AP	0	0	0	0	102	101	0	0	1
AQ	13961	12587	31099	30565	164	-251	-17138	-17978	415
AR	7015	7419	4992	5079	2	248	2023	2340	
AS	0	0	0	50	749	747	0	-50	
AT	72911	61893	14811	9697	13535	5114	58100	52196	
AU	884	516	2945	3313	410	94	-2061	-2797	
AV	0	0	180	640	-1	-1	-180	-640	
AX	0	0	0	0	2	2	0	0	
OTHERS(C/S&F/R)	650829	603253	521306	508538	13235	5089	129523	94715	8146
TOTAL	1321387	1212834	1367226	1328749	96943	79079	-45839	-115915	17864
DORMANT DEALERS					12585	12585	0	-400	0
GRAND TOTAL	1321387	1212834	1367226	1328749	109528	91664	-45839	-116315	17864

AREA WISE SALES VOLUME AND VALUE FOR FIVE YEARS

AREA (Figs in Lacs)	2008 - 2009		2009 - 2010		2010 - 2011		2011 - 2012		2012 - 2013	
	Mtrs	NIV	Mtrs	NIV	Mtrs	NIV	Mtrs	NIV	Mtrs	NIV
Andhra Pradesh	0.44	98	0.37	87	0.40	91	0.64	125	0.19	50
Assam	0.37	90	0.50	114	0.47	107	0.96	223	0.61	123
Bengal	0.43	98	0.34	72	0.27	59	0.25	64	0.34	74
Bihar	0.19	55	0.54	110	1.59	287	2.72	551	2.17	446
Bombay	0.15	31	0.20	42	0.18	40	0.32	75	0.14	35
Brand House	0.61	115	0.86	176	0.72	145	0.47	103	0.29	73
Chhattisgarh	0.57	107	0.75	152	2.01	381	2.60	510	0.82	159
Delhi	1.32	202	0.87	156	1.56	283	1.69	334	1.17	225
East UP	0.30	63	0.33	70	0.47	87	0.39	81	0.25	49
Gujarat	0.06	14	0.23	55	0.33	77	0.25	61	0.13	36
Institutional	0.20	44	0.35	81	0.41	95	0.52	123	0.45	111
Jammu & Kashmir/HP	1.59	374	1.73	378	2.93	591	2.70	612	1.52	343
Jharkhand	0.98	246	1.24	295	1.40	307	1.64	385	1.17	304
Karnataka	1.22	236	1.59	316	3.45	669	4.04	892	1.14	290
Multiple Store 1	0.23	59	0.54	121	1.12	242	1.77	399	0.81	199
Multiple store 2	0.39	81	0.23	51	0.31	69	0.29	72	0.13	35
Madhya Pradesh	1.47	294	1.49	314	1.84	377	2.71	598	1.05	237
Nepal	1.40	295	2.93	567	3.53	682	5.03	1102	2.65	574
North Maharashtra	0.06	14	0.13	33	0.07	17	0.06	16	0.04	12
Orissa	0.03	6	0.02	4	0.00	1	0.01	1	0.00	0
Punjab/Haryana	0.94	218	0.89	208	1.56	314	1.16	276	1.12	292
Rajasthan	0.90	172	0.63	105	3.48	583	0.87	147	0.05	11
South Maharashtra / Goa	0.11	28	0.07	19	0.11	27	0.08	25	0.09	27
Tamil Nadu/Kerala/Pondicherry	0.01	4	0.01	3	0.00	0				
West UP	0.17	66	0.22	75	0.26	91	0.16	64	0.00	0
Total	14.12	3009	17.04	3604	28.45	5620	31.32	6835	16.33	3705
Less GR	0.54	129	0.38	86	0.39	86	0.74	159	0.97	217
Total Net of GR	13.58	2880	16.65	3519	28.06	5534	30.57	6677	15.36	3488
Avg.		212		211		197		218		227

*NIV- Net invoice value

CHANNEL WISE SALES DATA FOR FIVE YEARS

Example:

Category	Type	YEAR 1		2		3		4		5	
		Quantity Mtrs	Value Rs Lacs	Quantity Mtrs	Value Rs Lacs	Quantity Mtrs	Value Rs Lacs	Quantity Mtrs	Value Rs Lacs	Quantity Mtrs	Value Rs Lacs
MBO	Agent										
	Retailer										
	Wholesaler										
	Total										

EBO	Premium Store															
	Shop-In-Shop															
	Franchisee															
	Company Stores															
	Total															
INS	Institutional															
	Readymade Garments															
	Others															
	Total															
LFR	Large format Stores															
ML	Miscellaneous															
	Total															

*MBO-Multi brand outlet, EBO-Exclusive brand outlet, INS-Institutional, LFR-Large format retail outlets, ML-Miscellaneous

CATEGORY WISE DOMESTIC MARKET SALES DATA FOR FIVE YEARS

Example:

Domestic Market	2009-2010			2010-2011			2011-2012			2012-2013			2013-2014			Total		
Mkt Type	Qty	Value	Avg	Qty	Value	Avg	Qty	Value	Avg	Qty	Value	Avg	Qty	Value	Avg	Qty	Value	Avg
Regular																		
Regular Discounted																		
Monopoly																		
Regular Total :																		
Export Surplus																		
Stock Lot																		
Stock Yarn																		
SurplusYarn																		
Cuts																		
Others Total :																		
Export Surplus C&D																		
Regular C&D																		
Stock Yarn C&D																		
SurplusYarn C&D																		
C&D Total :																		
Grand Total																		
Goods Return																		
Net Sales																		

ANNUAL EXPORT SALES SUMMARY OF A 100% EXPORT ORIENTED WEAVING PLANT

Market	Customer	Quantity (Mtrs.)	Avg. Rate (USD)	Sale Value (Lac USD)	Avg. Rate (Rs.)	Sale Value (Lac Rs.)
ANNUAL BUSINESS PLAN FY -- SALES AT A GLANCE					USD = RS.	53.50
US/Canada	A	440,000	6.00	26.40	321.00	1,412.40
	B	125,000	6.98	8.73	373.43	466.79
	TOTAL	565,000	6.21	35.13	332.60	1,879
UK	C	50,000	8.40	4.20	449.40	224.70
	D	50,000	6.65	3.33	355.78	177.89
	TOTAL	100,000	7.53	7.53	402.59	403
Europe	E	90,000	8.00	7.20	428.00	385.20
	F	90,000	8.00	7.20	428.00	385.20
	TOTAL	180,000	7.77	14	428.00	770
Middle East	G	75,000	7.00	5.25	347.75	280.88
	H	10,000	6.50	0.65	347.75	34.78
	I	10,000	6.75	0.68	361.13	36.11
	J	10,000	7.20	0.72	385.20	38.52
	K	10,000	7.10	0.71	379.85	37.99
	L	5,000	7.10	0.36	379.85	18.99
	M	10,000	6.75	0.68	361.13	36.11
	TOTAL	130,000.00	6.95	9.04	371.83	483.37
Australia	N	35,000	6.50	2.28	347.75	121.71
	O	20,000	6.25	1.25	334.38	66.88
	P	10,000	6.45	0.65	345.08	34.51
	TOTAL	65,000	5.84	3.85	343.22	223
Deemed	Q	5,000	8.50	0.43	454.75	22.74
	TOTAL	5,000.0	8.6	0.4	454.75	22.7
Institutional	R	50,000	7.50	3.75	401.25	200.63
	M	25,000	8.00	2.00	428.00	107.00
	L	10,000	7.00	0.70	374.50	37.45
	BI	30,000	8.50	2.55	454.75	136.43
	BS	5,000	7.00	0.35	374.50	18.73
	AR	50,000	7.50	3.75	401.25	200.63
	LMD	75,000	7.50	5.63	401.25	300.94
	TOTAL	245,000	7.64	18.73	408.89	1,001.79
Domestic Market		20,000	9.25	1.85	494.88	98.98
		15,000	8.75	1.31	468.13	70.22
		15,000	7.75	1.16	414.63	62.19
		5,000	8.80	0.44	470.80	23.54
		5,000	8.50	0.43	454.75	22.74
	TOTAL	60,000.00	8.65	5.19	2,303.18	277.67
SUMMARY						
TOTAL	USA/Canada	565,000	6.18	35.13	332.60	1,879.19

	UK	100,000	8.00	7.53	402.59	402.59
	Europe	180,000	7.77	14.40	428.00	770.40
	Middle East	130,000	6.95	9.04	371.83	483.37
	Australia	65,000	6.41	3.85	343.22	223.10
	Deemed	5,000	8.50	0.43	454.75	22.74
	Institutional	245,000	7.64	18.73	408.89	1,001.79
	Domestic	60,000	43.05	5.19	2,303.18	277.67
	TOTAL	1,350,000.00	6.98	94.27	374.88	5,060.83

SALES AT A GLANCE					USD = RS.	53.50
Market	Customer	Quantity	Avg. Rate	Sale Value	Avg. Rate	Sale Value
		(Mtrs.)	(USD)	(Lac USD)	(Rs.)	(Lac Rs.)
USA/Canada	A	440,000	6.00	26.40	321.00	1,412.40
	B	125,000	6.98	8.73	373.43	466.79
	TOTAL	565,000	6.22	35.13	332.60	1,879.19
UK	C	50,000	8.40	4.20	449.40	224.70
	D	50,000	9.75	4.88	521.63	260.81
	TOTAL	100,000	9.08	9.08	485.51	485.51
Europe	E	90,000	7.75	6.98	414.63	373.16
	F	90,000	7.75	6.98	414.63	373.16
	TOTAL	180,000	7.75	13.95	414.63	746.33
Middle East	G	75,000	7.00	5.25	374.50	280.88
	H	10,000	6.50	0.65	347.75	34.78
	I	10,000	6.75	0.68	361.13	36.11
	J	10,000	7.20	0.72	385.20	38.52
	K	10,000	7.10	0.71	379.85	37.99
	L	5,000	7.10	0.36	379.85	18.99
	M	10,000	6.75	0.68	361.13	36.11
	TOTAL	130,000	6.95	9.04	371.83	483.37
Australia	N	35,000	6.50	2.28	347.75	121.71
	O	20,000	6.26	1.25	334.38	66.88
	P	10,000	6.45	0.65	345.08	34.51
	TOTAL	65,000	6.42	4.17	343.22	223.10
Deemed	Q	5,000	8.50	0.43	454.75	22.74
	TOTAL	5,000	8.50	0.43	454.75	22.74
Institutional	R	50,000	7.50	3.75	401.25	200.63
	M	25,000	8.00	2.00	428.00	107.00
	L	10,000	7.00	0.70	374.50	37.45
	BB	30,000	9.00	2.70	481.50	144.45
	BS	5,000	7.75	0.39	414.63	20.73
	AR	50,000	7.65	3.83	409.28	204.64
	LMD	75,000	7.40	5.55	395.90	296.93
	TOTAL	245,000	7.72	18.91	412.99	1,011.82
Domestic		20,000	9.25	1.85	494.88	98.98
		15,000	8.75	1.31	468.13	70.22
		15,000	7.75	1.16	414.63	62.19
		5,000	8.80	0.44	470.80	23.54

		5,000	8.50	0.43	454.75	22.74
	TOTAL	60,000	8.65	5.19	462.78	277.67
	SUMMARY					
	USA/Canada	565,000	6.22	35.13	332.60	1,879.19
	UK	100,000	9.08	9.08	485.51	485.51
	Europe	180,000	7.75	13.95	414.63	746.33
	Middle East	130,000	6.95	9.04	371.83	483.37
	Australia	65,000	6.42	4.17	343.22	223.10
	Deemed	5,000	8.50	0.43	454.75	22.74
	Institutional	245,000	7.72	18.91	412.99	1,011.82
	Domestic	60,000	8.65	5.19	462.78	277.67
	TOTAL	1,350,000	7.10	95.88	379.98	5,129.71

COUNTRY WISE EXPORT SALES DATA FOR FIVE YEARS

Country	2013 - 14		2014 - 15		2015 - 16		2016 - 17		2017 - 18	
	Nett Mtrs	Amount Rs.	Nett Mtrs	Amount Rs.	Nett Mtrs	Amount Rs.	Nett Mtrs	Amount Rs.	Nett Mtrs	Amount Rs.
Abu Dhabi										
Australia										
Bahrain										
Belgium										
Cambodia										
Canada										
China										
Cyprus										
Dubai										
Ethiopia										
Germany										
India										
Indonesia										
Israel										
Italy										
Japan										
Kenya										
Kuwait										
Lebanon										
New Zealand										
Pakistan										
Poland										
Portugal										
S. Korea										
Saudi Arabia										
Singapore										

Slovakia												
Spain												
Sri Lanka												
Sweden												
Thailand												
Turkey												
UK												
USA												
Grand Total												
Average												

AREA AND AGENT WISE MONTHLY SALES TARGET AND DAILY FOLLOW UP REPORT

Buyer No	Area	TARGET				ACTUAL						Balance To Achieve	
		For Jan 18		Per Day		Today			To Date				
		Qnt Mtrs	Value (Rs. Lacs)	Qnt Mtrs	Value (Rs. Lacs)	Qnt Mtrs	Value (Rs. Lacs)	Price/ mtr	Qnt Mtrs	Value (Rs. Lacs)	Price/ mtr	Qnt Mtrs	Value (Rs. Lacs)
	Andhra Pradesh	16000	107	696	4.65	1738	13.60	783	9476	64.06	676	6524	42.98
	Assam	20000	134	870	5.82	240	0.74	311	4028	17.41	432	15972	116.39
	Bengal	15000	100	652	4.36	130	0.67	515	7897	44.28	561	7103	56.07
	Bihar	12000	80	522	3.49	825	7.24	877	5178	34.64	669	6822	45.64
	Chhattisgarh	4000	27	174	1.16			0	589	4.07	692	3411	22.69
	East Uttar Pradesh	15000	100	652	4.36	39	0.22	561	8320	55.74	670	6680	44.61
	Gujarat	25000	167	1087	7.27	64	0.21	322	1016	3.89	383	23984	163.36
	Jammu & Kashmir	3000	20	130	0.87			0	2015	9.96	494	985	10.11
	Jharkhand	3000	20	130	0.87			0	1816	10.33	569	1184	9.74
	Karnataka	10000	67	435	2.91			0	4215	28.50	676	5785	38.40
	Madhya Pradesh	8000	54	348	2.33	227	1.21	533	4115	21.60	525	3885	31.92
	Mumbai	5000	33	217	1.45			0	10374	55.41	534	-5374	-21.96
	Nepal	16000	107	696	4.65	376	3.02	803	12307	82.48	670	3693	24.56
	New Delhi	20000	134	870	5.82	1078	4.98	462	15090	86.33	572	4910	47.47
	North Maharashtra	10000	67	435	2.91	627	5.81	927	3342	25.75	770	6658	41.15
	Orissa	10000	67	435	2.91			0	3414	24.61	721	6586	42.29
	Punjab & Haryana	10000	67	435	2.91			0	5394	38.92	721	4606	27.98
	Rajasthan	25000	167	1087	7.27	1977	5.13	259	67259	237.39	353	-42259	-70.14
	South Maharashtra	15000	100	652	4.36	471	4.21	894	6667	51.58	774	8333	48.77
	Tamil Nadu	2000	13	87	0.58			0				2000	13.38
	West Uttar Pradesh	3000	20	130	0.87			0	2293	14.72	642	707	5.35
	Institutional	0	0	0	0.00			0				0	0.00
	Exclusive	3000	20	130	0.87			0				3000	20.07
	TOTAL	**250000**	**1673**	**10870**	**72.72**	**7791**	**47.03**	**604**	**174808**	**911.67**	**522**	**75192**	**760.83**

DOMESTIC SALES PAYMENT TERMS AND AGE WISE OUTSTANDING – DAILY FOLLOW UP REPORT

To continue the company's operation cycle i.e. to cover up expenses for production and to provide goods for sales, regular fund flow is required There should be a balance between spendings and earnings. As much as sales is important for generating revenues, timely collection of funds for sold goods is equally important for the continuity of business. Age wise outstanding report should be available with sales representatives and be compared with sales policy. Further, delivery of required goods should be controlled or restricted before the situation goes out of hand and converts to bad debt. Respective sales representatives can control number of outstanding days with consciousness and in a smart way. Being a seasonal business, outstanding may increase during dull periods but efforts must be to bring it down to as minimum as possible during peak season.

Example of format:

Party Name	Pymt Term in Days	Credit Limit Rs Lacs	Outstanding 02-Nov-2017 Rs Lacs	Amt Not Due Rs Lacs	Out Standing Rs Lacs							
					0-15 days	16-30 days	31-45 Days	46-60 Days	61-75 Days	76-90 Days	Above 90 Days	Total Overdue
	75	30	6.18	6.18	-	-	-	-	-	-	-	-
	75	20	2.39	2.39	-	-	-	-	-	-	-	-
	75	75	12.50	2.78	-	-	-	4.85	-	3.09	1.77	9.71
	75	50	24.08	-	-	-	14.99	-	-	8.99	0.10	24.08
	75	5	(0.86)	(0.86)	-	-	-	-	-	-	-	-
	75	10	6.08	6.08	-	-	-	-	-	-	-	-
	75	50	(0.12)	(0.12)	-	-	-	-	-	-	-	-
	75	10	1.88	1.88	-	-	-	-	-	-	-	-
	75	30	10.21	7.73	-	-	-	2.48	-	-	0.00	2.49
	75	20	1.75	1.75	-	-	-	-	-	-	-	-
	75	20	3.88	0.53	-	3.35	-	-	-	-	-	3.35
	75	60	58.03	29.23	-	4.80	0.39	13.66	3.85	6.11	-	28.80
	75	40	(0.03)	(0.03)	-	-	-	-	-	-	-	-
	75	10	0.22	0.22	-	-	-	-	-	-	-	-
	75	5	1.36	-	-	-	-	-	-	-	1.36	1.36
	75	10	16.42	16.42	-	-	-	-	-	-	-	-
	75	10	1.14	1.14	-	-	-	-	-	-	-	-
	75	20	8.89	2.08	-	0.87	1.13	3.51	1.07	-	0.24	6.81
	75	10	5.50	5.50	-	-	-	-	-	-	-	-
	75	20	0.57	0.57	-	-	-	-	-	-	-	-
	75	20	5.49	3.88	-	-	-	0.33	-	1.28	-	1.61
	75	30	(0.50)	(0.50)	-	-	-	-	-	-	-	-
	75	20	2.22	-	-	-	2.22	-	-	-	-	2.22
	75	50	3.40	-	-	3.40	-	-	-	-	-	3.40
	75	10	2.98	-	-	-	2.98	-	-	-	-	2.98
	75	20	4.45	4.45	-	-	-	-	-	-	-	-
	75	20	3.32	-	-	3.32	-	-	-	-	-	3.32
	75	10	1.64	1.64	-	-	-	-	-	-	-	-
	75	75	61.76	27.58	18.03	-	16.16	-	-	-	-	34.18
	75	10	2.17	-	-	-	-	2.17	-	-	-	2.17

			75	10	3.91	3.91	-	-	-	-	-	-	-	-
					(10.38)	(10.38)	-	-	-	-	-	-	-	-
					4.30	4.30	-	-	-	-	-	-	-	-
			75	40	20.72	10.99	-	3.40	-	6.33	-	-	-	9.73
			75	30	27.94	5.58	-	-	-	5.93	7.73	8.70	-	22.36
			75	75	16.53	2.91	-	4.57	-	-	-	9.05	-	13.62
			75	2	9.79	-	-	9.79	-	-	-	-	-	9.79
					319.82	137.83	18.03	33.50	37.86	39.25	12.64	37.24	3.48	181.99

EXPORT SALES OUTSTANDING BILLS – DAILY FOLLOW UP REPORT

	Inv. No.	Inv. Date	Party	F/C	Sales			INR Revaluation	BL Recvd date	BL sent date	Due date
					F/C Value	Rate	INR Value				
A											
		20-Aug		GBP	83,009.38	84.05	69,76,938	76,32,712	30-Aug	3-Sep	10-Sep
		23-Aug		Euro	45,992.70	77.80	35,78,232	36,46,761	30-Aug	13-Sep	18-Sep
		23-Aug		Euro	34,649.53	77.80	26,95,733	27,47,361	30-Aug	13-Sep	18-Sep
		23-Aug		Euro	14,178.60	77.80	11,03,095	11,24,221	30-Aug	13-Sep	18-Sep
		31-Aug		GBP	90,747.22	84.05	76,27,304	83,44,207	13-Sep	19-Sep	25-Sep
		31-Aug-17		Euro	1,28,595.49	77.80	1,00,04,729	1,01,96,336	13-Se	19-Sep	25-Sep
		31-Aug-17		Euro	31,495.24	77.80	24,50,330	24,97,258	13-Sep	19-Sep	25-Sep
		31-Aug-17		Euro	30,431.60	77.80	23,67,578	24,12,922	13-Sep17	19-Sep	25-Sep
		13-09-201		Euro	26,593.10	77.85	20,70,273	21,08,567	20-Sep	24-Sep	25-Sep
		13-Sep-17		Euro	48,877.79	77.85	38,05,136	38,75,520	20-Sep	24-Sep	25-Sep
		19-Sep-17		Euro	69,444.12	77.85	54,06,225	55,06,224	27-Sep	4-Oct	9-Oct
		19-Sep-17		Euro	26,466.37	77.85	20,60,407	20,98,518	27-Sep	4-Oct	9-Oct
		21-Sep-17		GBP	78,581.27	87.35	68,64,074	72,25,548	4-Oct	5-Oct	9-Oct
		16-Oct-17		Euro	1,30,909.91	76.65	1,00,34,247	1,03,79,849	25-Oct	31-Oct	Collection under process
		18-Oct-17		Euro	1,07,765.68	77.65	83,68,005	85,44,741	25-Oct	31-Oct	Collection under process
	TOTAL						7,54,12,306	7,83,40,745			
B	B/L yet to be sent										
		26-Oct		GBP	99,725.43	90.00	89,75,289	91,69,753	8-Nov		5-Nov
		31-Oct		Euro	1,03,574.58	77.65	80,42,566	82,12,428	8-Nov		Not Due
		31-Oct		Euro	51,715.13	77.65	40,15,680	41,00,493	8-Nov		Not Due
		31-Oct		Euro	94,362.50	77.65	73,27,248	74,82,003	8-Nov		Not Due
	TOTAL						2,83,60,783	2,89,64,677			
	G TOTAL						10,37,73,089	9,14,47,422			

*FC-Foreign currency, INR-Indian Rs, Inv-Invoice, BL-Bill of lading

SALES HISTORICAL DATA IN GRAPHICAL FORM

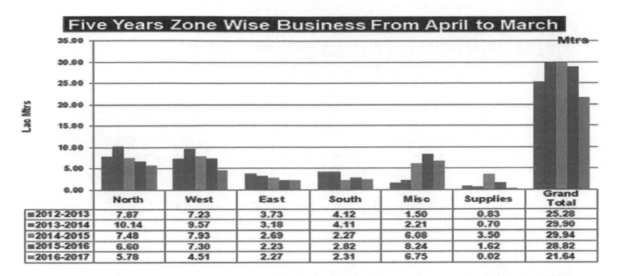

Five Years Zone Wise Business From April to March — Mtrs

	North	West	East	South	Misc	Supplies	Grand Total
2012-2013	7.87	7.23	3.73	4.12	1.50	0.83	25.28
2013-2014	10.14	9.57	3.18	4.11	2.21	0.70	29.90
2014-2015	7.48	7.93	2.69	2.27	6.08	3.50	29.94
2015-2016	6.60	7.30	2.23	2.82	8.24	1.62	28.82
2016-2017	5.78	4.51	2.27	2.31	6.75	0.02	21.64

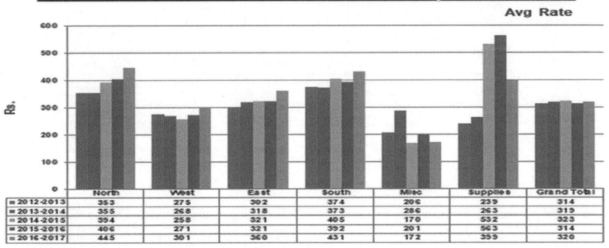

Five Years Zone Wise Business From April to March — Avg Rate

	North	West	East	South	Misc	Supplies	Grand Total
2012-2013	353	275	302	374	206	239	314
2013-2014	355	268	318	373	286	263	319
2014-2015	394	258	321	405	170	532	323
2015-2016	406	271	321	392	201	563	314
2016-2017	445	301	360	431	172	353	320

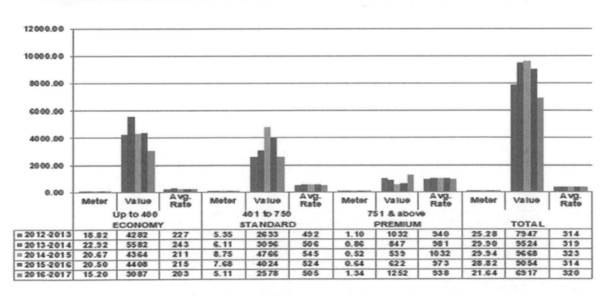

Five Years Segment Wise Business From April to March

	Meter	Value	Avg. Rate	Meter	Value	Avg. Rate	Meter	Value	Avg. Rate	Meter	Value	Avg. Rate
	Up to 400 ECONOMY			401 to 750 STANDARD			751 & above PREMIUM			TOTAL		
2012-2013	18.82	4282	227	5.35	2633	492	1.10	1032	940	25.28	7947	314
2013-2014	22.92	5582	243	6.11	3096	506	0.86	847	981	29.90	9524	319
2014-2015	20.67	4364	211	8.75	4766	545	0.52	539	1032	29.94	9668	323
2015-2016	20.50	4408	215	7.68	4024	524	0.64	622	973	28.82	9054	314
2016-2017	15.20	3087	203	5.11	2578	505	1.34	1252	938	21.64	6917	320

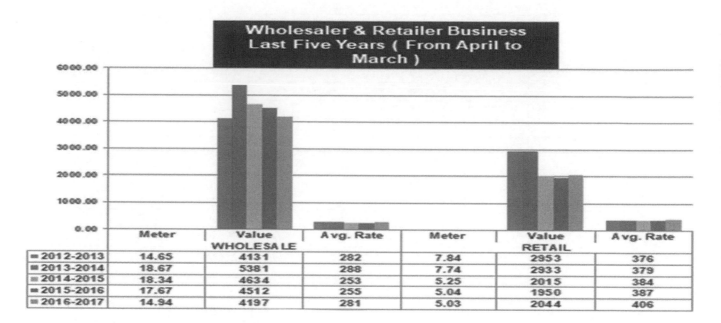

	Meter	Value WHOLESALE	Avg. Rate	Meter	Value RETAIL	Avg. Rate
2012-2013	14.65	4131	282	7.84	2953	376
2013-2014	18.67	5381	288	7.74	2933	379
2014-2015	18.34	4634	253	5.25	2015	384
2015-2016	17.67	4512	255	5.04	1950	387
2016-2017	14.94	4197	281	5.03	2044	406

RECORD OF SALES AGENTS

Sr. No	Agent Name & Address	Area	Contact Person	Contact Nos.	E-mail address
1					
2					
3					
4					
5					
6					
7					
8					
9					
10					

TENTATIVE SALES PROFIT AND LOSS STATEMENT

	SALES, EXPENSES AND P&L						
	Budget				Actual		
PARTICULARS	July		Today	To date	To day	To date	Balance to achieve
SALES (Rs, Lacs)							
Domestic –PW	1672.50		61.94	1610.56	47.06	912.49	760.01
Domestic –PC	378.00		14.00	364.00	7.34	199.52	178.48
Export PW	95.20		3.53	91.67	-	59.52	35.68
Export DEPB	1.90		0.07	1.83	-	2.98	(1.07)
TOTAL INCOME	2147.60	100%	79.54	2068.06	54.40	1174.51	973.10
EXPENSES							
Raw material consumption-PW	634.00	29.52%	23.48	610.52	18.34	355.63	278.37
-PC	203.00	9.45%	7.52	195.48	2.86	77.76	125.24

Total	837.00	38.97%	31.00	806.00	21.20	433.39	403.61
MANUFACTURING EXPENSES Rs Lacs							
Power Cost/ Utility	97.00	4.52%	3.59	93.41	3.59	93.41	3.59
Dyes & Chemicals	43.00	2.00%	1.59	41.41	1.59	41.41	1.59
Stores & spares	37.00	1.72%	1.37	35.63	1.37	35.63	1.37
Other Manufacturing expenses	31.00	1.44%	1.15	29.85	1.15	29.85	1.15
Wages	179.00	8.33%	6.63	172.37	6.63	172.37	6.63
Personnel Cost	106.00	4.94%	3.93	102.07	3.93	102.07	3.93
Administration cost	42.33	1.97%	1.57	40.76	1.57	40.76	1.57
Corporate Administration cost	60.00	2.79%	2.22	57.78	2.22	57.78	2.22
Selling Expenses	36.00	1.68%	1.33	34.67	1.33	34.67	1.33
TOTAL OPERATING COST	631.33	29.40%	23.38	607.95	23.38	607.95	23.38
TOTAL COST	1468.33	68.37%	54.38	1413.95	44.58	1041.34	426.99
GROSS MARGIN	679.27	31.63%	25.16	654.12	9.82	133.17	1400.09
SALES VARIABLE COST							
Publicity Charges	97.00	4.52%	3.59	93.41	3.59	93.41	3.59
Bonus & Commission	238.00	11.08%	8.81	229.19	4.08	88.09	149.91
TOTAL SALES VARIABLE	335.00	15.60%	12.41	322.59	7.67	181.50	153.50
PBDIT	344.27	16.03%	12.75	331.52	2.14	(48.33)	1553.59
Average Selling Price Rs/Mtr							
Domestic-PW	669				603.23	521.56	1012.71
Domestic-PC	252				267.41	231.96	278.94
Export-PW	476				#DIV/0!	374.83	865.83
Qty in Lac Mtrs							
Domestic-PW	2.50		0.09	2.41	0.08	1.75	0.75
Export-PW	0.20		0.01	0.19	-	0.16	0.04
Total PW	2.70		0.10	2.60	0.08	1.91	0.79
Domestic-PC	1.50		0.06	1.44	0.03	0.86	0.64

*P&L–Profit & Loss, PW-Poly/Wool, PC-Poly Cotton, Qty-Quantity

MARKET COMPLAINTS ANALYSIS BY QUALITY ASSURANCE

Month	Complaint Date	Customer Details	Quality /Shade No.	PC NO.	Nature of complaint	Defective sample Received	Defective Qty	Analysis	Final Remarks
April	10th				Sticker Mark	One Trouser	1.30	Sticker not used by company	Not accepted
	11th				Pilling	One Trouser	1.30	Rough Handling	Not accepted
	26th				Pilling	One Jacket	2.00	Rough Handling	Not accepted
							4.60		
May	12th				Contamination	3 Mtr Than	600.00	Hair line contamination	Accepted
	20th				Weaving Defects	11.15 Mtr Than	11.15	Weft bar due to poly filamen	Accepted
	22nd				Pilling & Weaving defects	11 Jacket +3 Trouser	28.65	Rough Use	11 Jacket+1 Trouser not accepted 2 Trouser accepted
							639.80		

MARKET COMPLAINTS ANALYSIS AND ITS VALUATION BY SALES AND ACCOUNT

Recd. dt	Buyer name	Buyer no	LR No.	LR dt.	Transport Market Related causes	Manufacture Related causes	No. of bales	GR Qty (mtr)	GR Value (Rs.)	PARTY REMARKS Reason for return by party		Qlty/shade	Old Pc. no.	Pc. length	QC inspection remarks	Prod Year	Des. Date
										Market Related causes	Manufacture Related causes						

Summary								
Market Related			Manufacture Related					
			Non defective		Actual Defects found by Q.A		
Mtr	Value		Mtr	Value	Mtr	Value	
Total Mtr			Total Value				

*LR- Lorry Receipt, Qty-Quantity, GR-Goods Receipt No, Pc-Piece, Prod-Production, Des-Dispatch, QC-Quality Control

11

Case Studies & Suggestions for Indian Textile Industry

CASE STUDY I

TURN AROUND OF A SICK COMPANY

In the late 90s, in an African country with an oil-based economy, most consumer products were imported and it mostly had a trading-based business environment. However, this led to country-wide unemployment and other social issues. To address this alarming issue of national interest, the government at the time launched attractive schemes in collaboration with the World Bank to encourage investment in manufacturing areas, including the textile sector, to make textile goods locally available for the public. Many foreign investors also invested at the time and installed textile units with European technology and state- of- the- art machineries. Since it was a seller's market, manufacturing units made handsome amounts of profits for several years but failed to reinvest sufficient amounts on maintenance of plant and machinery, design, development and innovation, which resulted in monotonous products, lower productivity levels and quality issues. Additionally, after the WTO treaty, when the markets opened up, a wide range of fashion textile options were made available to consumers at competitive prices. The local textile industries were unable to maintain and upgrade their manufacturing facilities. They were not sufficiently equipped to face new challenges, produce new varieties or meet quality standards. As a result, a majority of textile units began declining sharply and were compelled to close their operations within a period of few years.

My involvement: During this scenario I accepted a challenge to join a medium-sized composite textile company as a Business Head (in 1998).

STATE OF THINGS IN THE BEGINNING

OVERALL SITUATION

- Company was in loss since last three consecutive years and was converted to a sick company

- Liquidity crisis

- No product identity and goodwill in the market

- Low productivity levels

- High levels of breakdowns in machineries

- Quality issues and market complaints

- High waste generation and defective production

- High raw material and manufacturing costs

- Late deliveries of finished goods

- Dissatisfied suppliers & customers

- Dissatisfied employees due to more breakdowns, market complaints and no appreciation since last two-three years

- Previous management had presented a scheme for modernization with huge amount of Capex to the management to recover from this situation
- Production of most preferable product was very low in volume due to capacity constraints
- Previous management had made plans and projected to install new machineries to increase the capacity of this product, which needed huge capital investment
- Market scenario was negative and uncertain due to economic recession in the country and so the management was not keen on investing Management lost confidence in the business, confused about the future
- Ultimately, the Board decided to change the leader

WORKFORCE

- Entire workforce was unmotivated and demoralized
- Negative attitude
- Pessimistic thinking for future
- Lethargy amongst the work force
- Lack of loyalty for the company
- No teamwork
- No defined target and direction
- Low productivity levels
- No production targets / incentive schemes
- No training programs
- No welfare schemes

MACHINERIES

- Most machines were old/outdated with low productivity levels, poor maintenance
- High rates of breakdown of machineries
- Several quality issues and market complaints
- Management was not convinced for investment in spare parts or replacement of old machineries
- As a result, widespread frustration amongst workers and staff members

MATERIALS

- Vendors had long outstanding dues (> six months)
- They were not willing to supply on credit
- They were offering higher rates and the management was not in a position to negotiate
- Management had no choice of vendors
- Company was restricted to procure cotton yarn and polyester yarn from group company at a higher price compared to market rates and quality issues were not entertained
- Compromise on raw materials

- Higher rejection percentages of finished goods
- Higher percentage of rework

INVENTORY STATUS

- Dead Inventory of RM (Yarn and dyes & 4O MT of remnant yarn in dead stock)
- High stock of finished fabrics
- High waste generation and defective percentage of fabrics
- Import of dyes on higher price
- Sudden changes in fashion caused high stock of non-regular dyes/expired life

SALES AND MARKETING

Customers' trust was fragile due to:

- Quality issues and market complaints
- Unattractive packaging
- Untimely, unassorted and short delivery
- Limited designs, not matching market trends
- A practice of pushing materials to dealers due to pressure on sales team
- Longer credit days (3-6 Months) and higher credit limits
- Limited number of dealers resulted in longer business cycles and affected cash flow

MONEY (Liquidity)

- Liquidity Crisis
- Poor cash flo
- Financial institutions were not willing to increase credit limit due to poor performance of the business and lack of concrete future business plans

STEPS TAKEN TO ADDRESS THE ISSUES

ADDRESSED MANAGEMENT POLICY AND INTERACTION WITH WORKFORCE

- Held discussions with Chairman & Board of Directors and understood their point of view
- Agreed to avoid immediate capital investment to save huge initial investment, repayment, interest and depreciation cost. Convinced them for phase-wise, small and essential investments to improve machinery conditions
- Targeted stopping the bleeding ASAP
- Started addressing the staff and workers regularly
- Started visiting the plant, machine to machine on a daily basis
- Listening process included a two-way interaction through meetings with staff members and functional heads
- Allowed them to open up and share their personal and plant related problems and other grievances which helped ease their built-up frustrations

- Regular meetings with union leaders, group of workers to make them aware about the financial situation of the company and asked them to share their views and suggestions

- Developed personal interactions with individual staff members

- Daily factory rounds along with HODs to cover every corner

- Motivational speeches for rebuilding trust levels amongst entire workforce to revive the business

- Focused on culture building exercises

MESSAGE TO THE TEAM

As a team leader, I made immediate efforts to take command of the situation. All functional heads and senior staff members were taken into confidence while trying to convey a philosophy of 'Nothing is impossible' and that if we worked collectively as a team, one could surely turn around the business. However, the task was not simple; it was a long path ahead, and simply conveying such messages to the team is never enough.

A leader has to prove his acceptability, create followers, build a dedicated team for proactive participation and establish a safe environment to enable everyone to share their ideas towards improving the business. It is a leader's past record of accomplishment, good will, conviction, knowledge about the subject, approach towards problems, attitude for the team members, etc., which makes a strong team leadership.

A clear message was communicated to the team that, it is possible to revive the business if, we accept it as a challenge to save the company's image and if we work consistently with:

- Honesty

- Sincerity

- Belongingness

- Focus on work

- Transparency

- No groupism

- Rewards on the basis of performance (Develop an impartial culture)

MACHINERIES

- Conducted audits of individual machines through respective maintenance teams and department heads. Prepared details of individual machines' conditions

- Prepared a list of required spare parts and its estimate based on immediate, short term and long term requirements

- Prepared customized preventive maintenance schedules and checklists for individual machines in the plant

- Prepared maintenance record files for each machine and placed them on individual machines so it was easier to refer to the file instantly in case of any breakdown and investigate the reason of breakdown with the concerned technician who attended to the machine last

- Fixed accountability of individuals, from HODs to shop floor   workers (by name)

- Focused on energy savings by arresting air, water, steam leakages in the entire plant and their consumption related to production on a daily basis

- Conducted energy audits in-house with the Chief engineer and his team

- Began providing spares priority wise

- Began analysing the reasons for each break down & identified the root cause/responsible person(s)

- Began maintaining machine history records

- Began monitoring, record keeping and analysing the various reasons for machine stoppages and corresponding production loss

- Began identifying category wise reasons for stoppages such as manual fault, material fault or machine fault and ways to minimize them

- Introduced production based incentive schemes to motivate the workforce. Prepared process control check list and SOP for each process

MATERIAL

- After trust levels of the workforce began improving, convinced the management to release funds for spare parts in phases

- Convinced management to clear restrictions for buying raw material from the group company due to quality issues and higher prices of raw materials supplied by them (cotton yarn and polyester texturized yarn) Comparative quality parameters, market prices and economy between the group company and market rates and quality was presented to the board. Began smart buying (better quality of cotton yarn from open market) and started importing polyester texturised yarn in much competitive prices

- Changed the procurement policy. Stopped bulk buying of dyes and chemicals due to price benefits. Instead negotiated prices and quantity for the year and asked the local supplier to supply as per the plant's needs with credit facilities of 90-120 days without interest. No excess inventory in the plant. Supplier was assured about his market and we got good prices

- Suppliers began regaining their trust and started supplying on credit due to assured payments

- Slowly, credit for 100% purchases was increased upto 90-120 days

- During off season, procured 100 MT of one variety of cotton yarn in a very competitive price on credit which increased sales margin during season

FOCUSED ON PRODUCT QUALITY AND ENHANCING PRODUCTIVITY LEVELS

- Introduced production-based incentive schemes to motivate the workforce

- Prepared process control checklist and SOP for each process

- Strengthened online quality assurance systems in each department from raw material to finished product

- Strengthened system for testing of raw materials, dyes, chemicals, lubricants, etc. to ensure right quantity and quality

- Instant corrective actions to address market complaints

- Increased interaction with market to understand their issues

- Focused on product development

- All joint efforts resulted in improvement in productivity and quality of product

- After improved working, weaving machines allocation increased from - 6 to 8 & 8 to 10

- In view of high demand in the market in 2001-02, decided to procure three lacs meter grey fabric to increase sales volume and fast delivery

- Introduced a fabric-dyeing machine to increase fabric-dyeing capacity

- Started a new line of business of dyed swing thread and synthetic hairs for local market
- Also diversified the business into consumable plastic items of high demand (on cash and carry).

AGGRESSIVE SALES AND MARKETING

- Frequent market visits in different areas; interacted with dealers to understand their issues
- Collected samples of fabrics of all competitors, compared their qualities and identified shortcomings of products and packaging
- Involved concerned department heads to understand quality issues and the way to address them
- Improved packing materials for finished products
- Regular comparison of product with competitors & continuous improvement
- Frequent meetings with customers and implementation of their suggestions
- Focused on branding
- Assured the dealers with respect to consistent improvement in quality & faster delivery
- After few months invited the dealers once again in the plant and shared the steps taken by the management to address the situation, displayed improved products and packaging, and gave an assurance that consistency will be maintained
- Reviewed sales policy, introduced performance-based incentive schemes, etc. in consultation with dealers
- Introduced medium and small size, new customers instead of depending on few numbers of big size customers
- Close monitoring of customer performance

After addressing quality issues, the trust level of dealers improved. This conveyed a positive message to all stakeholders. It not only increased confidence levels of the management but also of suppliers and energized the entire workforce.

IMPLEMENTATION OF INNOVATIVE IDEAS

- Due to small but regular investment in spare parts, motivational programs, interaction with dealers, implementation of process control, quality assurance, and preventive maintenance schemes, there was a significant cultural change in the company. The challenge was to achieve consistency in quality of products, improve the production, and ensure timely delivery.
- There was a limited production capacity for the most preferred quality - printed material. Set of machines included Mercerizer, Dryer, Print Washers, Printing Machines, Curing Machine, Stenter and Calendering Machine. Main limitation was Mercerizer, Print washer and Curing capacity.
- Since there was a limitation towards adding new machines, out of the box ideas were adopted to almost double their production by adopting non-conventional methods.
- **Mercerizer:** Increased capacity of mercerizing by super imposed feeding (double fabric feeding), reduced 15% speed of machine to maintain caustic soda treatment to ensure proper dye pick up. Hence its capacity was increased by 85%
- **Print Washer:** An additional water pump was installed in one of the tanks to increase water force of spray (idea adopted from a reputed machine manufacturer). Although water consumption increased, 90% printed fabrics except black and blue, passed in one passage as a result of which, Print Washer capacity almost doubled

- **Curing Machine:** Additional guide rollers for fabric passage were fitted inside the chamber between the existing rollers. This increased no. of loops formation of fabric and created almost double capacity of curing

- Stenters and Printing machines were overhauled. Thus, the capacity for printed fabric was almost doubled without adding machines, saving huge capital investment, repayment burden, interest and depreciation.

RATIONALISATION

- After getting fully settled in about two years' time the trust level of workforce was improved, as a result of which I was able to initiate interactions with union leaders and concerned workers to convince them to implement rationalisation schemes in the best interest of all the stake holders

- Reduced 220 workers and staff member by stopping 220 slow speed conventional looms

- Removing 220 workers was a SERIOUS CHALLENGE!

- However, the exercise was conducted smoothly, with amicable settlements with union leaders and workers after a series of meetings

- 220 old looms were replaced with 22 high speed wider looms (three width) (A good amount of Capex was realised through the sale of looms scraps)

COST REDUCTION DRIVE

- Various steps were taken towards energy savings by critical monitoring of generation, transmission losses and consumption

- Minimised waste generation - shift wise and section-wise monitoring of waste generation

- Reduction of defective production - better RM, strict process control and online quality assurance

- Higher productivity levels - better maintenance, minimum machine stoppages, better quality of raw material, etc.

- Hands reduction – Because 220 Looms stopped

- Smart and transparent procurement policy

- Minimum inventory of raw material, WIP and finished goods

- Took a bold decision and converted 40 MT of mixed remnant yarn (dead inventory) in a consumable quality and dyed in black shade, which was sold in season

- After improved working, weaving machines allocation increased from - 6 to 8 & 8 to 10

- Rejection and rework percentage minimised as a result of better quality of RM, better machine conditions, strict process control and overall improvement in work culture.

OPERATION COMPARATIVE RESULTS

FABRIC SALES AND GREY STOCK

Lacs Mtrs	March 98	March 99	March 2000	March 2001	March 2002
Fabric Stock	16.7	31.28	17.61	8.9	12.97
Average Sales /month	6.96	6.78	8.6	9.9	11.81
Stock Equivalent Months	2.4	4.6	2.0	0,9	1.0

WEAVING PRODUCTION (Since March 2000- 220 old looms were stopped) 2002: Weekend working suspended due to bad market to avoid high stock.

Annual	1997	1998	1999	2000	2001	2002
Weaving Production Lacs Mtrs	97.7	102	105	92.8	133	144

PRINTING PRODUCTION (Ave/Month -Highest in a month- 12.6 Lacs, Mtrs)

Ave /Month	1996	1997	1998	1999	2000	2001	2002
Lacs Mtrs	5.22	5.5	4.63	5.9	7.77	9.63	9.06

FINISHING PRODUCTION (Highest in a month -17 Lacs, Mtrs)

Production Lacs, Mtrs	1998	1999	2000	2001	2002
Ave /Month	7.45	9.06	9.59	12.3	13.1

SALES

Sales for the Year	1998	1999	2000	2001	2002
Mtrs Lacs	83.52	81.36	103	119	142
Value M N =	62	60	75	94	127

INCREASED INPUT COST DURING THE PERIOD

- Wages Increased @   10% CAGR

	2001	2002
LPFO Rates/Ltrs	11.50	13.18
Power Rates/Unit	4.74	8.50
Interest Mn Naira	42.00	69.00
Depreciation Mn N	27.50	38.00

SIUATION AFTER FOUR YEARS

- Bad debt - Nil
- Receivable Days < 30 Days (Sales outstanding)
- Payable days > 90 days (Suppliers outstanding)
- Competition among vendors, hence increased negotiating power and credit facilities
- Introduction of small customers with no credit (cash customers)
- Performance Incentives to staff & workers
- Happy and motivated work force with high moral
- Dedicated teamwork
- Focused and result oriented team

- Enhanced brand image of the products and company

- High trust level of trade due to improved quality, timely delivery and product developments

- High trust level of suppliers due to assured payments. Emerged as one of the most profitable companies in the field

CASE STUDY II

STEEP DECLINE AND CLOSURE OF A GIANT SIZE PROFITABLE TEXTILE COMPANY

Size: A large size fully integrated (composite) textile company, approx. 70 years of age in 1970

Product range: 100% Cotton poplin, markin, drill, voil, dyed, and printed dress material, etc. PV and PC plain and dobby design shirting, PV suitings, soddy cotton blankets

Yarn Count Range: 10'S Ne TO 2/80 Ne, 75 -150 denier polyester filament yarn

Quality of Products: Product quality was good and well accepted in the market. PC and PV shirting quality was excellent. It was a classic example of producing dobby designs and leno shirtings upto 16 Shaft in conventional shuttle looms, with loose reed motion, side weft fork and without warp stop motion.

Quality assurance, process control, and preventive maintenance system were well in place.

Volume: More than 1 lac Mtrs /Day

Machineries Setup: The Company had a blend of conventional and modern machineries at important places in spinning, weaving preparatory and process house

Market Goodwill: Very Good

Financial Health: Sound

Management: The Chairman was deeply rooted, had a strong control, and was involved in the business. Unfortunately, due to serious illness his involvement from the business reduced slowly. Since any immediate family members, including his son were not interested in day-to-day affairs of the business, powers were transferred to the MD who had been associated with the company for a long time and had almost family-like relations with the Chairman. With time, the MD introduced his son as the President of the company who had built an image as a strong administrator, like an autocrat.

Sales Policy: This was the age of a 'Seller's Market' and there weren't too many challenges related to sales or marketing. The company had a monopoly sales agent and entire volume was sold through him to his dealers throughout the country. Company was not dealing with multiple numbers of dealers directly. Hence, essentially there were no major efforts required for sales and revenue collection.

Diversification of Products:

- During the 70s, there was a new trend of synthetic fabrics production in central and western Indian textile companies. The concerned company also accepted this challenge and following market trends, decided to switch over from 100% cotton products to synthetic products. Technicians and managers were trained, appropriate and experienced senior technicians were hired and introduced to share technical know-how and build a 'synthetic' culture.

- With vigorous and sincere efforts of the team, the company was successful in introducing quality products in the synthetic field too and was fully prepared to compete with the markets.

- The company achieved its share and established an identity in the synthetic markets.

- Business was good and the company was running smoothly except shortage of power (dependence on generator sets), irregular supply of cotton (previously), on and off union and labour related issues as this was a common phenomenon in those days in Indian textile mills.

Technological Advancement:

- Meanwhile, there were revolutionary changes in the global textile industries. Technologically advanced and high speed spinning machines, non-conventional shuttle-less Projectile and Air-jet and Rapier looms, Jet Dyeing, Rotary Printing, and advanced finishing machineries were introduced in the market.

- The Indian textile industry was also looking towards modernisation and technological upgradations.

- The company also took a bold step in adopting these modern technologies and introduced imported high speed shuttle-less Projectile and Air-jet looms during the late eighties.

Major Mistakes of Management:

- Huge capital investments in imported shuttle-less weaving machineries

- No new product mix was planned to produce in high-speed capital intensive machines

- Fully depreciated and interest-free machines were replaced with high cost machines

- High interest and depreciation cost

- Heavy burden of interest cost and repayment to the banks

- Due to heavy burden of interest and depreciation cost, the new technology proved unviable for the present product mix

- Union's adamant attitude opposed implementation of workforce rationalisation, which was essential due to automation and management failed to handle the situation

- Since product mix remained same, no increase in sales realisation

- No market research and advance preparation by sales and marketing team to develop high value new product mix

- Present sales policy did not allow introduction of new dealers or for exploration of new markets

- In view of modern technology, management decided to recruit over-qualified machine operators who became a liability over a period of time (proved to be a wrong decision)

- Overheads increased but realisation did not increase proportionately

- As a result, company started heading towards a financial crisis

- Meanwhile, few years ago, two members of third generation from the Chairman's family, entered the management as Chairman and Co-chairperson and started penetrating the system. Being qualified and belonging to a new generation, they believed in a system driven management, and started scrutinizing all the functions of the business. Post audit, loopholes were detected at various places including financial discrepancies

- There was a question mark on the credibility of the management. This led to ego clashes and created an uncomfortable situation for the established President who had built a strong hold in the business over a long period and had also in-parallell built his own business with the support of the present company's infrastructures

- Ultimately, differences escalated, situation deteriorated and ended with the separation of the President

- However, allowing the President to quit, who had deep roots and a strong hold in the business, especially in sales and marketing, before complete preparation - was a hasty and immature way of handling the situation

- The credibility of the monopoly sales agent was also questioned, and as a result, he too lost interest and decided to quit, giving a huge financial hit by default in dues from the markets

- The decision of the monopoly sales agent was disastrous for the company

- In absence of advance planning for developing alternative sales channels, it was a serious threat to the company

- The company was not in a position to bear the loss of a monopoly sales agent and failed to recover

- Shortly, the company reached a serious liquidity crisis and started defaulting in repayments to the banks and financial institutions

- In late 90s, the company was declared closed.

LESSONS TO LEARN

- Board of directors should be vigilant, keep a close eye on the functioning of the management, and provide directives to the management regarding its policies.

- The management should have proper systems in place along with transparency in every corner of the business.

- Experts of the field and skilled professionals must be involved in the strategic planning and policy decisions of the company.

- There must be internal and external auditors monitoring day-to-day operations of the company who can inform the board of directors about the current situation of the company, and raise alarms on time in case any serious issues are detected.

- The company should have a transparent sales and marketing policy and should not be dependent on few large size agents/dealers for business and revenue collections. All sales channels should be explored including wholesale, retailing, EBO's, MBO's, franchise, institutional, export, e-business, etc.

- There should be critical and regular monitoring of dealer-wise, area-wise and quality-wise sales, revenue collection, and credit balances. This should be analysed for evaluating performances of individual dealers.

- A SWAT analysis exercise, break-even analysis, and pay-back period calculations must be conducted and discussed with all functional heads before deciding on any expansion plans, modernisation or technological upgradations in the company.

- Since the present company was more than eight decades old, there were several union issues, high overheads, old working culture, and overall adamant attitude from shop floor workers to the top management and even upto sales agents.

- It would have been a wiser decision to establish a new set up of shuttle-less looms, away from the existing company as an independent identity, in an interior location. Thus, there would be freedom to recruit a new workforce as per new conditions, train them and develop an entirely new culture along with an independent profit center head accountable for its operations. After permanent closure of the mother company, this new set up would still be available for revival with different options (outsourced yarn, process house and new dealers).

- The opinion and consent of all stakeholders including union leaders, technicians, sales and marketing team and dealers are important for such key decisions. The company is liable for taking the interest of all the stakeholders into account and hence, an individual like President, COO or CEO of the company should not be disproportionately empowered without any accountability.

- Directly or indirectly, every business is done in the country by using public resources in terms of finance from banks, subsidies from government, land and infrastructure from government or through a private person, indigenous work force, energy, etc. Therefore, it is the moral responsibility of a management to be liable not only to the company but also to the society as a whole.

- Any person in a key position must not be allowed to satisfy their ego at the cost of the company's interest.

SUGGESTIONS TO IMPROVE OVERALL PERFORMANCE OF INDIAN TEXTILE INDUSTRIES

In the author's opinion, to achieve this target, the Indian textile & apparel industry and Ministry of Textiles will need to work closely and in conjunction, with focus on the following:

1. The Textile & Apparel industry is divided into many sectors and often, there is a clash of interest among them. For example,

   ■ **Cotton merchants,** who buy cotton from the farmers or ginners, wish to free the export of cotton and discourage the import of cotton or raise demands to increase import duty.

   ■ **Spinning companies,** in yarn business, oppose the free export of cotton and import of yarn or demand reduction in yarn import duty. They demand either a ban on cotton export or imposing higher import duty on yarn import.

   ■ **Synthetic Filament/Texturized yarn manufacturers** demand imposition of high import duty on synthetics yarn.

   ■ **Fabric Manufacturers** demand reduction of import duty on yarn to check the prices of local spinners. They also oppose the importation of fabrics or reduction in import duty and demand to increase it. While the demand of Garment manufacturers remains reverse of it.

   ■ **All above players** demand for export incentives from the government to promote export.

   There are different associations for these sectors – each representing its members interests independently.

   ### Sales Policy of Above Businesses In Indigenous Markets

   Cotton Fibres: Rates are uniform, fixed by CCI and traders based on crops, demand, and its quality.

   Payment terms: Normally cash against sale or 15-30 days credit based on size and credibility of the buyers

   Cotton and Blended yarn: Rates fixed and declared by the spinning companies, remains almost uniform based on count, blends, and its quality parameters against the international standards. However, high value superfine yarn from imported fibres and speciality yarn prices are decided by individual manufacturers based on demand and supply.

   Synthetic Fibres and Yarn: Rates are declared by manufacturers based on international price of crude oil

   Payment Terms: Either advance payment or within 15-30 days based on size and credibility of business

   Fabrics: Yarn once converted into the fabrics changes its form in numbers of varieties

   Type of Fabrics: 100% Cotton, synthetics, woollen, silk, linen, or blended fabrics in plain, dobby, or jacquards structure, white, dyed or printed fabrics in course, medium, fine and superfine range, etc. with various combinations. Further, there may be categories such as fibre dyed, melange, yarn dyed, fabric dyed, and fabrics with speciality finishes, etc.

   End use: Different sectors of society in domestic or international markets uses fabrics for a wide variety of different end uses such as consumers textiles, lower, middle, and higher segments, premium, luxury, super luxury textiles, ethnic wear, fashion wear, special textiles for work wear and uniform materials for different sectors such as armed forces, paramilitary forces and technical textiles, etc.

There are fast changing fashion and trends in the fabric business. Usually a team of creative members in different parts of the world are involved in research and development towards developing shades, designs, and styles and providing varied choices and consequently also setting new trends in the society. Obviously these expensive exercises.

Prices: Except the rates of consumer textiles, which are mostly uniform, prices of other fabrics are decided based on its creative designs, quality and end use and its brand image, etc. There are no fixed parameters to measure the value of creativity and brand value. Therefore there are wide variations in the prices of such products as they are decided based on its real cost and extra margins for its creativity, satisfaction of the users and special efforts to produce the same.

Payment terms: In domestic markets, fabric sale is normally seasonal and business is mainly on credit except in retail sectors. The fabrics are sold through regional agents in wholesale and retail. In a majority of cases, the payment terms are 30 to 90 days credit or more depending on seasons and demand in the markets. Additionally, there are always chances of bad debt, hence provisions are made in budgets and added in the costing to avoid any losses on this account.

Business Cycles

Procurement of raw material for sale and tentative days of revenue collection

	Production Cycle Days	Revenue Collection Days	Total Cycle Days
Cotton planting to harvest ready	130-160	15-30	145-190
Cotton fiber	5-10 (Preparation)	15-30	20-40
Synthetic fibres and Yarn	10-15	15	25-30
Cotton or Blended yarn	30	15	45
Yarn to fabric	15-30	60-90	75-120
Yarn to Knitted Fabric	10-15	30-90	40-105
Fabric in a fully integrated mill From fibre to fabric production (Domestic Market)	45-90	60-90	105-180
Export	45-90	30-60	75-150
Garmenting: Fabric to Garment production	15-30	60-90	75-120
Export	15-30	30-60	45-90

It is evident that there is a wide difference in the business cycle of fibre, yarn and fabric businesses and garmenting.

2. In view of long business cycles, it is essential to provide competitive resources to the industries, such as low cost finance for working capital, relaxation in various taxes and availability of low cost energy to be viable and competitive in the global market.

3. There should be attractive schemes to encourage investments in modern and fully integrated manufacturing facilities to produce designed quality with minimum cost and within shortest lead-time. Allocation of funds needs to be increased under TUFS and other schemes of governments to expedite investments.

4. To produce value added products with specialized designing and finishes. Currently Indian exports mainly cater to the low to medium price segments. There is a huge scope to cater to the fashion market with higher margins compared to discount segments.

5. Areas of major scope for the Indian textile industry are industrial work wears, inner wear, sleepwear, formal suits, technical textiles, home textiles and fashion fabrics, all of which have tremendous opportunities.

9. There is a need to restructure policies to further encourage fabric & garment production and avoid export of low margin raw & semi raw materials to discourage export of fabrics by our competitors after value addition to our own country.

10. Duty draw back/export incentive schemes must be based on value addition and should be segregated for fibers, yarn, fabric and garments as a scheme of incremental export incentive on value addition.

11. Need to facilitate areas of in-house production of technologically advanced machineries from fibre to garment. Until then all textile and clothing, machineries must be exempted from custom & excise duty.

12. To increase investment in the areas of research and development, with world class facilities for the development and production of fine and super fine cotton and wool fibres (still dependent on import), product design developments and textile & clothing manufacturing processes.

13. To focus on brand building in the domestic and global markets instead of selling goods in low margins and doing job work for international brands who outsource materials from our country, converting garments and selling apparels with high margins worldwide, including in our own country.

14. Active and effective participation of expert technocrats and professionals in policy decisions with the government as well as private entrepreneur activities.

15. Quick attention & response by government agencies for timely attention and action on the industry's issues to face global challenges.

16. To develop a mechanism of 'Watch on Industries' for continuous upgradation/ modernization instead of siphoning out profits and having a tendency towards converting/declaring a company sick/insolvent.

17. Appropriate norms may be set up to allocate requisite land(s) to new industries in synchronization with the capacity installed. Extra, un-utilized land by the existing industries must be surrendered if not used within few specified years, say 5 to 10 years .

18. It must be the last priority to convert agricultural land for industrial use. Land is the most important asset of the country. There is a need for economical and optimum utilization of available land. Government may decide on a formula for allocation of land according to the type and size of the industry to be installed. A possibility may be explored for construction of multi-layered buildings for small/medium industries in new industrial parks, especially for knitting, embroidery, digital printing, narrow weaving looms and garmenting machineries, etc.

19. A promoter uses national resources like land, finance from bank (public money), manpower, energy etc. Hence, top management and chairperson's earning may be restricted to reasonable levels and their attitude towards workforce needs to be more responsive and constructive.

20. Continuous drives of skill development programs for developing human assets, not only to retain them but also to keep them happy - by creating belongingness, implementing performance-based incentives schemes and treating them as a profit partner.

21. Labour unions also need to be educated and motivated equally for overall improvement in the work culture. The efficiency of workers needs to be improved to globally competitive and accountability should be fixed. A message should be conveyed that both management and labourers are responsible equally to run the industry.

12

Statutory Compliance and Auditing

'Statutory compliance' refers to fulfilment of all obligations by an individual, company, or firm, as are prescribed under different laws that are applicable to them (including all acts such as the Companies Act, rules, regulations, notifications like CST/GST, labour laws, and laws related to employees provident fund, ESI, etc.)

'Audit' is the process of examining such compliances and relevant documents thoroughly and ascertaining whether the entity has fulfilled all applicable tax and legal compliances. It is an independent inspection of the financial and legal information of any organization, whether profit-oriented or not profit-oriented, irrespective of its legal form, status or size. The statutory auditor appointed by a company in its annual general meeting carries out the audit of a company.

AUDITOR: An auditor is an independent professional who possesses the requisite qualifications to conduct an audit. The role of a financial auditor in accounting is to evaluate the validity and reliability of a company or an organization's financial statements and further to evaluate whether all compliances under various laws have been met by a company and the relevant documents are in order.

THE OBJECTIVES OF AN AUDIT: The objective of a financial audit is to verify the books of accounts of a business by an independent professional to ensure that the accounting has been carried out as per the relevant regulatory requirements and to check the veracity of transactions and provide an objective analysis on whether the books of accounts show a true and fair view of financial transactions by the business.

A limited company has to close its accounts every financial year and prepare financial statements as per the books of accounts depicting a true and fair view of the affairs of the company. Then, the financial statements shall be audited by the statutory auditor and have to be placed before the board members for approval.

Every company has to get its accounts audited by its statutory auditor irrespective of its size and turnover and file the same with the Registrar of Companies.

APPOINTEMENT OF AUDITOR: As per the prescibed rules, it is mandatory for every company to appoint an auditor (who can either be an individual or a firm) at its first annual general meeting. The first auditor shall be appointed within 30 days from the date of incorporation of the company by the board of directors. If the board fails to appoint one, the members of the company have to appoint the first auditor at an extraordinary general meeting within 120 days from the date of incorporation.

TENURE OF AUDITOR: A company can appoint an individual as an auditor for more than one term of five consecutive years and an audit firm as an auditor for more than two terms of five consecutive years.

QUALIFICATION AND ELIGIBILITY OF AN AUDITOR: Only a chartered accountant, who holds a valid certificate of practice, authorised under the Chartered Accountants Act, can become an auditor of the company. An employee of a company or a person whose relative is a director or the employee of the company or a person who has been convicted by a court of an offence involving fraud will be disqualified as an auditor.

REMOVAL AND CHANGE OF AUDITOR: The auditor appointed may be removed from his office before the expiry of his term only by a special resolution of the company.

RESIGNATION OF AUDITOR: If an auditor resigns, he needs to submit the statement in a prescribed form to the company and the Registrar of Companies within thirty days from the date of resignation.

POWERS AND DUTIES OF AUDITORS AND AUDITING STANDARDS: Every auditor of a company shall have the right to access the book of accounts and vouchers of the company at all times. The auditor shall be entitled raise requisitions from officers of the company demanding such information and explanations as he may consider necessary for performance of his duties.

AUDITOR REPORTS: The auditor shall make a report for members of the company on accounts examined by him with every financial statement. The auditor report shall also state:

- Whether he has sought and obtained all the necessary information and explanations

- Whether proper books of account have been kept

- Whether company's balance sheet and profit and loss statements are in agreement with books of accounts and returns

AUDITOR TO SIGN AUDITOR REPORTS: The statutory auditor shall sign the auditor's report or certify financial statements and other documents as required in accordance with the provisions of law.

COMPANY ANNUAL FILING: Auditing is a continuous process, which involves examining various compliances, risks and accounting practices. At the end of a financial year (31st March, the company shall prepare its financial statements as per the books of accounts maintained and the board of directors shall approve it for the audit process.

The auditor shall then make a detailed report disclosing the accuracy of transactions whether reflected accurately or not.

The Company has to conduct its Annual General Meeting (AGM) within the specified time limit as per the Companies Act, from the closure of a financial year, and all members have to be notified a certain prescribed number of days before such AGM. A copy of the Director's Report and audited financial statement along with the Auditor's Report has to be distributed amongst the members along with the notice of AGM. So, the Auditor's Report must be made available before sending the notice.

The auditor of the company shall sign the Auditor's Report and can sign or certify any other document of the company. Any financial transactions or matters, which have any adverse effect on the functioning of the company mentioned in the Auditor's Report shall be read before the company in the general meeting and shall be open to inspection by any member of the company.

There are certain fixed guidelines related to auditors. For examples:

- Right to remuneration

- Consent of auditor

- Auditing standard

WINDING UP: As per specified rules, at the time of voluntary winding up of a company it is mandatory for an auditor to attach a copy of the audits of the company prepared by him.

STATUTORY REGULATIONS: All companies operating in India have to ensure compliance with certain categories of statutory regulations and mandates. For example,

- Employment laws

- Tax laws

- Labour laws

- Competition and antitrust laws

- Environmental regulations

- Financial and corporate laws

- Industry-specific laws

- Advertising laws

- Data privacy laws
- Intellectual property laws
- Foreign exchange laws, etc.

Thus, all Indian companies need professionals who can ensure statutory compliance with the ever-changing environment of regulations in the country.

EXAMPLES OF STATUTORY COMPLIANCES

- Payment of Wages Act ,1936
- Minimum Wages Act, 1948
- The Payment of Bonus Act, 1965
- Tax Deduction at Source (TDS)
- Professional Tax
- Maternity Benefit Act, 1961
- Equal Remuneration Act,1976
- Shops & Establishments Act
- The Employees' State Insurance Act (ESIC)
- The Payment of Gratuity Act
- Employees Provident Fund (PF) Act, 1952
- Labour Welfare Fund Act (LFW), 1965
- Sexual Harassment of Women at Workplace (Prevention, Prohibition & Redressal) Act, 2013
- Industrial Establishment (N&FH) Act, 1973
- Interstate Migrant Workmen (Regulation of Employment and Conditions of Services) Act, 1979
- Child Labour (Prohibition and Regulation) Act, 1986
- Other laws providing for annual leave and holidays, compensation, fair treatment, health and safety, trade union etc.

IMPORTANCE OF STATUTORY COMPLIANCE FOR PAYROLL

If a company in India does not comply with the extant central and state labour laws, it can land into serious trouble, which can make its growth stagnant. Depending on the seriousness of defaults, following are some common risks of statutory non-compliance:

- Work stoppage
- Loss of the company's integrity, reputation, and goodwill
- Cancellation and suspension of operational licenses
- Loss of faith in the company's stakeholders and investors
- Imposition of heavy fines
- Loss of customer loyalty
- Civil and criminal liabilities
- Negative impact on the company's productivity
- Penal liabilities against the Directors or officers of the company
- Withdrawal of fiscal benefits
- A shutdown of the company, in extreme cases

Also, statutory compliance must be looked at positively - as critical checks in place by the government for a smooth, equitable, transparent and safe work environment for all involved.

Therefore, since all labour and taxation laws of India prescribe a wide range of statutory compliances, in order to avoid extreme situations like fines, penalties or even complete shutdown, a company should spend a good amount of resources to ensure statutory compliance. The rules keep evolving from time to time, and the laws are dynamic. The concerned departments must understand all the current labour and taxation laws of India to stay updated about any changes they need to incorporate within their organizations. A good compliance officer can help a company manage a demanding regulatory environment and avoid non-compliance risks and penalties.

EXAMPLE OF STATUTORY COMPLIANCE FORMATS:

Monthly GST Compliance Chart (Subject to change time to time as per Govt notifications									Remarks
Name of Returns	Particulars	April		May		June			
		Due Date	Complied on	Due Date	Complied on	Due Date	Complied on		
GSTR - 3B	Summary of Sales, Purchase & Expenditure								
GSTR - 1	Outward supplies made by taxpayer								
GSTR - 2	Inward supplies received by a taxpayer								
GSTR - 3	Monthly return								
GSTR - 7	Return for Tax Deducted at Source								
GSTR - 9	Registered Taxable Person								
TRANS - 1	Goods held in stock as on 30.06.17 (Transitional Credit)								
RFD - 11	LUT (for Export) (Renewal Once in year)								
ITC 04	Goods sent for job work outside factory. (Quarterly)								

MONTHLY INCOME TAX COMPLIANCE REPORT

Particulars		Payment of TDS/ TCS in Previous month	Quarterly TDS Return (Other than Govt.) Examples: Form Nos 24Q, 26Q, 27EQ, 27Q	Issue of TDS Certificates Form : 16/ 16A	Filing of Income Tax Return	Advance Tax Payment	Annual Information Return
January	Due Date						
	Complied on						
February	Due Date						
	Complied on						
March	Due Date						
	Complied on						
April	Due Date						
	Complied on						

MONTHLY ESIC / PF / PROFESSIONAL TAX COMPLIANCE REPORT

Particulars		Online ESIC Payment for previous month	Online PF Payment for previous month	Professional tax for previous month to be deposited with Form No--	Professional Tax Annual Renewal
January	Due Date				
	Complied on				
February	Due Date				
	Complied on				
March	Due Date				
	Complied on				
April	Due Date				
	Complied on				

FACTORY ACTS AND STATUTORY COMPLIANCES

There are specific rules and regulations prepared by governments to provide directions to manufacturing factories depending on the type and size of businesses and the same are implemented through the different laws. The objective of these laws is to monitor the factories and regulate safety, security & health issues, work environment, basic facilities and overall welfare of the employees engaged in such factories. These laws also cover the interest of employers. There are specific forms, registers, records, billboards, etc. designed and approved by the government to be maintained and followed by factories as statutory requirements as per the respective laws. These laws may be amended by the government from time to time in the best interest of employees and employers.

Sometimes factories do not follow these laws strictly and try to compromise or dilute their compliance. Such attitude towards statutory compliances not only invites legal complications but also negatively affects the overall performance and image of the factories or companies.

Timely compliance of prescribed laws directly or indirectly also provides additional benefits to the factories such as improved business opportunities, higher employee retention, prevention of legal troubles, Improvement in quality, improvement in business operations, brand loyalty, boosting employee morale, improved public relations and helps in the development of good will and brand image.

It is observed that except concerned departments, other staff members in the factory, who deal the workforce directly or indirectly, are often ignorant about many important rules, regulations and factory laws. As a result of such ignorance, they invite problems for the factory management. For example, in most cases disciplinary actions against employees are initiated and recommended without proper charge and evidence. Hence, it is advisable that not only the Labour or Administration Department but all concerned in the factories should be educated about basics of extant laws and encouraged to adhere to them strictly.

Examples:

Statue/Nature of Compliance			Provision Complied or not (mention Y or N)
Payment of Wages, Register/Records to be maintained (As per Acts)			
1. Muster Roll			
2. Register of Wages			
3. Register of Deductions			
4. Register of Overtime			
5. Register of Fines			

6. Register of Advances			
7. Wage Slip			
As Per Factories Act			
1. Annual Return			
2. Half Yearly Return			
3. Factory License Renewal			
4. Stability Certificate (– years validity)			
5. Annual Register of Accident in specific Form no. __			

Minimum Wages Act
1. Abstract of minimum wages
2. Minimum rate of wages fixe
3. Annual return
4. Notices required to be displayed at work site, under minimum wages act.
Payment of Wages Act
1. Annual return - for calendar year
Payment of Bonus Act
1. Register of bonus
2. Annual return
Payment of Gratuity
1. Notice of opening
2. Declaration
3. Abstract of payment of gratuity under the relevant Act
Other Employee related laws
1. Returns under Maternity Benefits Act
2. Employment exchange –
3. Employment exchange –
4. Half yearly return – Labour welfare fund
Environment Laws
1. Pollution control board consent (water & air)
2. Pollution control board consent (H.waste)
3. Water cess return
4. Environment statement to state govt. body for last FY
5. Return regarding disposal of hazardous waste to the State Govt body
6. Batteries Management and handling rule
7. Return regarding disposal of e-waste

Employees State Insurance Act	Yes/No
1. Issue of ESI Cards to workers	
2. Register of employees (specified form No.–)	
3. Timely report of accidents (specified form No.–))	
4. Register of accidents (specified form No.–))	
The Contract Labour (Regulation & Abolition) Act	
1. Applications to be made to registering office	

2. For registration of establishments employing contract labour	
3. For licence to contractor - specified form No. –	
4. Half Yealy return – (contractor)	
5. Annual return – (principal employer)	
Registers/Records to be maintained	
1. Register of contractors (principal employer)	
2. Register of contract labour (contractor)	
3. Employment card to every labour (contractor)	
4. Certificate to contractor by principal employer	
5. Notice of commencement and completion of contract work	
Contractor and Principal Employer	
1. Muster roll-Specified form No.– (contractor)	
2. Register of wages – (contractor)	
3. Register of deductions – (contractor)	
4. Register of overtime – (contractor)	
5. Register of fines – (contractor)	
6. Register of advances – (contractor)	
7. Wage slip – (contractor)	

STATUTORY AMENITIES IN THE FACTORY

Statutory amenities are to be provided in the factories as a compliance of welfare measures **(Applied As per Acts)**

Example:

1	Drinking water points			Yes/No
2	Cooled supply of water			
3	Latrine accommodation			
4	Urinal accommodation			
5	Washing facility			
6	First aid appliances			
7	Ambulance room			
8	Rest room/shelter/lunch room			
9	Creche			
10	Canteen			
11	Occupational health centres (in respect of hazardous process industries)			

STATUTORY OFFICERS IN THE FACTORY (As per specified rules, may be amended by central or state government)

Example:

S. No.	Particular of Appointments	Nos Required	Qualificatio	Yes/No
1	Safety Officers (for factories employing more than (as per rule) workmen and notified factories)			
2	Welfare Officers (for factories employing more than (as per rule) workmen)			
3	Factory Medical Officer in respect of industries involving hazardous processes			

4	Supervisors (in factories involving hazardous processes)			
5	First Aid trained personnel			
6	Qualified nursing staff (wherever creche facilities are provided)			

REGISTERS MAINTAINED BY MANAGEMENT UNDER THE FACTORIES ACT AND RULES (As per specified rules, may be amended by central or state government)

(In prescribed form

1	Muster Roll
2	Register of adult workers
3	Register of leave with wages
4	Register of accident or dangerous occurrence
5	Inspection book
6	Overtime register in respect of exempted workers and corresponding copy of the overtime slip
7	Health register in respect of factories involving hazardous processes and dangerous operations

STATUTORY NOTICES TO BE DISPLAYED BY THE FACTORY (As per specified rules, may be amended by central or state government)

STATUTORY NOTICES TO BE DISPLAYED
Notice of periods of work along with weekly holiday, name and address of the inspector and the medical office
Abstract of the Act and rules made thereunder
Relevant cautionary notices wherever required
Safety and health policy of the factory
Relevant safety posters to create awareness amongst the workers
STATUTORY RETURNS TO BE FIELD UNDER THE FACTORIES ACT (AS AMENDED FROM TIME TO TIME) WITH DUE DATES
The Factories Manager of all the registered factories are required to submit the following returns to the area inspector.
1. Annual Return
2. Half yearly return
3. Monthly accident return
STATUTORY PERIODICAL EXAMINATIONS OF PLANTS AND EQUIPMENT TO BE CARRIED OUT
Factories Act envisages provisions for periodical tests and examinations of the following equipment and plant used in the factory.
1. Lifting machine and connected tackles
2. Overhead and pendent operated cranes
3. Hoists
4. Pressure vessels and pressure plants
5. Power presses
6. Ventilation system
7. Solvent extraction plants
8. Dusts Extraction system
The result of test and examinations shall be maintained in prescribed forms issued by the competent authorities.

EXAMPLE OF AUDIT REPORT OF A COMPANY FOR ONE QUARTER

The auditing committee normally chooses certain areas at a time and scrutinizes them in depth.

The Management Audit work for 1st quarter has been conducted for the following areas:

A. Observation report on

(a) Raw material (b) Logistic management

B. Taxation

(a) Tax deduction at source (TDS) (b)Tax collected at source (TCS) c) Goods and Services Tax (GST)

C. Vouching

(a) Major Expense vouching (b) Journal Vouching (c) Purchase and sales vouching

Content of Reports

Sr No	Particulars	Annexure
1	Letter for Management	
2	Gist For Management attention	
3	Report on Raw Materials and Stores	
	Summary - Raw Materials	
	Master List of Suppliers	
	Scrutiny Of Records	
	Issue of Materials To Departments	
	Physical Verification By Management	
	Waste Generation Slip	
	Waste Stock Register	
	Issue/Dispatch of waste materials	
	Physical Verification of Raw Material Stock by Audit Team	
	Physical Verification of Waste Material Stock by Audit Team	
	Making Top Yield Report	
	Summary of Top Produced Yield	
	Detailed Stores Dead Stock Report	
4	Report on Logistics	
	Area-wise Sales	
	Turnaround Time	
	Pricing Policy	
	Other Points	
5	Taxation summary	
	Report on Taxation- TDS on Regular Contractor	
	Report on Taxation- TDS on Service Contractor	
	Report on Taxation- TDS on Professional Services	
	Report on Taxation- TDS on Rent	
	Report on Taxation- TDS on Salary	
	Report on Taxation- TDS Discrepancies	
	Report on Taxation- Tax Collected at Source	
	Report on Taxation- Sales Working	

	Report on Taxation- Purchase Working	
	Report on Taxation- GST Under Reverse Charge Mechanism	
	Vouching	
	Other Discrepancies	
6	Other Discrepancies - Scrutiny of Accounts	
	Other Discrepancies - Loans and Advances	

Detailed findings for each point submitted in an elaborated form as annexures

GUIDELINES FOR GENERAL CHECK POINTS IN THE BUSINESS

For the smooth running of any business, it is essential to adopt strong operating systems and cross verification. A company has internal and external auditors but their role is to examine the system and records after the completion of jobs. However, before this stage, it is essential for an effective management to implement online checking and verification systems for critical processes and records so that any non-compliances, malpractices, pilferage or fraud may be arrested on the spot and can prevent unexpected losses to the company.

Following are the examples for daily, weekly and monthly checkpoints:

A. DAILY VERIFICATION

1. Verify all cash and bank payments.

2. Verify all sales receipts and ensure that official receipts are issued, and Cash/Cheques are lodged in the bank without any delay. Also, verify that the bank tellers are tallying with the receipts issued. At the end of the day, confirm with the CFO as to the confirmation of actual lodgments into the Bank.

3. Verify all Journal Vouchers, Debit Notes, and Credit Notes, and ensure that the CFO/COO have approved these documents.

4. Petty Cash and Main Cash are to be verified physically, tallied on a daily basis, and signed by both the Cashier and Internal Auditor. Any variance detected, should be brought to the notice of the CFO/COO.

5. Verify all invoices, gate passes, and delivery waybills, and ensure that proper system for delivering goods has been followed and invoice references are available on all delivery waybills.

6. Verify and ensure that all deliveries to Customers/Distributors/Branches are made through registered transporters with whom the Company has a written agreement in place. Also ensure that Vehicle particulars including Registration Certificate, Insurance Certificate, Driver's License, etc. are duly verified by the unit and photocopies are taken before goods are delivered.

7. Verify the receipts of goods delivered to bonafide Customers/Branches (in cases of Stock Transfer).

8. Verify that invoices are being raised for all services provided and ensure that no service is provided free of cost (To Sister Company).

9. Verify all deliveries leaving the factory for Port for stuffing into containers for exports.

10. Ensure that all export invoices are issued and they conform with the quantity dispatched to the Port and in conformity with L/Cs opened by overseas customers. In case of open account transactions, check the prices and quantity as per existing contract.

11. Verify if all the purchases have been made with proper procedures, approvals and ensure that indents are raised for all purchases and duly taken to stock (as per approved Procurement Policy).

B. WEEKLY VERIFICTION

1. Verify and review all returnable and non-returnable gate passes and ensure that goods sent out under returnable gate passes are brought back or recorded for regular monitoring.

2. Verify receipt of all export proceeds and its utilization and the rates at which export proceeds are being sold.

3. Verify the reconciliation of all export sales and Overseas Debtors and report any delays in realization of export sales to the CFO/COO.

4. Travel tickets are purchased by approved agent and supported by approved procedure.

C. MONTHLY VERIFICTION

1. Verify attendance and that the wage payment is being made as per the attendance evidenced from the clocking machine. Daily attendance to be verified and the timings checked. In case of any variations, appropriate authority must certify and check the variances and approve the payment accordingly.

2. Overtime payments to be checked and ensured that approvals have been taken and calculations are done based on the actual hours worked as per clocking cards.

3. Verify approved credit limits for customers and any deviations to be brought to the notice of the concerned Business Head.

4. Conduct RM, WIP and finished goods and stores physical stock taking exercise and reconciliation at the end of the month and financial year. Any variations to be reported with justification by concerned department heads to CFO.

5. Verify the quantities of Scrap/Waste generated and report any variance between the scrap generated and sold, and adherence to scrap disposal procedure.

6. Verify and get the CFO's confirmation on all bank reconciliation statements. Any unknown debits or credits should be immediately brought to the knowledge of the CFO.

7. Verify interest and other charges like remittance charges, Forex commissions and L/C opening and negotiating charges charged by all Banks. Verify these charges against the agreement with the Bank and report any abnormal charges to the CFO.

8. Verify if monthly reconciliations are done with the Group Units.

9. Verify monthly report on sale of waste and scraps.

10. Review of staff advances report and recovery status.

11. Report of all goods for which L/Cs are opened and any shipments delayed or any other abnormal situation on a monthly basis.

12. Verify if the consignments are being cleared on time, if not, report the days for clearance for each consignment and expenditure on demurrage.

13. Verify if all capital expenditure incurred is as per approval.

14. Ensure that Assets Register is updated immediately on purchase and payment is effected.

15. Verify and ensure that all statutory payments are made on time and report compliance/default on a monthly basis giving details of the amount paid/outstanding.

16. Verify and ensure that all telephone bills and vehicles' expenses are received and paid as per the current policy.

17. Ensure that reconciliation of accounts with overseas companies (if any) is done once every quarter.

18. Physical verification and reconciliation of all the Fixed Assets in the Fixed Assets Register on a yearly basis including vehicles, furniture and fittings, machineries and all other assets.

GENERAL CHECK LIST FOR SALES AND FINANCE

ITEMS	REMARK
1. Cash and Bank Payments a) CPV issued during the week CPV No. _____ to CPV No._____ b) BPV issued during the week BPV No. _____ to BPV No._____ c) Details of CPV/BPV issued during the week but not pre-audited d) Ensure that the entries are booked under proper heads of accounts and fall under the right accounting code as per the coding given in the accounting package. Instances of errors observed and corrections recommended by the audit to be noted for audit records. Recommendations should be in a separate sheet of paper attached to the payment voucher. Also mention the cases where the management has not made the necessary corrections even after being pointed out by the audit. e) Report the cases where payment for purchases is not as per the terms, rate and quantity mentioned in the LPO f) The accepted quantity as per the GRN g) Auditor's observations on any specific Payment Voucher * (CPV-Cash purchase voucher, BPV-Bank Payment voucher, LPO-Local purchase order)	
2. Cash/Cheque receipts and Lodgements Ensure computerized receipts are issued for all cash/cheque receipts. Cash Receipts issued during the week CR. Nos_____ Bank Receipts issued during the week BR. Nos_____ Deviations from daily lodgement of cash/cheque into bank Re-confirmation from CFO as to the actual lodgements to bank Auditor's observations on any specific receipts/lodgements *(CR- Cash receipt, BR-Bank receipt)	
3. Journal Vouchers, Debit Notes and Credit notes Ensure computerized JV, DN and CN are Issued in all the cases and are approved by a proper Authority. a) Journal Vouchers issued during the week JV Nos_____ b) Debit notes issued during the week DN Nos_____ c) Credit notes issued during the week CN Nos_____ d) Auditor's observations on any of the above document *(JV-Journal Voucher, DN-Debit note, CN-Credit note)	
4. Daily physical verification of cash a) Exceptions to daily physical verification of cash, if any b) Difference found on physical verification of cash, if any	
5. Invoices, Gate Passes and Way Bills a) Gate passes issued during the week No._____ to No._____ b) Invoices issued during the week No._____ to No._____ c) Details of all export sales invoices (date, customer, quantity, quality, rate, value, etc) d) Whether invoices have been raised for all sales, deliveries including export sales with reference of the waybill on the invoice. Exception, if any, to be reported e) Whether invoices have been raised for the entire quantity of material as shown in the waybill. Any short invoicing to be reported separately. f) Whether the price reckoned for invoicing is as per approved priced list (if no price list is available, please state so and also state the source from which the rates have been confirmed g) Ensure PCH/Marketing manager/CFO approves all the invoices. Deviations to be reported, if any. h) Ensure invoices are raised for all the services provided by any group company	
6. Sales Deliveries a) Whether there is a written agreement with the transporter b) The company retains photocopies of vehicle particulars and driver's particulars. c) Check the quantity leaving the factory for domestic market or to port for export d) Waybill Nos. quantity and quality of goods dispatched for domestic market or to port for export e) Auditor's observations on deviations, if any	

7. **Confirmation of receipt of goods at the destination** a) To confirm that the goods have reached the destination b) Auditor's observations on non-confirmation, abnormal delays etc to be reported	
8. **Purchases** a) LPO issued during the week LPO No.to LPO No._____ b) Whether proper procedures are complied with before LPO is raised c) Details of all LPO issued with approval from appropriate authority d) Details of materials received for which GRN not prepared e) Details of payments made for purchases for which there is no LPO or GRN f) Details of non-recording of purchases in stock records g) Whether purchase bookings have done for all GRN generated. Give a listing of GRN cut off number with cross reference to the relevant purchase voucher. # (LPO-Local purchase order, GRN-Goods received note)	
9. **Returnable and Non-Returnable gate passes** a) Abnormal transactions if any to be reported b) List of all outstanding items at the end of the week	
10. **Export proceeds** a) Details of receipts of export proceeds during the week b) Rates at which export proceeds are sold c) Utilization of export proceeds during the week	
11. **Overseas Debtors** a) Details of non-booking of export sales after the delivery b) Schedule of invoice wise break up of overseas debtors with the due date for receipts of export proceeds c) Delays in realization/repatriation of export proceeds to be listed	
12. **Overseas Travel** a) Details of payments for overseas travel, approval with appropriate authority and objective of travel and its outcome. b) Details of overseas travel expenses and verification if it is as per policy	
13. **Nominees Account** Details of abnormal transactions if any Nominees Account	
14. **IOU (I owe You)** a) List of IOU as at the close of business on all weekends b) Explanation	
15. **Any other matter of Audit relevance** a) Pending audit observations from the earlier report b) Other issues, if any	

VARIOUS LICENCES, REGISTRATION, CERTIFICATES, PERMISSION, CONSENT FOR THE COMPANY

From incorporation and throughout their existence, companies or factories have to take memberships of different organisations to update their knowledge, seek help or get current information for the benefits of their business. They also have to deal with different private agencies and organisations, various offices of local, state and central government and fulfil their obligations in terms of fees or taxes, etc. For timely compliances, it is important to prepare a list of such agencies or departments and an annual calendar - indicating amounts to be paid, timing, etc. that will provide relevant information at a glance to the management for considering in the annual budget.

Example:

RECORD AND FOLLOW CHART

Items	Authority	Certificate No.	Commencement Date	Validity/ Expiry Date	Fees Rs.	Frequency
Incorporation Certificate	Govt of India-Ministry of Corporate Affairs, Registrar of Companies					
Commencement of Business Certificate	Registrar, Joint Stock companies					
Factory Licence	Municipal Corporation					
Factory Licence	Chief Inspector of Factories, State Govt					
Consents from State Electricity Board	Environmental Engineer, State Pollution Control Board					
Central Sales Tax Registration	Sales Tax Office - Local					
Composite Textile Mills Certificat	Govt. of India, Ministry of Textiles, Mumbai					
GST Registration	Government of India					
Central Excise Registration	Dy Commissioner Central Excise					
Service Tax Registration	Superintendent of Service Tax					
Permanent A/C No	Commissioner of Income Tax					
Professional Tax Registration Company's	Municipal Corporation					
Professional Tax Registration Employees	Municipal Corporation					
Exemption from power Restriction Permission for 7 days working	Energy & Petrochemicals Dept., State Govt					
Chamber of Commerce & Industry	Local Chamber of Commerce & Industry					
DG set Registration	Office of the commissioner of Electricity					
Certificate for use of Boiler	Boiler Inspector, State Govt					
Contract Labour Act Registration	Dy Commissioner of Labour & Registering Office, Contract Labour					
Employees State Insurance Registration						
Provident Fund Registration						
Common Hazardous Waste dumping Certificate	State Govt					
Gun Licence	District Magistrate -Local					
License for working of lift	Inspector of Lifts, state Govt					
Certificate for stamping for weighing Scales	Weights & Measures Dept-Local					
Diesel Storage Permission	Jt. Chief Controller of Explosives					

Power Trading Agreement	Agreement With Pvt Supplier					
Tax Deduction Account Number (TAN)	National Securities Depository Limited (Income Tax Department)					
Director General Tender & Deposit	Ministry of Commerce & Industry dept of commerce, Directorate General of Supplies & Disposals					
Any other Government Supply Registration	Ministry of Defence - Ordnance Clothing Factory Shahjahanpur					
Trade Mark	Government of India, Registrar of Trade Marks					
Shop & Establishment No.	Inspector of State Govt, Shop & Establishment					
Vehicle Registration Certificat	RTO -Local					
Woolmark Blend Licence Agreement	Woolmark Company					
Quality System Certificate (License ISO)	Bureau of Indian Standards, Mumbai					
OEKO-TEX Certificate (for export)	Hohenstein Textile Testing Institute Gmbh & Co. KG, Germany					
Export Promotion council, Registration Cum Membership	Wool Industry Export Promotion Council, Mumbai					
Export Promotion council, Registration Cum Membership	Federation of Indian Export Organisation					
Indian Woollen Mill's Federation(for worsted/woolen companies)	Indian Woollen Mills Federation, Mumbai					
IEM	Ministry of commerce & Industry, New Delhi					
Textile Mills Serial No. for statistical return	Textile Commissioner, Ministry of Textiles, Govt of India, Mumbai					
Import Export code	Jt, Director General of Foreign Trade, Regional Office					
Textile committee Registration	Textile Commissioner, Ministry of Textiles, Govt of India, Mumbai					

13

Import and Export

Import and export of goods or services are the primary means of foreign trade. Exporting is selling of goods and services from the home country to a foreign nation whereas, importing refers to the purchase of foreign products and bringing them into one's home country.

Every nation has some strengths with certain resources, assets, and abilities. For instance, a few nations are rich in wool, cotton, petroleum products, timber, fertile soil, or valuable metals, etc. while other nations have deficiencies of these resources.

GENERAL TERMS USED IN INTERNATIONAL TRADE (IMPORT AND EXPORT)

In sea transport, it is often difficult to precisely determine the arrival and departure of a ship on account of various reasons with which ships come into contact during their voyage. At times the same can happen with air travel too. The date and time at which a ship/airplane is expected to depart from a certain sea port/airport is commonly known. Some terms related to sea/air transport are:

ETA: Expected time of arrival (at destination)

ATD: Actual time of arrival

ETD: Expected time of departure

ATD: Actual time of departure

FCL: It refers to a shipment where all the goods in a container belong to one party

LCL: It involves multiple party's goods packed together

FOB (Free on Board or Freight on Board): This means that the cost of delivering goods to the nearest port is included but the buyer is responsible for shipping expenses and all other fees associated with getting the goods to his country/address. Seller delivers the shipment to the carrier nominated by the buyer.

Ex Works: The seller makes the item available for pick up at the factory and is not responsible for the product once it leaves the factory premises. The buyer is responsible for transporting it from the factory and covers all export or import clearances as well as insurance costs.

CIF (Cost, Insurance, and Freight): In this case, the price also includes sea freight charges and insurance to deliver the goods to buyer's nearest port. From the port onwards, all expenses are borne by the buyer.

CNF (Cost & Freight , no Insurance): Similar to CIF, except insurance is not included.

DDP (Delivered duty paid): Seller has the responsibility to deliver goods at the buyer's premises or any other place as agreed. It means that from seller's premises to buyer's premises or any other agreed place, all the expenses are borne by the seller.

DAP (Delivered at place): This means seller delivers when shipment arrives at final destination, ready for unloading from the arriving mode of transport. Seller bears all the costs and risks in bringing the goods to this place.

DDU (Delivered Duty Unpaid): It means the seller is responsible for ensuring that the goods arrive safely to a destination; the buyer is responsible for import duties. By contrast, Delivered Duty Paid (DDP) indicates that the seller must cover duties, import clearance, and any taxes.

Shipper or Consignor: A shipper (also known as a consignor) is a person or a company responsible for organising and transporting goods from one point to another.

When goods are sent by the manufacturer or the producer to the buyer, the act is referred to as consignment.

Consignee: In a consignment, the receiver of the goods is termed as the consignee. A consignee is only a receiver and not the owner of the goods. The ownership is transferred only when the consignee has paid the consignor.

AD Code of Bank: An Authorised Dealer Code (AD Code is a numerical code provided by a bank with which the business has a current account. The company will need to register an AD Code at every port from where its goods are cleared by customs. At the time of customs' clearance the company's Customs House Agent (CHA) will ask you to provide the AD Code for that particular port.

Customs House Agent (CHA: Most companies who are involved in import and export business, appoint a CHA for fast and smooth customs clearance of goods. The CHA is like a legal adviser in that area and handles all the documentation and complications if any on behalf of the company.

IMPORT PROCEDURES

A. Trade Enquiry

The first stage in an import transaction is making trade enquiries. An enquiry is a written request from the intending buyer or his agent for providing information regarding the price and the terms on which the exporter will be able to supply goods. The importer should mention in the enquiry all details such as the goods required, their description, catalogue number or grade, size, weight and the quantity required. Similarly, the time and method of delivery, method of packing, terms, and conditions about payment should also be indicated. In reply to this enquiry, the importer will receive a quotation from the exporter.

The quotation contains the details as to the goods available, their quality, etc., the price at which the goods will be supplied and the terms and conditions of the sale.

B. Procurement of Import Licence and Quota

The importer consults the Export Import (EXIM Policy in place all together to know whether the merchandise is subject to import license or not. Import trade in India is subject to the Imports and Exports, (Control Act, 1950. A person or a firm cannot import goods into India without a valid import licence. An import licence may either be a general licence or specific licence. Under a general licence, goods can be imported from any country, whereas a specific or individual licence authorises one to import only from specific countries.

For issuing licences, importers are divided into three categories:

(a) Established importer,

(b) Actual users, and

(c) Registered exporters, i.e., those importing under any of the export promotion schemes.

In order to obtain an import licence, the intending importer has to make an application in the prescribed form to the licensing authority. If a person imports goods belonging to the class in which he is interested during the period prescribed for such class, he is treated as an established importer. An established importer can make an application to secure a Quota Certificate. The certificate specifies the quantity and value of goods, which the importer can import. For this, he furnishes details of the goods imported in any one year in the basic period prescribed for the goods, together with documentary evidence for the same, including a certificate from a chartered accountant in the prescribed form certifying the CIF value of the goods imported in the selected year. The CIF value includes the invoice price of the goods and the freight and insurance paid for the goods in transit. The Quota Certificate entitles the established importer to import up to the value indicated therein (called Quota) which is calculated on the basis of past imports.

If the importer is an actual user and wants to import goods for their own use in industrial manufacturing process, they have to obtain a licence through the prescribed sponsoring authority.

The sponsoring authority certifies their requirements and recommends the grant of licence. In case of small industries, they have to apply for licences through the Director of Industries of the state where the industry is located or some other authority expressly prescribed by the government.

Registered exporters importing against exports made under a scheme of export promotion and others have to obtain licence from the Chief Controller of Exports and Imports. The Government issues a list of commodities and products from time to time, which can be imported by obtaining a general permission only, known as the OGL or Open General Licence.

C. Obtaining Foreign Exchange

After obtaining the relevant licence (or quota, in case of an established importer), the importer has to make arrangements for obtaining necessary foreign exchange for making payments for imports in the currency of the exporting country.

The foreign exchange reserves in many countries are controlled by the government and are released through its central bank. In India, the Exchange Control Department of the Reserve Bank of India deals with foreign exchange. For this, the importer has to submit an application in the prescribed form along with the import licence to any exchange bank as per the provisions of Foreign Exchange Management Act, 1999.

The exchange bank endorses and forwards applications to the Exchange Control Department of the Reserve Bank of India. The Reserve Bank of India sanctions the release of foreign exchange after scrutinizing the application based on the exchange policy of the Government of India in force at the time of application.

The importer gets the necessary foreign exchange from the exchange bank concerned. It is to be noted that whereas import licence is issued for a particular period, exchange is released only for a specific transaction.

D. Placing the Indent or Order

After the initial formalities are over and the importer has obtained a licence quota and the necessary amount of foreign exchange, the next step is the placing of order. This order is known as indent. An indent is an order placed by an importer to the exporter for the supply of certain goods. It contains instructions by the importer about the quantity and quality of goods required, method of forwarding them, nature of packing, mode of payment and the price etc. An indent is usually prepared in duplicate or triplicate. The indent may be of several types like open indent, closed indent and confirmatory indent.

In an open indent, the necessary particulars of goods, price, etc. are not mentioned. The exporter is not certain on completing formalities, at his own end.

On the other hand, if complete particulars of goods, the price, the brand, and packing, shipping, insurance, etc. are mentioned clearly, it is called a closed indent.

In a confirmatory indent, an order is placed subject to a confirmation by the importer. The importer puts in an import request or indents to the exporter about the supply of merchandise. The request contains information regarding cost, quality, quantity, size, instructions about packaging, delivery shipping, and a method of payment etc.

E. Dispatching a Letter of Credit

Generally, foreign traders are not familiar with each other at the primary stage or even later and do not have control in other countries. Therefore, an exporter wants to be sure about the credit worthiness of an importer before shipping goods, to be sure that there is no risk of non payment. Usually, for this purpose, he asks importers to send a letter of credit to him. A letter of credit, popularly known as 'L/C' or 'L.C' is an undertaking by its issuer (usually importer's bank) that the bills of exchange drawn by the foreign dealer, on the importer will be honoured on presentation up to a specified amount. Once payment terms are agreed between the importer and the exporter, the importer gets the letter of credit from its banker and forwards it to the overseas provider.

The importer arranges for money in advance to pay the exporter on arrival of goods at the port. This empowers the importer to avoid penalties on imported goods lying uncleared at the port for need of payment.

F. Obtaining Necessary Documents

After dispatching a letter of credit, the importer hasn't much left to do. On receipt of the letter of credit, the exporter arranges for the shipment of goods and sends an Advice Note to the importer immediately after the shipment of goods. An Advice Note is a document sent to a purchaser of goods to inform him that goods have been dispatched. It may also indicate the probable date of ship's arrival at the port of destination.

The exporter then draws a bill of exchange on the importer for the invoice value of goods. The shipping documents such as the bill of lading, invoice, insurance policy, certificate of origin, consumer invoice etc., are also attached to the Bill of Exchange. Bill of Exchange with all these attached documents is called a Documentary Bill. Documentary Bill of Exchange is forwarded to the importer through a foreign exchange bank, which has a branch or an agent in the importer's country for collecting payment of the bill.

There are two types of Documentary Bills:

(a) D/P or D.P. (or Documents against payment) bills.

(b) D/A or D.A. (or Document against acceptance) bills.

If the bill of exchange is a D/P bill, then the documents of title of goods are delivered to the drawee (i.e. importer) after payment of the bill in full.

D/P bill may be sight bill or usance bill. In case of sight bill, the payment has to be made immediately on the presentation of the bill. However, usually a grace period of 24 hours is granted.

Usance bill is to be paid within a particular period after sight. If the bill is a D/A bill, then the documents of title of goods are released to the drawee on his acceptance of the bill and is retained by the banker till the date of maturity.

Usually, 30 to 90 days are provided for the payment of the bill.

G. Customs Formalities and Clearing of Goods

After receiving the documents of title of the goods, the importer's only concern is to take delivery of the goods after the ship arrives at the port and reaches its place of business. The importer has to comply with many formalities for taking delivery of goods. At the point when the goods arrive in the importer's nation, the concerned person conveys to the officer in control at the dock or the airport about it. The individual responsible for the ship or airway gives a report about it.

Imported merchandise or goods are subjected to customs which is an extensive process and takes a considerable time to complete. The importer often appoints a C&F operator for completing these custom formalities. Essentially, the merchant acquires a delivery order, which is called an endorsement for delivery. This order allows the importer to take the delivery of merchandise subsequent to paying the cargo charges. Importer likewise needs to pay dock dues for getting port trust dues receipts for which he submits two duplicate, filled-in forms known as "application to import" to the Landing and "Delivering Dues Office".

After paying dock dues, the importer gets back one copy of the application as a receipt, which is called as 'Port Trust Levy Receipts'. At the end, the importer fills a form known as 'bill of entry' for appraisal of customs import duty. An inspector inspects the merchandise and gives his report regarding the bill of entry. This bill is then introduced to the port administration which issues discharge arrangements.

DOCUMENTS USED IN IMPORT TRANSACTIONS

Proforma Invoice: A record containing points of interest about the quality, design, weight, etc. of the exported merchandise and the terms and conditions of transaction.

Import order or Indent: It is a document in which the importer orders for supply of imperative merchandise to the supplier. The order contains data such as amount and nature of merchandise value, a technique for sending the merchandise, packing process, method of payment, etc.

Shipment advice: The exporter sends a shipment advice to the importer for informing him that the merchandise has been dispatched. It contains invoice number, bill of lading/airway bill number and date, the name of the vessel to date, the port of export, description of products and amount, date of cruising of the vessel, etc.

Bill of lading: It is readied and marked by the captain of the ship recognizing the receipt of merchandise on board. It contains the terms and conditions on which the products are to be taken to the destination.

Bill of entry: It is a form provided by the customs office to the importer to be filled at the time of getting the merchandise. It is usually in triplicate and is to be submitted to the customs office

Letter of credit: It is a document that contains a certification from the importer bank to the exporter's bank - an assurance to respect the payment up to a specific sum of the bills issued by the exporter for transportation of products to the importer.

Trade Enquiry: It is a written request made by a logistic firm or giving data regarding the cost and different terms and conditions for trading merchandise.

IMPORT FLOW CHART

1. Purchase Order / Contract Scrutiny
2. Search and fix Freight Forwarder & CHA (Custom house agent)
3. Delivery Terms Ex works /CIF / CFR / FOB (CIF-Cost, Insurance and Freight, CFR - Cost and Freight, FOB-Free on Board)
4. Material Availability & ETD from Port (Estimated time of departure)
5. Shipping - Copy of Documents (Docs)
6. Shipping Original Docs
7. Docs Submit To CHA / Verify ETA (Estimated time of arrival)
8. Shpping Line Payment & Delivery Order
9. CHA to Pre Check and confirm of Duty Tariff
10. Customs Clearance, Duty Payment
11. Material Transportation upto Plant
12. Service Payment Transporter / CHA
13. Foreign Supplier Payment
14. Stock Entry in the plant, Purchase department

IMPORT DOCUMENTATION AND SOP

1. Ascertain & scrutinise Purchase Order given by the Purchase department and provide logistics support.
2. Search suitable freight forwarder and CHA for customs clearance of Container / Box cargo by Sea / Air

3. Check Delivery terms in P.O such as Ex-works / FOB / CFR / CIF / DDP / DDU and Payment terms to be verified

4. Coordinate with supplier for availability of material to be shipped from origin country

5. Check & verify copies of documents received by the Purchase Department and send to the concerned CHA

6. Coordinate with the Purchase Department or Accounts Department for original documents

7. Find ETA of cargo at Port / Airport and arrange to submit documents to the concerned CHA

8. Coordinate with Shipping Line or their Agent in India for Delivery Order & statutory charges invoice

9. Follow up with CHA for Bill of Entry and check all details and particulars about Customs Duty and tariff

10. After arrival of material at Port, pay related customs duty or arrange MEIS Licence to free Duty (If applicable)

11. With the help of CHA arrange transport to bring material to the plant from Port / Airport

12. Make payment to service provider and supplier after receiving material in the plant, as per agreed terms

13. Submit copy of Bill of Entry to Bank for supplier payment

14. Send documents, Bill of Entry, etc. to Purchase Dept for entry of material in stock books

IMPORT DOCUMENTS BY SEA / AIR: RAW MATERIAL

1.	Purchase Order / Sales Contract duly signed by Seller & Buyer both.	1 Original + 1 copy
2.	Letter of Credit (If under LC)	1 Original + 1 copy
3.	Commercial Invoice	2 Original + 1 copy
4.	Packing List Summary / Weight List / Detailed Packing List	1 Original + 1 copy
5.	Bill of Lading (Master & House BL if HBL)	3 Original + 3 Non Negotiable
6.	Freight Certificate from Shipping Company	1 Original + 1 copy
7.	Mill Test Report / Technical Analysis Report / Weight etc	1 Original + 1 copy
8.	Certificate of Origin	2 Original + 1 copy
9.	Insurance Certificat	1 Original + 1 copy
10.	Certificate of Origin - MICECA if applicable (import from Malaysia)	1 Original + 1 copy
11.	Health Certificate confirming non-hazardous & no animal residues	1 Original + 1 copy
12.	Beneficiary Certificat	1 Original
13.	Beneficiary s Certificate Shipping Advise	1 Original
14.	Vessel Certificate from Shipping Company	1 Original + 1 copy
15.	Bill of Entry – Customs Cleared	1 Original + 1 copy

IMPORT DOCUMENTS: MACHINERY & THEIR SPARE PARTS

1.	Purchase Order / Sales Contract duly signed by Seller & Buyer both.	1 Original / 1 copy
2.	Letter of Credit (If under LC)	1 Original / 1 copy
3.	Commercial Invoice	2 Original + 1 copy
4.	Packing List / Weight List / Detailed Packing List	1 Original + 1 copy

5. AWB or Bill of Lading (Master & House BL if HBL)	3 Original + 3 Non-negotiable copy
6. Freight Certificate from Airline or Shipping Company	1 Original + 1 copy
7. Mill Test Report / Technical Analysis Report	1 Original + 1 copy
8. Certificate of Origin	2 Original + 1 copy
9. Insurance Certificate	1 Original + 1 copy
10. Technical Write up or Catalogue	1 Original + 1 copy
11. Fitness Certificate and Manufacturing date of material if second hand	1 Original + 1 copy
12. Certificate for need & usage of specified material with age & expiry date	1 Original
13. Bill of Entry – Customs Cleared	1 Original + 1 copy

EXPORT PROCEDURES:

1. Firstly, an exporter gets a request from the potential buyer asking for data with respect to the cost, standard and different terms & conditions for transportation of merchandise. The exporter responds with a citation known as a Proforma Invoice.

2. Once the purchaser approves of the terms and conditions, he raises an 'indent' for the merchandise.

3. After receiving the indent or before, the exporter tries to inquire about the financial condition and credibility of the importer to evaluate the risk of default.

4. As indicated by customs laws, the exporter or the export firm should have a fare permit before proceeding for export. The following steps are taken for acquiring the export license:

   ■ Opening account in any approved bank

   ■ Apply to acquire import export code (IEC) number from Directorate General Foreign Trade (DGFT) or Regional Import Export Licensing Authority (RIELA).

   ■ Register with suitable export promoting committee.

   ■ To be enrolled with Export Credit and Guarantee Corporation (ECGC).

5. After getting the export license, the exporter meets with his banker to sanction pre dispatch finance for carrying out production. (If required)

6. Exporter, after getting the pre shipment fund from the bank, starts production as per the requirements of the importer.

7. As per Indian laws, selective products are permitted for export. The exporter needs to introduce pre-shipment examination report along with other required papers at the time of dispatch.

8. Excise duty on the material is to be paid, as prescribed by the Central Excise Tariff Act. Exporter applies to the concerned Excise Commissioner in the area with a receipt.

9. To get tariff concessions or diverse exclusions, the importer may ask the exporter to send an authentication of origin.

10. The exporter applies to the logistics organization for planning for transportation space. He needs to give full information as for the merchandise to be dispatched, conceivable date of shipment and port of destination. The logistics organization issues a transportation plan, which is a guideline to the captain of the ship, after accepting an application for dispatch.

11. The merchandise is stuffed and set apart with crucial data like name and address of the importing person, gross and net weight, port of shipment and destination etc. After this, the exporter makes the strategy for the transportation of merchandise to the port.

12. To protect the merchandise amid ocean travel, the exporter gets the material insured through an insurance agency.

13. Before stacking the merchandise on the ship all documents are to be submitted by the client. The exporter makes the bill and submits 5 copies of the bill along with:

   ■ Certificate of origin

   ■ Commercial Invoice

   ■ Export Order

   ■ Letter of credit

   ■ Certificate of Inspection, wherever required

   ■ Marine Insurance Policy.

   On presenting the mentioned documents, the director of the concerned port issues guideline to the staff at the entryway of the port to allow the cargo in the dock.

14. After the merchandise is stacked onto the ship, the captain issues mate's receipt (a document originally issued by first mate of the ship) to the port administrator which contains the vessel no., bill, information about the merchandise, date of shipment, the state of the merchandise.

15. The clearing and forwarding specialist (C&F operator) hands over the mate's receipt to the transportation organization for analysing the cargo. On accepting the cargo, the transportation organization issues a Bill of Lading. The bill of lading is a legal document between the shipper of a particular good and the carrier detailing the type, quantity, and destination of the goods being carried. This also serves as a receipt of shipment when the good is delivered to the predetermined destination. This document must accompany the shipped goods irrespective of mode of transportation, and must be signed by an authorised representative from the carrier, shipper, and receiver.

16. The exporter releases a receipt for the outgoing merchandise. The receipt contains the quantity of merchandise sent and the amount to be paid by the importer. It is confirmed by the customs.

17. After dispatching the merchandise, the exporter forwards reports to the importer along with an attested duplicate copy of the receipt, bill of lading packing list, insurance arrangement, certificate of origin, and letter of credit through his bank. These records are required by the importing merchant for getting the products cleared from customs.

DOCUMENTS USED IN EXPORT TRANSACTIONS

A. Documents related to Goods

Seller's Bill: Contains data about products like amount, number of packages, blemishes on packaging, the name of the ship, port of destination, terms of delivery and payment and so on.

Certificate of Inspection: For giving guarantee about quality, the government has made inspection of specific products compulsory by some approved organizations like Trade Inspection Board, Export Inspection Council of India (EICI, etc. Post reviewing the merchandise, the organization issues a certificate of inspection.

Packing List: This document shows the number of cases or packs and the details of products contained in these packs.

It gives finish insights with respect to the products sent out and the condition in which they are being sent.

Testament of Origin: This authentication indicates the nation in which the merchandise is produced. This authentication empowers the importer to claim levy concessions or different exemptions.

B. Documents related to Shipment

Transportation Bill: It is the basic document based on which consent is accorded for the export of merchandise by the customs office. It contains details of as to whom the merchandise is being sent to, the name of the vessel, exporter's name, and address, a destination country.

Mate's Receipt: This receipt is issued by the captain or mate of the ship to the exporter after the merchandise is stacked onboard the ship. It contains the name of the vessel, quantity, marks, condition of the packaging at the time of receipt on board or ship, etc.

Bill of lading: It is a record issued by the shipping organization. It is a proof of acknowledgment of the delivery organization to convey the merchandise to the port of destination.

Airway Bill: Similar to a shipping bill, it is a record issued by the airline organization on getting the products onboard.

Cart Ticket: Also known as cart chit or gate pass, the exporter establishes it. It contains insights with respect to sending out pay load like number of items, shipping charge number, port of destination, etc.

Marine Insurance Policy: It is a contract between the exporter and the insurance company to safeguard goods against the risks of ocean /air travel. The amount paid to cover the insurance is called premium and in case of any damage of goods, the insurance company is liable to reimburse the loss as per the terms and conditions of the policy.

C. Documents related to Payment

Letter of credit: It is an assurance letter issued by the importer's bank expressing that it will respect the export bills to the bank of the exporter up to a specific sum.

Bill of exchange: In export and import exchange, exporter draws the bill on the importer requesting that he will pay pre-determined monies to someone in particular or the owner of the goods. The records required by the importer for guaranteeing the title of exported merchandise, are passed on to him just when the importer acknowledges this bill.

Bank Certificate of Payment: It is a declaration that the required documents identifying the specific e port deal have been arranged and payment is received as per the exchange control regulations.

EXPORT FLOW CHART

1. Compliance with sales Department / Legal framework

2. Concluding an Export Deal

3. Arranging Export Finance

4. Producing / Manufacturing Goods

5. Appointing C&F Agent (CHA)

6. Arranging Cargo Insurance

7. Negotiating Ocean / Air Freight Charges with C&F Agent

8. Booking Shipping Space

9. Dispatch of goods to C&F Agent (at Port/Airport)

10. Sending Documents to C&F Agent (CHA)

10. Receipt of documents & cargo by C&F Agent (CHA)

12. Customs Clearance, completion of port/Airport formalities, and cargo loading by C&F Agent CHA

13. Receiving Docs from C&F Agent / Shipping Line or Airline

14. Sending Shipping advice to Buyer

15. Claiming Export Incentives DBK + MEIS

16. BANK Presentation of documents to negotiating Bank

17. Receiving Export Incentives DBK, as per tariff

18. Receiving payment from Buyer

19. Check BRC with Bank / Submit POF

20. Receiving Export Incentive MEIS on FOB

EXPORT DOCUMENTATION AND SOP

1. Coordinate with Sales department, Send samples of product to Buyer as per their advice

2. As soon as order confirmed by Sales department, ask for sales contract and Letter of Credit (if required)

3. Coordinate with Accounts department, Export Advance booking with Bank (PCFC) (Packing credit loan in foreign currency a form of pre-shipment finance

4. Coordinate with the Production Department for availability of produced material to be dispatch on due date

5. Identify suitable CHA for customs clearance of Container / Box cargo from sea port / air port

6. Coordinate with Insurance Agency or their agent for Export Insurance of material (by road / sea / air)

7. Negotiate with Shipping Line for Ocean (Sea) Freight / Air Freight Charges up to Destination (CIF/C&F)

8. After freight is finalised, obtain booking slip of space allocation with concerned shipping line or airline

9. Dispatch Material at required port of loading such as JNPT / NSICT / GTIL or SAHAR Air Cargo Mumbai etc

10. After material dispatch or stuffed inside container, send documents to related CHA for Customs Clearance

11. Confirm with CHA / freight forwarding agent for receiving of documents & Cargo / Material at site

12. Coordinate with CHA & get confirmation of proper Customs Clearance of material and loading of container / cargo at allocated space provided by shipping line or air line

13. Check and receive documents from CHA shipping / airline, shipping bill, bill of lading, etc.

14. Send cargo-loading information to buyer through shipping advice and prepare post shipment documents15-Follow up with CHA for export incentives and submit documents if any query raised by customs

16. Submit all post shipment documents of export against LC or other payment terms to bank for CAD/LC and direct to buyer in case of TT

17. Check with Accounts Department if DBK claim received, if not follow up with CHA

18. Check with Accounts Department if payment from buyer is received vide LC or by TT, if not follow up with Sales Department

19. Check with Documents Negotiating Bank for BRC (Bank realisation certificate) through Account Dept.

20. Check if MEIS Licence is issued by the concerned agency, and is available online, if not follow up with Agent or Dealer

EXPORT DOCUMENTS – SEA / AIR

PRESHIPMENT DOCUMENTS – DIRECT EXPORT

1. Purchase Order / Sales contract duly signed by Seller & Buyer both.   –   1 Original + 1 copy
2. Customs Invoice   –   4 Original + 4 copies
3. Packing List and Detailed Packing List   –   2 Original + 2 copies
4. GST Invoice   –   1 Original + 1 copy
5. Test Report / Technical Analysis Report / Quality Certificate   –   1 Original + 1 copy
6. DBK Declaration (as per the case)   –   1 Original + 1 copy
7. SDF/EDF Form / GR form deceleration   –   1 Original + 1 copy
8. Container Examination Report (Self Sealing)   –   1 Original + 1 copy

POST SHIPMENT DOCUMENTS – DIRECT EXPORT

1. Letter of Credit (if required)   –   1 Original
2. Commercial Invoice   –   4 Original + 4 copies
3. Packing List / Detailed Packing List   –   2 Original + 2 copies
4. Bill of Lading   –   3 Original + 3 Non-negotiable
5. Test Report / Technical Analysis Report   –   1 Original
6. Certificate of Origin   –   2 Original + 1 copy
7. SDF/EDF Form   –   1 Original + 1 copy
8. Insurance Certificate   –   1 Original + 1 copy
9. Certificate of Chamber of Commerce   –   2 Original + 1 copy
10. Beneficiary Certificate   –   1 Original
11. Beneficiary s Certificate Shipping Advise   –   1 Original
12. Vessel Certificate from Shipping Company   –   1 Original + 1 copy
13. Freight Certificate from Shipping Company   –   1 Original + 1 copy
14. Bill of Exchange   –   1 Original
15. Bank Forwarding Letter   –   1 Original + 1 copy
16. GSP Certificate (for Europe + USA) (If asked)   –   1 Original + 1 copy

   (this certificate is issued from DGFT / EPC - Export Promotion Council)

   In addition, all other documents mentioned in Letter of Credit.

PROOF OF EXPORT DOCUMENTS LIST

1. Shipping Bill Copy
2. Customs Attested Invoice Copy
3. Customs Attested Packing List Copy

4. BRC to be linked online by Bank for MEIS Copy

5. Bill of Lading Copy

6. Mate Receipt issued by Shipping Line Copy

EXPORT INCENTIVES

1. DBK - Duty Drawback Claim after 7 days from Let Export Date. To Check and follow weekly

2. MEIS Licence after BRC Linked by Bank. To Check and follow weekly

EXPORT DOCUMENTS CHECKLIST

Export Shipments Documents Checklist	
Invoice No	Sea /Air
Invoice date	
Documents	No. of Copies
Invoice Cum Packing List 1st & 2nd	Three Copies
Detailed Packing List	One Copy
GST Invoice	Two Copies
Shipping Bill	Three Copies
Accepted Copy of Sale Note	One Copy
Samples	
Bill of Lading / Air Way Bill /(AWB)	
Certificate of Origin /GSP	
Insurance of Certificate and Declaration	
Letter of Credit	
Mill Test Certificate	

EXPORT DOCUMENTS TO CHECK AND VERIFY

L/C Checklist

1. Beneficiary / shipper as per invoice

2. LC expiry date

    (a) LC amount - not to exceed invoice amount

    (b) Tolerance for quantity & value (5 to 10 %) (As per Invoice)

    (c) Documents presentation time - 21 days

    (d) Bank name for bill of exchange, reimbursement bank etc

    (e) Latest date of shipment

3. Consignee as per invoice or as per LC (in case of LC)

4. Notify party as per invoice or to order (in case of order) or as per LC (in case of LC)

5. Notify as per invoice

6. Port of loading as per shipping company's loading

7. Port of discharge as per invoice

8. Final destination as per invoice

9. Description of goods as per invoice, gross weight, net weight, no. of cartons etc

   As per LC (in case of LC)

10. Trade terms, freight collect/freight payable at destination

    (in case of FOB shipment). Freight prepaid (in case of C&F and CIF shipment).

    Inland haulage prepaid in case of C&F and CIF basis.

11. "Shipped on Board" stamp with date and signature

12. Type of movement - FCL/LCL in case of full container covering only one invoice

13. Vessel name as per planning or actual loading of the container.

14. Container no. as per invoice

15. Others:

    (a) Name, address, contacts no and fax no./email ID of the shipping company agent at the port of destination.

    (b) Invoice number and date on the bill of lading (as per invoice)

    (c) Shipping bill No, Date and GR No., Date on the B/L

    (d) Three originals B/L with revenue stamp

INVOICE CUM PACKING LIST CHECKLIST

1. Consignee: As per sale note / letter of credit

2. Notify party: As per sale note / letter of credit

3. Also notify as per sale note / letter of credit

4. Buyer: As per sale note / letter of credit

5. Port of loading: As per sale note / letter of credit

6. Final destination: As per sale note / letter of credit

7. Price term: As per sale note / letter of credit

8. Description of goods: As per sale note / letter of credit

9. Payment terms: As per sale note / letter of credit

10. Mode of dispatch: As per sale note / letter of credit

11. H.S. code: As per description of goods and tariff

12. Container no.: As per shipping line

13. Number of cartons: Total cartons as per description of goods

B/L CHECKLIST

1. Shipper: As per invoice

2. Consignee: As per invoice or To order (in case of order) or as per LC (in case of LC)

3. Notify party: As per invoice or to order (in case of order) or as per LC (in case of LC)

4. Also notify as per invoice

5. Port of loading: As per advised by shipping line

6. Port of discharge : As per invoice

7. Final destination: As per invoice

8. Description of goods: As per invoice. Gross weight, net weight, no.of cartons as per LC (in case of LC)

9. Trade terms : Freight collect/freight payable at destination (in case of fob shipment) Freight prepaid

   (in case of C&F and CIF shipment) .Inland haulage and destination of prepaid in case of C&F and CIF basis

10. Shipped on board : "Shipped on board" stamp with date and sign

11. Type of movement: FCL/LCL in case of full container covering only one invoice

    FCL/LCL in case of full container covering one or more invoice. FCL/LCL in case of loose shipment12. Vessel name: As per planning or actual loading of the container13. Container no : As per invoice

14. Others :

    A. Name, address, Contact No and Fax no. /email ID of the shipping company's agent at the port of destination

    B. Invoice number and date on the Bill of Lading (as per invoice)

    C. Shipping Bill No. date and GR No, Date on the B/L as per GR.

    D. Three originals B/L with revenue stamp must be received.

    E. 14 days free time detention & demurrage at port of destination

INSURANCE DOCUMENTS CHECKLIST

1. Shipper: As per invoice

2. Consignee: As per invoice or to order.(in case of order) or as per LC (in case of LC)

3. Notify party: As per invoice or to order.(in case of order) or as per LC (in case of LC)

4. Also notify: As per invoice

5. Port of loading: As per shipping company's loading

6. Port of discharge: as per invoice

7. Final destination: As per invoice

8. Description of goods: As per invoice , carton nos., gross wt, net wt, no.of cartons as per l/c (in case of l/c)

9. Trade terms: Freight collect/freight payable at destination. (in case of FOB shipment) ,Freight prepaid (in case of C&F and CIF shipment)

   (in case of FOB shipment), Freight prepaid (in case of C&F and CIF shipment). Inland haulage prepaid in case of C&F and CIF basis

10. Shipped on board: "shipped on board" stamp with date and signature

11. Type of movement: FCL/LCL in case of full container covering only one invoice

    FCL/LCL in case of full container covering one or more invoice, LCL/FCL in case of loose shipment.

12. Vessel name: As per our planning or actual loading of the container.

13. Container no : As per invoice

14. Others:

    A. Name, address, Contact No. and Fax no. of the shipping company's agent at the port of destination.

    B. Invoice number and date on the bill of lading (as per invoice)

C. Shipping bill no.date and GR no. date on the BL as per GR

D. Three originals BL with revenue stamp must be received.

CERTIFICATE OF ORIGIN CHECK LIST

1. Exporter's name, address, and country as per invoice or as per LC

2. Importer's name, address, and country as per invoice or as per LC

3. Producer's name, address, and country as per invoice or as per LC

4. Transport details (means and route) ,mode of shipment from Indian port to destination port

5. Item number, marks, packages description as per invoice or as per LC, Goods HS Code

6. Preference criteria: "B" in case of Japan, CTH+IVC in case of Korea, etc.

7. Quantity: Gross weight and net weight.

8. Invoice number and date as per invoice

9. Remark: Third party name and address and tick in case of third party invoice

10. Issued retroactively- tick and sealed issued retroactively in case of issued

     after B/L date

11. Sign and place of issue: Signed by authorised signatory of the company

12. Sign of authority : Sign and seal of issuing authority

SHIPPING BILL CHECKLIST

1. Exporter: As per invoice

2. Consignee: As per invoice

3. Invoice no & date: As per invoice

4. Port of discharge: As per invoice

5. Country of destination: As per invoice

6. Description of goods: As per invoice

7. Currency: As per invoice

8. Amount : FOB, freight and insurance in the currency of invoice, total amount as per invoice amount9. Fob value: FOB value in Indian Rupees.

10. Nature of contract: FOB / C&F / CIF, as per invoice

11. Name of the bank: Name of the bank through which the documents will be routed

12. AD code of the bank: AD code of the bank through which the documents will be routed13. Quantity and packages: Gross weight, net weight, number of cartons, etc.

14. Duty drawback scheme: If under duty drawback scheme then check as per tariff

13. Signature: Duly signed by the customs official on the backside of the GR14. Container number: As per invoice15. Export incentives scheme: Check if MEIS scheme is enabled & mentioned as yes on shipping bill, if applicable

PACKING LIST CHECKLIST

1. Check packing list : Lot number and number of cartons.

2. Check total gross weight, net weight and invoice weight.

3. Check total cartons, total gross weight, total net weight.

FORMAT FOR IMPORT CONSIGNMENTS STATUS REPORT

P.O. No	S.O. No	Name of supplier	Items	Payment Terms	Value FC	Value FC/ INR	Loading/ Port	Container No	SOB DT.	E.T.A Port	Sum of Inv. Qty (Tons)	Container size	Gross Qunt	NO. Of Bale	Shipping Line	Dt.Of Rcvd.in Plant	CHA Name	Remark

*PO-Purchase order , SO-Supplier order ,PT-Payment terms, FC-Foreign Currency, SOB-Shipped on board, ETA-Estimated time of arrival,

Inv-Invoice, Ship-Shipping line, Recd-received, CHA-Custom house agent, LC 38- Letter of credit /38 days, Cash against documents, TT-Telegraphic transfer

FORMAT FOR EXPORT CONSIGNMENTS STATUS REPORT

Inv No	Inv date	Consg	Dst	Country	Container size	SL	Cont No	B/L NO.	BL Date	ETD India	ETA Destination	DOC NEG. DT.	DAYS FROM Inv to Loading	Days from Inv to ETA	Currency	Inv Value	Inv wt Kg	DHL Detail	CHA
	3rd Feb		Laspezea	Prato, Italy	20 GP	Hapag			14 Feb	14 Feb	Ist March	17th Feb	11	26	EURO	9,458.46	4,777.00		
	6th Feb		Genoa	Biella, Italy	20 GP	Hapag			14 Feb	14 Feb	2nd March	17th Feb	8	24	USD	84,101.40	5,236.70		
	6th Feb		Genoa	Biella, Italy					14 Feb	14 Feb	2ndMarch	17th Feb	8	24	USD	12,679.00	1,283.30		
	7th Feb		Nagoya	Japan	20 GP	Interasia			11Feb	11Feb	8th March	13th Feb	4	29	USD	38,341.80	2,142.00		
	7th Feb		Nagoya	Japan					11Feb	11Feb	8th March	13th Feb	4	29	USD	46,332.60	4,929.00		
	11th Feb		Busan	Korea	20 GP	Wan Hai			21 Feb	21 Feb	15Tth March	26th Feb	10	32	USD	91,931.91	7,256.30		

*Inv-Invoice , Consg-Consignee(Buyer), SL-Shipping line, B/L-Bill of Lading, ETA-Expected time of arrival, CHA-Custom house agent, DOC-Documents, NEG-Negotiation

FOREIGN TRADE PROMOTIONS INCENTIVE, TRADE PROMOTIONS INCENTIVES AND ORGANISATIONAL SUPPORT

Foreign trade leads to division of labour and specialization at a global level. There is sufficient availability of labour in India. That is one of the reasons the Indian government promotes and stimulates policies and schemes for expanding foreign trade. The government takes various measures to support foreign trade, such as initiating different incentives and plans to help business firms enhance the competitiveness of their exports. The Government has also established a number of institutions to offer infrastructural and marketing help to organizations conducting international business.

FOREIGN TRADE PROMOTION MEASURES AND SCHEMES

1. **Duty Drawback Scheme:** The Duty Drawback Scheme provides exporters a refund of customs duty paid on unused imported goods, or goods that will be treated, processed, or incorporated into other goods for export. On showing verification of export of these products to the concerning authority, charges paid as excise, levy charges, and customs duties are refunded. Such refunds are 'Duty Drawbacks'.

2. **Export manufacturing under the Bond Scheme:** Through bonded manufacturing, all types of businesses can avail exemption on customs duty on imported inputs used in the production of finished goods to be exported.

3. **Advance License Scheme:** An Advance licence or Authorisation is issued to allow duty free import of inputs, which are physically incorporated in export product (making normal allowance for wastage). In addition, fuel, oil, energy, catalysts which are consumed/ utilised to produce export products, may also be allowed. DGFT, by means of Public Notice, may exclude any product(s) from purview of Advance Authorisation.

4. **Export Processing Zones:** An export-processing zone, or EPZ, is an area set up to enhance commercial and industrial exports by encouraging economic growth through investment from foreign entities. EPZ zones can accommodate both domestic and foreign firms. Incentives such as tax exemptions and a barrier-free environment are the main attractions of an EPZ. They intend to provide an internationally competitive duty free environment for export production at low cost. There are different measures, for example, availability of export fund, export promotion, capital merchandise scheme is in use for foreign trade promotion.

5. **Organizational Support:** The government has set up from time to time various institutions in order to facilitate the process of foreign trade. Following are some of them:

   (a) **Department of Commerce:** This is the most authoritative body responsible for the country's international trade and all jurisdictions linked with it. This might be in the form of expanding business relations with other nations, state trading, export promotional measures and development or regulation of certain export oriented industries and commodities. The Department of Commerce formulates policies in the sphere of foreign trade. It also frames the import and export policy of the country in general.

   (b) **Export Promotion Councils:** Export Promotion Councils are non profit institutions registered under the Companies Act or the Societies Registration Act. The fundamental objective of these councils is to market and produce the nation's exports of particular products falling under their jurisdiction.

   (c) **Commodity Boards:** Commodity Boards are establishments of the Government of India for development of manufacturing of traditional merchandise and their exports.

   (d) **Export Inspection Council:** Export Inspection Council of India is also an establishment by the Government of India. The council aims at the sound development of export trade through quality control and pre shipment inspection. The council is a vital body for managing the operations with standard control and pre shipment inspection of merchandise for export. Barring a few exceptions, all merchandise destined for exports must be passed by EIC.

   (e) **Indian Trade Promotion Organisation:** The Indian Trade Promotion Organisation works under the Ministry of Commerce, Government of India. Its objective is to support export organizations engaged in international trade fairs and exhibitions, etc. Also, developing exports of new items, providing support and updated commercial business information.

   (f) **Indian Institute of Foreign Trade (IIFT):** Indian Institute of Foreign Trade has been established under the Government of India as an autonomous body. The prime objective is to professionalise the country's foreign trade management. It provides training in international business, conducts research in areas of international business, and analyses and disseminates information relating to international trade and investments scenario.

   (g) **State Trading Organisation:** The main purpose of STC is to promote trade, primarily export trade among different trading partners of the globe.

   (h) **Indian Institute of Packaging (IIP):** The Indian Institute of Packaging is a national institute mutually run by the Ministry of Commerce, Government of India, and the Indian Packaging industry and allied interests.

14

Utility

The textile industry is supported by many service departments to run operations smoothly. In a textile mill, the core operation is that of production of yarn in spinning, production of fabric in weaving or knitting, and dyeing and finishing of fabrics in processing departments. These operations are conducted independently in large volumes. A number of small size factories run in decentralised sectors independently while a fully integrated textile mill consists of fibre to finish or even garmenting facilities within the same premises.

These operations are dependent on energy in the form of electricity, compressed air, water, steam, gas, etc. These are basic requirements to run any operation. The Utility or Engineering Department manages a combination of such services. This department is also responsible for maintaining the climatic conditions mainly within the spinning, weaving, knitting, folding and clothing departments by keeping humidity and temperatures under control. Their main job is to not only ensure and maintain uninterrupted supply of the above utilities, but also ensure that corresponding costs stay under control.

The head of the Engineering Department is normally known as the 'Chief Engineer'. The responsibilities of a Chief Engineer include carrying out regular preventive maintenance of the plant and equipment like transformers, power house, generator sets, compressors, boilers, water pumps, RO plants, captive power plants (if installed, humidification plant, air conditioning and maintenance of buildings, etc. Some of the tasks of a Chief Engineer include:

- Prepare a comprehensive preventive maintenance schedule and checklist for all machinery and equipment and implement them holistically
- Maintain transformers, generator sets, power station, etc.
- Maintain air compressors and the quality of air released within the departments
- Manage fuels – gaseous, liquid and solid
- Manage the quality of water and its uninterrupted supply for various equipment and services
- Maintain boilers and quality of steam
- Maintain heating media – various polymer liquids
- Maintain refrigeration units generating chilled water
- Maintain safety and security instruments in the factory
- Maintain effluent treatment plant, etc.
- Get clearance from the Pollution Control Board, etc.

ENERGY CONSERVATION

There is a significant contribution of energy cost in the total cost of the end-product. The need for energy conservation is of paramount importance in the textile industry on account of rising energy costs on the one hand and severe energy shortage on the other hand.

Out of total energy consumption in the manufacturing of yarn and fabric, a major share of power consumption pattern in spinning and weaving, dyeing and finishing consume a majority of thermal energy, largely in two operations i.e. in heating of water and drying of water for evaporation of moisture from fabrics at various stages of wet processing and also for heating of process chemicals.

Modernisation and technological upgradation of the plant and machinery could be effective in reducing energy consumption. Some of the important factors for energy conservation are regular energy audits, machines' maintenance, instrumental control, rework and waste control and heat recovery, etc.

It is an utmost important task for the utility department to not only take necessary measures to control energy consumption but also be vigilant and put continuous efforts to save and reduce the energy cost by various measures such as :

1. Conduct of periodical energy audits by internal as well as external experts

2. Use of energy efficient illuminating devices and ensuring optimum utilization

3. Replacement of conventional street lights by Solar Powered LED lights (if viable)

4. Replace old motors and pumps with energy efficient motors

5. Replacement of undersized or oversized motors – saving depending on the percentage of loading on the motors

6. Use of highly recommended temperature grease according to insulation class of motors

7. Conduct no load power study of motor and replace the motors consuming high no load power

8. Investigate reasons of burning and rewinding as per original technical data

9. Conversion from DC to AC variable frequency drive in the machines to ensure saving of power

10. Critical monitoring and control of rework at different stages of manufacturing

11. Monitoring and control of leakages and wastage of compressed air

12. May replace low efficiency transformer with high efficiency transformer

13. Install and monitor heat recovery systems to recover heat from various machines through steam/hot water to the boiler feed

14. Steam Trap management to increase condensate recovery

15. May install special humidity sensors in air-conditioning systems to help in controlling the air washer units so as to automatically stop the air washer pump during rainy seasons

16. May fix sun control films and sun blinds on window glasses to reduce load on the air-conditioning system

17. May install the wind turbines to help minimize the use of ceiling and exhaust fans.

18. Efficient removal of water using heavy squeezing enables 15-20% reduction in energy requirement for drying

19. Continuous efforts for planation of maximum possible trees all around the campus in order to regulate temperature and maintain the ground water level

20. The possibilities of utilizing new energy resources like solar energy, wind power, tidal power, nuclear energy, etc. can be be explored – but should be implemented after considering the initial costs, commercial viability and payback period.

21. Install and use EMS (Energy management systems) for greater accuracy of energy consumption reports for utility departments and process machines so as to take effective measures for saving wherever necessary, etc.

22. Energy efficient layout to be prepared in the project stage itself so that the running costs after project implementation are maintained at an economic level considering the ever rising energy & maintenance costs.

ENERGY AUDIT

Energy conservation is a continuous process which needs regular and close monitoring and control at the generation, distribution as well as user end. For optimum utilization there needs to be high contributions at human level to stop wastage. In the case of capital investment, the management and Chief Engineer need to take decisions for its implementation after preparing a short term as well as long term plan depending on cost benefit analysis.

SCOPE OF ENERGY AUDIT (Example)

The following points may be covered under the scope of an audit :

Generation, distribution & utilisation of:

- Steam boilers, steam distribution systems, steam traps, condensate recovery systems and line sizing determination
- Substation, metering systems, power factor & demand control
- Thermic fluid heaters/furnaces, line sizing & distribution systems, tf oil analysis
- Thermal insulation, identifying heat loss areas & heat recovery areas
- Pumps & blowers
- Air compressors efficiency volumetric & isothermal & air dryer systems
- Refrigeration & air-conditioning plants, determining the thermal, refrigeration & electrical parameters based on ashrae charts
- Cooling towers efficiency, blow down%, make up water %, evap losses
- Air handling units efficiency determination, output computation (chilled water & dx type
- Transformers, determining efficiency at various loads and determining optimal loading
- Motors analysis for active, reactive & apparent power, current, voltage, p.f
- Cables loading and suitability
- Lighting systems, ups & stabilisers, rectifier
- Harmonic analysis
- Power plants , cogent systems, gas & steam turbines, comprehensive audits of power plants with various type of solid & gaseous fuel
- Textile mill audits : weaving, spinning, process house m/cs, humidification plants, power & lighting systems, steam and compressed air systems, etc.

Project management services

Electrical engineering:

1. Electrical distribution system design, eht/ht/lt
2. Substation design & layout preparation
3. Ht/lt metering system design
4. Pccs/mccs/lighting system design
5. Cables sizing & selection
6. Safety relays design
7. Earthing system design to comply with the Indian Electricity Act 1954 & 2005
8. Lightning arrestor design/ surge arrestors
9. Preparation of slds for above
10. Site visits

Thermal engineering:

1. Boilers selection design / piping of steam lines

2. Steam distribution systems/ condensate recovery

3. Insulation design

4. Thermopac capacity design/ thermopac selection & evaluation criteria design

5. Thermic fluid distribution design

6. Compressed air selection/piping design

7. Water distribution system design

8. Preparation of utility process flow sheets

9. Instrumentation services for utilities

10. Refrigeration plants design/cooling load calculations, plant selection

RECORDS OF UTILITIES IN A FACTORY

For effective control, it is important that the details of utilities in a factory is maintained for ready reference. This can be done in the following formats:

Examples:

LOAD CHART: MACHINE & UTILITY OF A WORSTED SPINNING UNIT						
MACHINE						
SR. NO.	MACHINES	MODEL	NO. OF M/C.	CONNECTED LOAD (KW) OF EACH M/C	TOTAL CONNECTED LOAD (KW)	% Age
	A. WOOL TOP MAKING					
1	OPENER		1	52.00	52.00	1.29
2	CARDING MACHINE CA-7		2	72.00	144.00	3.58
3	MATERIAL FEED UNIT		1	7.50	7.50	0.19
4	CARDING CONDENSOR		1	35.50	35.50	0.88
5	GILL BOX GC-13		1	12.50	12.50	0.31
6	GILL BOX GC-14		1	12.50	12.50	0.31
7	GILL BOX GC-15		2	12.50	25.00	0.62
8	GILL BOX GC-15		2	12.50	25.00	0.62
9	COMBER		11	5.00	55.00	1.37
	TOTAL (A)				369.00	9.16
	B. PREPARATORY					0.00
10	TT-11		1	25.00	25.00	0.62
11	GILL BOX GC-13		3	12.50	37.50	0.93
12	GILL BOX GC-14		8	12.50	100.00	2.48
13	GILL BOX GC-14		2	12.50	25.00	0.62
14	GILL BOX GC-14		3	12.50	37.50	0.93
15	GILL BOX GC-14		2	12.50	25.00	0.62
16	GILL BOX GC-15		1	12.50	12.50	0.31
17	GILL BOX GV-11		2	15.00	30.00	0.74
18	GILL BOX GV-20		1	15.00	15.00	0.37
19	COMBER		2	5.00	10.00	0.25

20	COMBER		15	5.00	75.00	1.86
21	COMBER		1	5.00	5.00	0.12
22	RUBBING FRAME		2	26.50	53.00	1.32
23	RUBBING FRAME		2	20.00	40.00	0.99
24	RUBBING FRAME		2	30.00	60.00	1.49
	TOTAL (B)				550.50	13.67
	C. SPINNING					0.00
25	RING FRAME		14	40.00	560.00	13.91
26	RING FRAME		8	48.00	384.00	9.53
27	RING FRAME (SIRO)		3	58.00	174.00	4.32
28	OVERHEAD CLEANER		14	2.20	30.80	0.76
29	OVERHEAD CLEANER		11	1.50	16.50	0.41
30	SUCTION FAN		1	3.50	3.50	0.09
	TOTAL (C)				1168.80	29.02
	D. POST SPINNING					0.00
31	AUTO CONER 238		8	32.00	256.00	6.36
32	AUTO CONER 338		2	22.00	44.00	1.09
33	ASSEMBLY WINDER		5	25.00	125.00	3.10
34	FADIS		2	12.00	24.00	0.60
35	AUTOCLAVE (ARNOLD)		2	15.00	30.00	0.74
36	T.F.O. VOLKMANN		14	60.00	840.00	20.86
37	T.F.O.MURATEC		10	60.00	600.00	14.90
38	LAB EQUIPMENT AND					
	M/C MAINTENANCE			20.00	20.00	0.50
	TOTAL (D)				1939.00	48.15
	TOTAL (A+B+C+D)				4027.30	100.00

UTILITIES

SR. NO.	MACHINES	MODEL	NO. OF M/C.	CONNECTED LOAD (KW) OF EACH M/C	TOTAL CONNECTED LOAD KW	% Age
1	AIR COMPRESSOR		7	45.00	315.00	10.27
2	AIR COMP. PUMP		3	5.50	16.50	0.54
3	AIR DRYER		2	7.50	15.00	0.49
						0.00
4	CHILLER 400 TR		2	202.00	404.00	13.18
5	CHILLER 525 TR		2	265.00	530.00	17.29
6	CHILLER PUMP		5	30.00	150.00	4.89
7	CONDENSER PUMP		5	30.00	150.00	4.89
8	COOLING TOWER		4	15.00	60.00	1.96
						0.00
9	CARDING AWT		1	68.00	68.00	2.22
10	PREPERATORY AWT (W.T)		1	90.00	90.00	2.94
11	SEMI CENTRAL UNIT		1	15.37	15.37	0.50
12	PREPARATORY AWT PHV		1	112.00	112.00	3.65
13	SPINNING AWT PHV		1	165.00	165.00	5.38

14	WINDING AWT PHV		1	74.00	74.00	2.41
15	T.F.O. AWT PHV		1	180.00	180.00	5.87
16	PREPARATORY AWT PHVI		1	87.00	87.00	2.84
17	SPG.+WINDING AWT PHVI		1	225.00	225.00	7.34
18	T.F.O. AWT PHVI		1	202.00	202.00	6.59
19	PACKAGE A/C LAB		1	13.00	13.00	0.42
20	PLANT LIGHTING			194.00	194.00	6.33
	TOTAL				**3065.87**	**100.00**

DETAILS OF UTILITIES IN A TEXTILE MILL AT A GLANCE (Example)

FIRE FIGHTING		
1	Hydrant/sprinkler/fire extinguishers	
2	The area(s) where sprinklers are installed	
3	Is hydrant system pressurised?	
4	Whether the sprinkler system installed is operative or not	
5	No. of times in a year fire incidents in plant	
6	Area / section of maximum fire incidents & reasons for it	
7	Actual loss due to fire in the year -	
8	Training by firefighting agency	
8	Month/ year of last fire incident	
9	No of firefighting members/team	

STORAGY CAPACITY		
1	Lignite + Coal (MT)	800
2	Water (KL)	2000
3	Diesel (Lt)	90000

INSULATION		
1	Insulation Details of Thermic Fluid Line	
	Type of insulation	Glasswool
	Thickness of Insulation MM	50
	Ambient Temperature Degree Centigrade (C)	30
	Temp. at Outer Surface of Insulation C	
2	Insulation Details of Steam Line	
	Type of insulation	Glasswool
	Thickness of Insulation MM	50
	Ambient Temp. Deg. C	30
	Temp. at Outer Surface of Insulation	

STEAM BOILER	
Type of Boiler	1) Water Tube 2) Smoke Tube With Water Wall
No. of Boiler	2
Total Capacity (Ton/Hour) of Boiler	7+6=13
Steam requirement Ton/Hour	4
% Loading of Boiler	60
Standby Boiler	1

No. of Boilers used at a time	1
Fuel Used In Boiler	Lignite + Coal (size 0-50)
Test report of fuel being used In Boiler	
Steam Meter / Water Meter installed, Location (Ideally at inlet of each department)	Water Meter on 7 TPH Boiler
Meter Measure Steam /Water (Ideally at inlet of user department)	Water
Steam Pressure In Boiler (KG/CM2)	7
Maximum Pressure required in Plant (KG/CM2)	6
Maximum Temperature to be maintained in Plant (Degree Cent)	
CO_2 % in Fuel Gases	7.535
Flue Gas Temp. (Degree Cent)	127
Efficiency of Boiler %	75
Auxiliary Electrical load (KW)	70
% Blow Down	1 Time/Shift for 60 Seconds
Boiler Feed Water Test Report	
Any treatment to Boiler Feed Water is given	Yes
Chemical Dosing to Boiler Feed Water	Yes (Mextreat 3222 & Mextreat 3200)
Quantity of Chemical Used/Day	2+1 = 3 kg
Cost of Chemical Used /Day (Rs)	465
Cost of Fuel (Rs)/Kg	4
Calorific Value of Fue l/Kg	AVG 3500
Cost of Steam /Kg (Rs)	1.69
Boiler Safety, Details & Present Status	Good
Condensate being recovered or bot	Yes/Partially

	AIR COMPRESSOR AND AIR DRYER									
		1	2	3	4	5	6	7	8	9
1	Details of Air Compressor									
	MAKE									
	MODEL									
2	Type of Air Compressor	Rotary Screw	Rotary Screw	Rotary Screw	Rotary Screw	Rotary Screw	Rotary Screw	Rotary Screw	Reciprocating	Reciprocating
	Capacity (CFM)	167	200	97	97	110	723	100	100	100
3	No. of Air Compressors	9								
4	Total Installed Capacity (CFM)	1694								
5	Loading /Unloading Pressure of	7/8	7/8	7/8	7/8	6/7	8.5/7.5	7/6	7/8	7/8
	Air Compressor (Kg/ Cm^2)									
6	Is Installation Centralised or Not	yes								
7	No of Compressors Run at a Time	Energy Efficient Comp No. 2 Atlas + Comp. No. 5 Atlas + Comp. No. 6 Kaeser (total CFM 1033) = 3 Nos								
8	Stand By Compressor (If Any)	6								
9	Details of All Compressors Used For	IR 7100 (44 CFM) High Pressure requirement for KD M/C 1 No.								

	Cleaning of M/cs	IR 253 (25 CFM)- 4 Nos +IR 234 (15 CFM) - 2 Nos.							
	Capacity & Nos. of Such Compressors								
10	Is Cleaning of M/cs Done Through A								
	Separate Low Pressure Air line	Yes/Partially							
11	Air Receiver Pressure Testing Done	Apr –							
	Last on								
12	Test Pressure (Kg/Cm²)								
13	Total load (KW) of Air Comp & Air Dryers	175							

WATER

1	Water Source Details		Borewell + Govt. Supply
2	Raw water Test Report		
3	Yield of each Borewell (Ave. Ltr/hour)		No 1) 20000, No.2) 25000,No.3) 20000
4	No. of Borewells		3
5	Total Yield (M3/hr)		21
6	Electrical Load of Borewell Pump (Kw)		7.5
7	Depth of The Borewell (Mtr)		150
8	Water Consumption (Kl/Day)		800
9	Details of water Treatment Plant		Water Softening Plant
10	Treated Water Test results		
11	Water Cost Rs/KL Including Power Cost & Chemical Cost		
12	Water Storage Capacity(KL)		2200
13	Is Water Fed to Plant Through O.H. Tank		Fresh Water-100 KL
14	Water Meter Installed or Not Used		Yes
15	M/c Cooling Water is being Re-used or Not		Yes
	Quantity of Cooling water (KL/Day)		

Sr. No.	STEAM Description	UOM	Boiler NO. 1	Boiler NO. 2 (Stand by)
1	Make		Thermax	Thermax
2	Capacity	TPH	7	6
3	Type		Solid fuel, water tube	Solid fuel, smoke tube with water wall
4	Year of installation			
5	Fuel		Lignite + Steam coal	Lignite + Steam coal
6	Cost			
	Lignite	Rs/Ton	3560	
	Coal (0-50)	Rs/Ton	4600	
7	Consumption			
	Lignite	MT	5170	
	Coal (0-50)	MT	805	
	Total		5975	
8	Steam generation	MT	24174	
9	Production (Own)	Meter	2429981	
10	Fuel consumption	Kg/meter	2.459	
		Rs/meter	9.098	

		per day	19.212	
11	Steam consumption	Kg/Meter	9.948	
	Scheduled caliberation		Steam flow meter &	
			Water flow meter	
	Department wise			
	Wool Scouring	%	16	
	VA Chiller	MT/Hour	1.5	
	Recombing	%	10	
	Dyeing	%	45	
	Finishing	%	40	
	Spinning	%	5	
12	No of Work force/shift			
	Boiler Attendant	No	1	1
	Fireman	No	1	1
	Helper	No	1/2	1

THERMIC FLUID HEATER

1	Make	–		
2	Capacity	Lac K.Cal/Hour	10	
3	Year of installation	–		
4	Type	–	Solid fuel, Manual firin	
			W/out Air/water pre heater	
			Size <25 MM	
5	Annual Consumption			
	Lignite	MT	1689	
	Coal (0-50)	MT	174	
	Total		1863	
6	Fuel consumption	Kg/meter	0.767	
		Rs/meter	2.804	
		per day	5.990	
7	Heat consumption		Stanter No. 1 with 5 Chambers	
			Stanter No. 2 with 5 chambers	
8	Work force/shift			
	Fireman	No	1	
	Coolie	No	1/2	

AIR COMPRESSURE

Sr. No.	Make	Capacity (cfm)	Type	Model	KW	Year of Installation	Pressure (bar)
1		730	Screw	SD 202	110		8.5
2		200	Screw	GA 37 AEL 2	37		10
4		150	Screw	GA 22 AEL 2	22		10
		1080	Running				
3		167	Screw	GA 37	37		10
5		97	Screw	GA 22	22		10
6		97	Screw	GA 22	22		10
7		100	Screw	SR 2025	18		7
8		100	Reciprocating	7 X 5	22		8
9		100	Reciprocating	7 X 5	22		8
		1644					

Moisture Seperature							
1	Puriflai	1000	Horizontal		0		
Air Dryers							
1	–	750	Refrigeration	2KD+ 075	3.8		
	Department wise air requirement (cfm)						
1	Combing	150					
2	Dyeing	100					
3	Spinning	450					
4	Weaving	150					
5	Finishing	150					
6	Boiler	50					
	Total	1050					

WATER					
Internal source					
Sr.No.	Description	KL/Day	Rate (Re/KL)		Water meter
1	JMC	35	55		Yes
2	Bore well	600			Yes
Consumption					
			KL/Day		Water meter
			Fresh	**Recycle**	
1	Wool Scouring			30	Yes
2	Dyeing		300		Yes
3	Finishing		2	200	Yes
4	AC Plant			200	Yes
5	Boiler		75		Yes
6	Domestic		200		Yes
7	Garding			100	Yes
8	Dack Wash		30	30	Yes
	Total		607	560	
Work force/shift					
	Operator			1	
	Helper			1	
Effluent Treatment plant					
	Capacity		KL/Day	900	
	Work force/shift				
	Operator			1	
	Helper			1	
Storage capacity					
Sr. No.	Description		KL		

WATER SOURCE AND STORAGE			
		Size:(Meter)	HP
Bore well:	1.	150 (Depth)	10
	2.	150 (Depth)	10
	3.	150 (Depth)	10

			KL
Well:		8 X 8 X 12 (Depth)	770
Water storage Tank:	RCC Tank:	12X15X3 (3Nos)	1890
	Fresh water O/H Tank	7.2X4.9X3	100
	Recycle water O/H Tank	6.1X6.1X2.7	100
	Fresh water Ground Tank	7.4X3.7X1.5	40
	Boiler O/H water tank	5X3.6X2.5	45
	Staff Colony RCC Tank	7.2X5.7X3.6	150
Softener	No 1	1.8 Dia X 2.5 H	6.3
	No 2	1.5 Dia X 2..05 H	3.6
	No 3	2.2 Dia X 2.40 H	9.1
Filter		2 Dia X 2.1 H	6.6
ETP	Plant Capacity		800

List of Abbrevations

A$/A$$ – Australian Dollar

A/C – Account

ACK – Australian Dollar cents

Act – Actual

Admin – Administration

Adv – Advance

AGM – Annual general meeting

AGO – Automotive gas oil (Diesel)

Amt   Amount

AP – Accounts Payable

AR – Accounts Receivable

ATD – Actual time of arrival

ATD – Actual time of departure

Avg – Average

AW – All wool

BD – Brought down

BEP – Break-even Point

BL – Bill of lading

BPV – Bank Payment voucher

BR – Bank receipt

BR – Blow room

C – Combed

C&D – Cuts and defectives

C&F – Cost and freight

CA – Current Assets

CAD – Cash against delivery

CAPEX – Expenditure for Capital items

CB – Credit balance

CC – Cash credit

CD – Carried down

CD – Cash discount

CDD - Carded

CCO – Chief Commercial Officer

CEO – Chief Executive officer

CFO – Chief Financial Officer

CIF – Cost Insurance and fright

CHA – Customs House Agent

Cont – Contribution

CL – Current Liabilities

CMO – Chief Marketing Officer

CN – Credit note

CNF – Cost & Freight, no Insurance

Co – Company

Commn – Commission

Comp – Compressor

Cons – Consumption

Cont – Contribution

Contrs – Contract

COO – Chief Operating Officer

C/COT – Cotton

CPV – Cash purchase voucher

Cı – Credit

CR – Cash receipt

CS – Closing stock

CS – Company secretory

CSMO – Chief sales and marketing office

CSO – Chief Sales officer

CSR – Corporate Social Responsibility

CST –

CTC – Cost to company

Ctns – Cartons

Cum – Commissioning

D/A or D.A. – Document against acceptance

D/P or D.P – Documents against payment

DAP – Delivered at place

DB – Debit Balance

DD – Direct documents

DDP – Delivered duty paid

DDU – Delivered Duty Unpaid

Dec/Decr – Decrease,

Def – Defective

Deg – Designing

Del – delivery

DEPB\ – Duty entitlement passbook scheme

DEPB – Duty belittlement passbook scheme

Dept – Department

Disc – Discount

Dist – Distribution

Divn – Division

DN – Debit note

DOP – Dioctyl Phthalate

Doz - Dozen

Dr – Debit

Drs – Debtors

Dyg – Dyeing

EBITDA – Earnings before interest, taxes, depreciation, and amortization

EBIT – Earnings before interest and taxes

EBO – Exclusive brand outlet

EDU,PTV,HTVS – Name of machine models

Eff – Efficiency

Eng – Engineering

EPF – It is the contribution of the employee to the provident fund

EPI – Early payment incentive

EPS – Earnings per share

ESIC – Employees state insurance scheme

ETA – Expected time of arrival (at destination)

ETD – Expected time of departure

ETP – Effluent treatment plant

EU – European Union

Exp/Exps – Expenses

Exptd – Expected

Fab – Fabric

FA – Fixed assets

FCL – Full container load

FCL – It refers to the shipment where all the goods in a container belong to one party

FD – Fixed deposit

FG – Finished goods

FOB – Free on board/Freight on Board

FOI – Free of interest

Fsg – Finishing

FY – Financial year

GB – Gill box

GBP – Great Britain pounds

GC – Group company

GH – Guesthouse

GLM – Gram per linear meter

GM – Grams

Govt – Government

GRN – Goods received note

GSTR – Goods and service tax return

HFO – Heavy fuel oil

HRD – Human Resource Development

HR – Human resource

HSD – High speed diesel

H plant - Humidification plant

Inc/Incr – Increase

Incld – Including

Inp – Inspection

Inst – Institutional

ITC – Input tax credit

IOU – I owe You

JV – Journal Voucher

K – Carded

KL – Kilo liters

KSH – Kenyan shilling

Lacs – Lakhs

LCL – It involves multiple party's goods packed together

LC – Letter of credit

LCL – Less container load

LFR – Large format stores

LPFO – Low pour fuel oil

LP – Low pill polyester

LPO – Local purchase order

LR – Lakshmi Reiter

LR – Loading Receipt

Ltg – Lighting

LUT – Letter of undertaking

MBO – Multi brand outlet

MD – Maximum demand

Mfg – Manufacturing

Min – Minute

Mktg – Marketing

M – Micron

M/c - Machine

MT – Metric ton

M/Mtr – Meter

N – Naira (Nigerian Currency)

NP – Normal polyester

NRC – Nomination and Remuneration Committee

O/S – Outsourced

OE – Open end

OH – Over heads

OS – Opening stock

P Count – Per count

P&L – PROFIT AND LOSS ACCOUNT

P/E – Price earnings ratio

PA – Per annum

PBDT – Profit before depreciation and taxes

PBT – Profit before taxes

PCH – Production Centre head

PC – Polyester/Cotton

Permt – Permanent

PE – Spun Polyester

PF – Power factor

Pkg – Packaging

PLEC – Peak load exemption charges

PM – Per month

POP – Point of purchase

PPC – Production, planning and control

PPI – Prompt payment incentive

PP – Polypropylene

Prep – Preparatory

Prod/Prodn – Production

PVC – Polyvinyl Chloride

PV – Polyester/Viscose

PW – Polyester/Wool

PY – Previous year

QA – Quality Assurance

QC – Quality Control

QN – Quality Number

Qunt – Quantity

Recd – Received

RF – Ring Frame

RF – Ring frame

RMC – Risk Management Committee (RMC)

ROA – Return on Assets

ROCE – Return on Capital Employed

ROI – Return on investment

Rs – Rupees

RFD – Refund

SF – Speed frame

Spdl – Spindle

SOP - Standard operating procedure

Specs – Specification

Spg – Spinning

SRC – Stakeholders' Relationship Committee

TBL – Trilobal polyester

TDS – Tax deduction at source

TCS – Tax collection at source

TD – Term deposit

Tex – Texturized

TFO – Two for (doubling cum twisting m/c) TFO –
Two for one Machine

TL – Term loan

TM – Twist multiplier

TOD – Turn over discount

TOI – Turn over incentive

TPI – Twist per inch

TPM – Twist Per meter

TW – Terry wool

US$ – American dollar

US – United states

Uti – Utilization

Utiy – Utility

VAT – Value added tax (Tax in the past)

VT – Vinyl tube

Wdg – winding,

Wevg – weaving

WH – Warehouse

WIP – Work in progress

Wkr – Worker

Wt – Weight

Wvg – weaving

Wvg – Weaving

W – Width

BRC – Bank relisation certificate – Issued by bank
based on realisation of payment against export by an
exporter

DBK – Duty drawback – Rebate of any duty that is
chargeable on imported or excisable materials used in
goods for export

EDF – Export deceleration form (RBI requirement

GR Form (in export reference – GR form is an
exchange control document required by RBI

GSP – Generalised system of preferences (Instituted in
1971 under the aegis of UNCTAD to create an
enabling trading environment for developing
countries

HS Code – Harmonised system code of tariff for
import of goods

MEIS – Merchandise export from India scheme – A
scheme to reward to exporters to offset infrastructural
inefficiencies and associated cost

SDF Form – Statutory deceleration form (RBI
requirement

TT – Telegraphic transfer (Electronic method of
transferring funds